D0148437

DIFFERENTIAL GEOMETRY AND ITS APPLICATIONS

John Oprea

Cleveland State University

PRENTICE HALL

Upper Saddle River, NJ 07458

Library of Congress Cataloging-in-Publication Data

Oprea, John.
  Differential geometry and its applications / John Oprea.
     p.   cm.
  Includes bibliographical references and index.
  ISBN   0-13-340738-1
     1. Geometry, Differential. I. Title
QA614.O67 1997   96-9009
516.3'6—dc20   CIP

Acquisitions editor: *George Lobell*
Editorial assistant: *Gale Epps*
Editorial director: *Tom Bozik*
Editor-in-Chief: *Jerome Grant*
Assistant vice president of production and manufacturing: *David W. Riccardi*
Editorial/production supervision: *Robert C. Walters*
Managing editor: *Linda Mihatov Behrens*
Executive managing editor: *Kathleen Schiaparelli*
Manufacturing buyer: *Alan Fischer*
Manufacturing manager: *Trudy Pisciotti*
Marketing manager: *John Tweeddale*
Marketing assistant: *Diana Penha*
Creative director: *Paula Maylahn*
Art director: *Jayne Conte*
Cover designer: *Bruce Kenselaar*
Cover photo: *Southeastern Massachusetts House by Steve Rosenthal*

© 1997 by Prentice-Hall, Inc.
Simon & Schuster/A Viacom Company
Upper Saddle River, New Jersey 07458

The author and publisher of this book have used their best efforts in preparing this book. These efforts include the development, research, and testing of the theories and formulas to determine their effectiveness. The author and publisher shall not be liable in any event for incidental or consequential damages in connection with, or arising out of, the furnishing, performance, or use of these formulas.

Printed in the United States of America

10   9   8   7   6   5   4   3   2   1

ISBN 0-13-340738-1

Prentice-Hall International (UK) Limited, *London*
Prentice-Hall of Australia Pty. Limited, *Sydney*
Prentice-Hall Canada, Inc., *Toronto*
Prentice-Hall Hispanoamericana, S.A., *Mexico*
Prentice-Hall of India Private Limited, *New Delhi*
Prentice-Hall of Japan, Inc., *Tokyo*
Simon & Schuster Asia Pte. Ltd., *Singapore*
Editora Prentice-Hall do Brasil, Ltda., *Rio de Janeiro*

TO MY MOTHER AND FATHER

JEANNE AND JOHN OPREA

CONTENTS

PREFACE

How and what should we teach today's undergraduates to prepare them for careers in mathematically oriented areas? Furthermore, how can we ameliorate the quantum leap from introductory calculus and linear algebra to more abstract methods in both pure and applied mathematics? There *is* a subject which can take students of mathematics to the next level of development and this subject is, at once, intuitive, calculable, useful, interdisciplinary, and, *most importantly*, interesting. Of course, I'm talking here about Differential Geometry, a subject with a long, wonderful history and a subject which has found new relevance in areas ranging from machinery design to the classification of four-manifolds to the creation of theories of Nature's fundamental forces to the study of DNA.

Differential geometry provides the perfect transition course to higher mathematics and its applications. It is a subject which allows students to see mathematics for what it is — not the compartmentalized courses of a standard university curriculum, but a unified whole mixing together geometry, calculus, linear algebra, differential equations, complex variables, the calculus of variations, and various notions from the sciences. Moreover, differential geometry is not just for mathematics majors, but encompasses techniques and ideas relevant to students in engineering and the sciences. Furthermore, the subject itself is not quantized. By this, I mean that there is a continuous spectrum of results which proceeds from those which depend on calculation alone to those whose proofs are quite abstract. In this way students gradually are transformed from calculators to thinkers.

Into the mix of these ideas now comes the opportunity to visualize concepts and constructions through the use of computer algebra systems such as MAPLE and MATHEMATICA. Indeed, it is often the case that the consequent visualization goes hand-in-hand with the understanding of the mathematics behind the computer construction. For instance, in Chapter 5, I use MAPLE to visualize geodesics *on surfaces* and this requires an understanding of the idea of solving a system of differential equations numerically and displaying the solution. Further, in this case, visualization is not an empty exercise in computer technology, but actually clarifies various phenomena such as the bound on geodesics due to the Clairaut relation. There are many other examples of the benefits of computer algebra systems to understanding concepts and solving problems. In particular, the procedure for plotting geodesics can be modified to show equations of motion of particles constrained to surfaces. This is done in Chapter 8 along with describing procedures relevant to the calculus of variations and optimal control. At the end of Chapters 1, 2, 3, 5, 7, and 8 there are sections devoted to explaining how MAPLE fits into the framework of differential geometry. I have tried to make these sections a rather informal tutorial as opposed to just laying out procedures. This is both good and bad for the reader. The good comes from the little tips about pitfalls and ways to avoid them; the bad comes from my personal predelictions and the simple fact that I am not a MAPLE expert. What you will find in this text is the sort of MAPLE that anyone can do. Also, I happen to think that MAPLE is easier for students to learn than MATHEMATICA and so I use it here. If you prefer MATHEMATICA, then you can, without too much trouble I think, translate my procedures from MAPLE into MATHEMATICA or you can look at A. Gray, *Differential Geometry of Curves and Surfaces*, for a huge number of MATHEMATICA geometry procedures and examples.

In spite of the use of computer algebra systems here, this text is traditional in the sense of approaching the subject from the point of view of the 1800's. What *is* different about this book is that a conscious effort has been made to include material that I feel science and math majors should know. For example, although it is possible to find mechanistic descriptions of phenomena such as Clairaut's relation or Jacobi's theorem and geometric descriptions of mechanistic phenomena such as the precession of Foucault's pendulum in advanced texts (see V.I. Arnol'd, *Mathematical Methods of Classical Mechanics* and J. Marsden *Lectures on Mechanics*), I believe they appear here for the first time in an undergraduate text. Also, even when dealing with mathematical matters alone, I have always tried to keep some application, whether mathematical or not, in mind. As an example of this, I would note that the Weierstrass-Enneper representation is not discussed merely for its own sake, but for its use in solving the problem of Björling *and* because it is essential to understanding why (a portion of) Enneper's surface is minimal, but not area minimizing. In fact, I think this last topic helps to show the boundaries between physics (e.g. soap films) and mathematics (e.g. minimal surfaces) as well as being amenable to the power of MAPLE (see Chapter 7).

This book originally began as an attempt to fashion a one-quarter course in differential geometry. In fact, I have taught such a course for mathematics, physics, engineering, chemistry, biology and philosophy majors and I have used topics from Chapters 1, 2, 3, 4, 5, 6 and 8. This does not mean that I have covered these chapters exhaustively in one quarter, but that I have chosen certain parts to emphasize and allowed students to do projects, say, involving other parts. For example, students have done projects on involutes and gear teeth design, re-creation of curves from curvature and torsion, Enneper's surface and area minimization (see above), geodesics on minimal surfaces, and the Euler-Lagrange equations in relativity. In many cases, students have gone way beyond what this book contains and I owe them my thanks for expanding my knowledge. The book, as it now stands, is suitable for either a one-quarter or one-semester course in differential geometry as well as a full-year course. In the case of the latter, all chapters may be completed. In the case of the former, I would recommend the chapters I've listed above, but there is a good choice of alternative material as well.

The reader should note two things about the layout of the book. First, the exercises are integrated into the text. While this may make them somewhat harder to find, it also makes them an essential part of the text. The reader should at least *read* the exercises when going through a chapter — they are important. Secondly, I have chosen to number theorems, lemmas, and the like separately from examples, definitions, and remarks. I did this to avoid having Lemma 27 followed by Theorem 82, but of course this makes it harder to find the examples. Therefore, I have included a list of examples directly before the solutions to problems at the back of the book.

There are several students who deserve special mention with regard to this text. Rob Clark first interested me in minimal surfaces and, together with Jack Chen, showed me the use of computers (e.g. Ken Brakke's Evolver program) in distinguishing "minimal" from "harmonic". Laszlo Ilyes provided many of the MAPLE procedures for optimal control while Carrie Kyser Heide took my original laughable "geodesic procedures" and transformed them wonderfully into one elegant procedure which does exactly what I want it to do. Sue Halamek did an excellent job on the first draft of the solutions to problems and any present errors are certainly due to my final editing. Thanks to you all and to all the students who watched me fumble my way to a book!

I would also like to acknowledge the contributions of my friend Allen Broughton. It was Allen who first taught a course from a sheaf of my handwritten notes and actually made sense of the notes and a success of the course. Allen also first explored the use of MAPLE for differential geometry and is responsible for producing the first procedures for calculating curvatures etc. Similarly, the handwritten notes referred to above would have remained just that without the TEXpertise of Joyce Pluth. Joyce typed the first draft of those notes and patiently tutored me in the intricacies of TEX until I stopped bothering her. Let me also thank Elaine Hoff and Dena Jones for helping me to photocopy, collate, cut, and paste to ready versions of this text for unsuspecting classes. I must also thank the members of the Cleveland Geometry/Topology Seminar for sitting through numerous lectures on various parts of this text.

Finally, the writing of this book would have been impossible without the help, advice, and understanding of my wife Jan and daughter Kathy. Thanks — the computer is now free!

John Oprea
oprea@math.csuohio.edu

Chapter 1

THE GEOMETRY
OF CURVES

1.1 INTRODUCTION

A *curve* in 3-space $\mathbb{R}^3$ is a continuous mapping $\alpha \colon I \to \mathbb{R}^3$ where I is some type of interval (e.g. $(0,1)$, (a,b), $[a,b]$, $(-\infty, a]$, $[0,1]$ etc.) on the real line $\mathbb{R}$. Because the range of α is $\mathbb{R}^3$, α's output has three coordinates. We then write, for $t \in I$,

$$\alpha(t) = (\alpha^1(t),\ \alpha^2(t),\ \alpha^3(t))$$

where the α^i are themselves functions $\alpha^i \colon I \to \mathbb{R}$. A useful way to think about curves is to consider t to be time and $\alpha(t)$ to be the path of a particle in space. We say α is *differentiable* or *smooth* if each coordinate function α^i is differentiable as an ordinary real-valued function of $\mathbb{R}$. The *velocity vector* of α at t_0 is defined to be

$$\alpha'(t_0) = \left(\left.\frac{d\alpha^1}{dt}\right|_{t=t_0},\ \left.\frac{d\alpha^2}{dt}\right|_{t=t_0},\ \left.\frac{d\alpha^3}{dt}\right|_{t=t_0} \right)$$

where $d\alpha^i/dt$ is the ordinary derivative and $|_{t=t_0}$ denotes evaluation of the derivative at $t = t_0$. We shall also write $d\alpha^i/dt(t_0)$ for this evaluation when it is convenient. In order to interpret α' geometrically, we introduce the

Example 1.1: First example of a curve: A line in $\mathbb{R}^3$.
We know that two points determine a line. In the plane $\mathbb{R}^2$, this leads to the usual slope-intercept equation which gives an algebraic description of the line. Unfortunately, in $\mathbb{R}^3$ we don't have a good notion of slope, so we can't expect exactly the same type of algebraic description to work. For two points, p and q, the line l joining them may be described as follows. To attain the line, add the vector p. To travel along the line, use the direction vector $q - p$ since this is the direction from p to q. A *parameter* t tells exactly how far along $q - p$ to go. Putting these steps together produces a curve which is a *parametrized line* in $\mathbb{R}^3$,

$$\alpha(t) = p + t(q - p).$$

For instance, if $p = (1, 2, 3)$ and $q = (-1, 4, -7)$, then $q - p = (-2, 2, -10)$, so the line through p and q is given by the curve

$$\alpha(t) = (1, 2, 3) + t(-2, 2, -10)$$
$$= (1 - 2t, 2 + 2t, 3 - 10t).$$

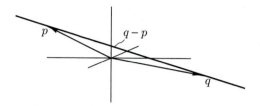

FIGURE 1.1.   A parametrized line

EXERCISE 1.1. What is the parametrization of the line through $(-1, 0, 5)$ and $(3, -1, -2)$?

EXERCISE 1.2. In $\mathbb{R}^4$, what is the parametrization of the line through $(-1, 6, 5, 0)$ and $(0, 1, -3, 9)$?

So we see that, given a line $\alpha(t) = p + t(q - p)$, $\alpha'(t) = q - p$. Since this *direction vector* $q - p$ is then the velocity vector, we often write

$$\alpha(t) = p + t\mathbf{v}.$$

Now that we understand lines, we can ask what the picture of the velocity vector is for any curve $\alpha(t)$? The definitions give

$$\alpha'(t_0) = \left(\frac{d\alpha^1}{dt}\bigg|_{t=t_0}, \; \frac{d\alpha^2}{dt}\bigg|_{t=t_0}, \; \frac{d\alpha^3}{dt}\bigg|_{t=t_0} \right)$$

$$= \left(\lim_{t \to t_0} \frac{\alpha^1(t) - \alpha^1(t_0)}{t - t_0}, \; \lim_{t \to t_0} \frac{\alpha^2(t) - \alpha^2(t_0)}{t - t_0}, \; \lim_{t \to t_0} \frac{\alpha^3(t) - \alpha^3(t_0)}{t - t_0} \right)$$

$$= \lim_{t \to t_0} \left(\frac{(\alpha^1(t), \alpha^2(t), \alpha^3(t)) - (\alpha^1(t_0), \alpha^2(t_0), \alpha^3(t_0))}{t - t_0} \right)$$

$$= \lim_{t \to t_0} \frac{\alpha(t) - \alpha(t_0)}{t - t_0}.$$

This looks much like our usual "slope" definition of the derivative. In fact, $\alpha(t) - \alpha(t_0)$ is exactly the vector shown below. Note that the approximating

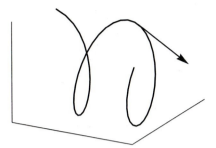

FIGURE 1.2.   A curve with tangent vector

vectors inside the limit always point in the direction of increasing t. For $t < t_0$, this follows since $t - t_0$ is negative.

Now, $t \to t_0$, so the vector moves closer and closer to the tangent vector to the curve at $\alpha(t_0)$. Hence, the velocity vector $\alpha'(t_0)$ to α at t_0 is exactly the tangent vector to α at $\alpha(t_0)$. Notice that we obtain a vector with a precise length, not a line. Remember that a vector $\mathbf{v} = (v^1, v^2, v^3) \in \mathbb{R}^3$, has a *length* given by the Pythagorean theorem,

$$|\mathbf{v}| = \sqrt{(v^1)^2 + (v^2)^2 + (v^3)^2}.$$

EXERCISE 1.3. Show that $|\mathbf{v}|$ is the distance of (v^1, v^2, v^3) from $(0,0,0)$ by two applications of the Pythagorean theorem.

In our situation, $|\alpha'(t)| = \sqrt{\left(d\alpha^1/dt\right)^2 + \left(d\alpha^2/dt\right)^2 + \left(d\alpha^3/dt\right)^2}$ is simply the *speed* of α. Again, thinking of α as the path of a particle and t as time, we see that the length of the velocity vector is precisely the speed of the particle at the given time. For a line, $\alpha(t) = p + t\mathbf{v}$, the speed is simply $|\mathbf{v}|$, the length of the direction vector.

If $\alpha(t) = \left(\alpha^1(t), \alpha^2(t), \alpha^3(t)\right)$ is a curve in $\mathbb{R}^3$, then its acceleration vector is given by

$$\alpha''(t) = \left(\frac{d^2}{dt^2}\alpha^1(t), \; \frac{d^2}{dt^2}\alpha^2(t), \; \frac{d^2}{dt^2}\alpha^3(t) \right).$$

EXERCISE 1.4. Newton asked the question, what is the curve which, when revolved about an axis, gives a surface offering the least resistance to motion through a "rare" fluid (e.g. air)? Newton's answer is the following curve (see Chapter 8).

$$(x(t), y(t)) = \left(\frac{\lambda}{2}\left[\frac{1}{t} + 2t + t^3\right], \; \frac{\lambda}{2}\left[\ln\left(\frac{1}{t}\right) + t^2 + \frac{3}{4}t^4\right] - \frac{7}{8}\lambda \right),$$

where λ is a constant and $x \geq 2\lambda$. Graph this curve and compute its velocity and acceleration vectors. What angle does the curve make with the x-axis at the intersection point $(2\lambda, 0)$?

We now come to the first (and simplest nontrivial) instance of calculus imposing a constraint on the geometry of a curve. In a real sense, this is what the differential geometry of curves is all about and we will see many more examples of it later.

Proposition 1.1. *The curve α is a straight line if and only if $\alpha'' = 0$.*

Proof. If $\alpha(t) = p + t\mathbf{v}$ is a line, then $\alpha'(t) = \mathbf{v}$ (which is a constant vector), so $\alpha''(t) = 0$.

If $\alpha''(t) = 0$ for all t, then $d^2\alpha^i(t)/dt^2 = 0$, for each coordinate function $\alpha^i(t)$. But a zero 2nd derivative simply means that $d\alpha^i(t)/dt = v^i$ is a constant. We may integrate with respect to t to obtain $\alpha^i(t) = p^i + v^i t$ where p^i is a constant of integration. Then

$$\alpha(t) = (p^1 + tv^1, p^2 + tv^2, p^3 + tv^3) = p + t\mathbf{v}$$

with $p = (p^1, p^2, p^3)$ and $\mathbf{v} = (v^1, v^2, v^3)$. Hence the curve α may be parametrized as a line. $\square$

This easy result indicates how we will use calculus to detect geometric properties. Just to see how far we can get using these elementary ideas, let's consider the following question. What is the shortest distance between two points p, $q \in \mathbb{R}^3$? We have been taught since we were children that the answer is a line, but now we can see why our intuition is correct. Calculus tells us that distance is simply the integral of speed. Therefore, the integral

$$L(\alpha) = \int_a^b |\alpha'(t)|\, dt$$

calculates the *arclength* of a curve $\alpha\colon [a, b] \to \mathbb{R}^3$ from $\alpha(a)$ to $\alpha(b)$. In order to see this, consider a small piece ("an infinitesimal") of arclength approximated as in Figure 1.3.

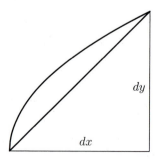

FIGURE 1.3.   An infinitesimal piece of curve

The Pythagorean theorem then says that $l = \sqrt{dx^2 + dy^2}$. Suppose both x and y are parametrized by t. Then $\alpha(t) = (x(t), y(t))$ and $dx^2 = (dx/dt)^2 \, dt^2$ and $dy^2 = (dy/dt)^2 \, dt^2$. Finally,

$$l = \sqrt{\left(\frac{dx}{dt}\right)^2 + \left(\frac{dy}{dt}\right)^2} \, dt$$

expressing the relation, distance equals rate times time. Now we add up all the pieces of l by integration to get,

$$L(\alpha) = \int_a^b \sqrt{\left(\frac{dx}{dt}\right)^2 + \left(\frac{dy}{dt}\right)^2} \, dt$$
$$= \int_a^b |\alpha'(t)| \, dt.$$

The three dimensional version of this intuitive description simply uses our earlier exercise on the Pythagorean theorem in three dimensions. Now, the length of the vector $\mathbf{v}$, $|\mathbf{v}|$, may be written $\sqrt{\mathbf{v} \cdot \mathbf{v}}$, where $\cdot$ denotes the *dot product* of vectors. Recall that, in general, $\mathbf{v} \cdot \mathbf{w} = v^1 w^1 + v^2 w^2 + v^3 w^3$, for $\mathbf{v} = (v^1, v^2, v^3)$ and $\mathbf{w} = (w^1, w^2, w^3)$. We also have the following

Proposition 1.2. *The dot product is computed to be $\mathbf{v} \cdot \mathbf{w} = |\mathbf{v}| \, |\mathbf{w}| \cos \theta$, where θ is the angle between the vectors $\mathbf{v}$ and $\mathbf{w}$.*

Proof. From Figure 1.4, we see that the usual properties of vectors give $\mathbf{u} = \mathbf{v} - \mathbf{w}$. Hence

$$|\mathbf{u}|^2 = \mathbf{u} \cdot \mathbf{u} = (\mathbf{v} - \mathbf{w}) \cdot (\mathbf{v} - \mathbf{w})$$
$$= \mathbf{v} \cdot \mathbf{v} - 2\mathbf{v} \cdot \mathbf{w} + \mathbf{w} \cdot \mathbf{w}$$
$$= |\mathbf{v}|^2 + |\mathbf{w}|^2 - 2\mathbf{v} \cdot \mathbf{w}.$$

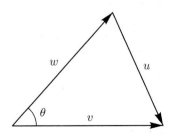

FIGURE 1.4

But the Law of Cosines for this triangle says $|\mathbf{u}|^2 = |\mathbf{v}|^2 + |\mathbf{w}|^2 - 2|\mathbf{v}|\,|\mathbf{w}|\cos\theta$. Equating the two righthand sides and eliminating like terms produces

$$\mathbf{v} \cdot \mathbf{w} = |\mathbf{v}|\,|\mathbf{w}|\cos\theta.$$

$\square$

Corollary 1.3 (Schwarz's Inequality). *The dot product obeys the inequality,*

$$|\mathbf{v} \cdot \mathbf{w}| \le |\mathbf{v}|\,|\mathbf{w}|.$$

Proof. Simply note that $|\cos\theta| \le 1$. $\square$

We note here that, with respect to the derivative, the dot product behaves just as ordinary multiplication — namely, the Leibniz (or product) rule holds. In particular,

Proposition 1.4. *If $\alpha(t)$ and $\beta(t)$ are two curves in $\mathbb{R}^3$, then*

$$\frac{d(\alpha \cdot \beta)}{dt} = \frac{d\alpha}{dt}\beta + \alpha\frac{d\beta}{dt}.$$

Proof. We use the ordinary product rule on the component functions,

$$\begin{aligned}
\frac{d(\alpha \cdot \beta)}{dt} &= \frac{d}{dt}(\alpha^1\beta^1 + \alpha^2\beta^2 + \alpha^3\beta^3) \\
&= \sum_i \left(\frac{d\alpha^i}{dt}\beta^i + \alpha^i\frac{d\beta^i}{dt} \right) \\
&= \sum_i \left(\frac{d\alpha^i}{dt}\beta^i + \sum_i \alpha^i\frac{d\beta^i}{dt} \right) \\
&= \frac{d\alpha}{dt} \cdot \beta + \alpha \cdot \frac{d\beta}{dt}.
\end{aligned}$$

$\square$

When the parameter t is understood, we will write the product rule as $(\alpha \cdot \beta)' = \alpha' \cdot \beta + \alpha \cdot \beta'$. We are now able to answer the question of which path between two points gives the shortest distance.

Theorem 1.5. *In $\mathbb{R}^3$, a line is the curve of least arclength between two points.*

Proof. Consider two points $p, q \in \mathbb{R}^3$. The line between them may be parametrized by $l(t) = p + t(q - p)$, where $q - p$ is the vector in the direction from p to q. Then $l'(t) = q - p$ and $|l'(t)| = |q - p|$, a constant. Therefore,

$$L(l) = \int_0^1 |l'(t)|\, dt = |q - p| \int_0^1 dt = |q - p|$$

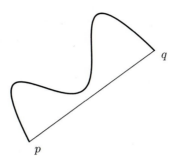

FIGURE 1.5.   Line and comparison curve

and the length of the line segment (or direction vector) from p to q is the distance from p to q (as of course we expected). Now consider another curve α which joins p and q.

We want to show that $L(\alpha) > L(l)$ and, since α is arbitrary, this will say that the straight line minimizes distance. Now, why should α be longer than l? One intuitive explanation is to say that α starts off in the wrong direction. That is, $\alpha'(a)$ is not "pointing toward" q. How can we measure this deviation? The angle between the unit vector in the direction of $q - p$, $\mathbf{u} = q - p/|q - p|$ and $\alpha'(a)$, the tangent vector of α at p, may be calculated by taking the dot product $\alpha'(a) \cdot \mathbf{u}$. The total deviation may be added up by integration to give us an idea of why $L(\alpha) > L(l)$ should hold. Precisely, we compute $\int_a^b \alpha'(t) \cdot \mathbf{u}\, dt$ in two ways to obtain the inequality. Now, we have $(\alpha(t) \cdot \mathbf{u})' = \alpha'(t) \cdot \mathbf{u} + \alpha(t) \cdot \mathbf{u}' = \alpha'(t) \cdot \mathbf{u}$ since $\mathbf{u} = $ constant vector. Also, the Fundamental Theorem of Calculus gives $\int_a^b df/dt\, dt = f(b) - f(a)$, so

$$\int_a^b \alpha'(t) \cdot \mathbf{u}\, dt = \int_a^b (\alpha(t) \cdot \mathbf{u})'\, dt = \alpha(b) \cdot \mathbf{u} - \alpha(a) \cdot \mathbf{u}$$

$$= q \cdot \mathbf{u} - p \cdot \mathbf{u} \qquad \text{since } \alpha(a) = p, \quad \alpha(b) = q$$

$$= (q - p) \cdot \mathbf{u}$$

(1)
$$= \frac{(q - p) \cdot (q - p)}{|q - p|}$$

$$= \frac{|q - p|^2}{|q - p|}$$

$$= |q - p|$$

$$= L(l)$$

the straightline distance from p to q. Consequently,

$$\int_a^b \alpha'(t) \cdot \mathbf{u}\, dt \leq \int_a^b |\alpha'(t)| \cdot |\mathbf{u}|\, dt \qquad \text{by the Schwarz inequality}$$

(2)
$$= \int_a^b |\alpha'(t)|\, dt \qquad \text{since } |\mathbf{u}| = 1$$

$$= L(\alpha)$$

Combining (1) and (2), we have $L(l) \leq L(\alpha)$. Note also that $\alpha'(t) \cdot \mathbf{u} = |\alpha'(t)| \cdot |\mathbf{u}|$ only when $\cos\theta = 1$, or $\theta = 0$. That is, $\alpha'(t)$ must be parallel to $q - p$ for all t. In this case α is the line from p to q. Therefore, we have the strict inequality $L(l) < L(\alpha)$ unless $\alpha =$ line.   □

So, using the first notions of calculus, we have seen how to identify lines in $\mathbb{R}^3$ and to determine exactly why lines give the shortest distance between two points. Of course, $\mathbb{R}^3$ is governed by Euclidean Geometry. One of the main goals of differential geometry is to develop analogous techniques for curved geometries as well.

Other Examples of Curves

While curves described parametrically may be defined at the whim of the definer, in fact they often arise from geometric or physical considerations. Several of the examples below illustrate this.

Example 1.2: The Circle of radius r (centered at $(0,0)$).
Of course, the question is, how can such a curve be parametrized? The definitions of $\sin\theta$ and $\cos\theta$ as the vertical and horizontal sides of the right triangle with angle θ depicted below give us a way of assigning coordinates to the point P.

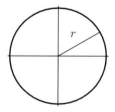

FIGURE 1.6.   Circle of radius r

Namely, the point P has coordinates $(\cos\theta,\ \sin\theta)$. For a circle of radius r, by the fundamental property of similar triangles, the vertical and horizontal sides of the triangle must be $r\sin\theta$ and $r\cos\theta$ respectively. Hence, the coordinates of the point on the circle are $(r\cos\theta,\ r\sin\theta)$. Therefore, the circle may be parametrized by the angle θ. Again thinking of the circle as the path of a particle, we write

$$\alpha(t) = (r\cos t,\ r\sin t), \qquad \text{for } 0 \le t < 2\pi.$$

There are two important observations to remember. First, $\alpha'(t) = (-r\sin t, r\cos t)$ so $\alpha(t) \cdot \alpha'(t) = 0$ for all t. In other words, the tangent vector is perpendicular to the radius. Second, $\alpha''(t) = (-r\cos t, -r\sin t) = -\alpha(t)$, so the acceleration vector points toward the center of the circle. Finally, we can compute the arclength (i.e. circumference) of the circle,

$$
\begin{aligned}
L(\alpha) &= \int_0^{2\pi} |\alpha'(t)|\, dt \\
&= \int_0^{2\pi} \sqrt{r^2 \sin^2 t + r^2 \cos^2 t}\, dt \\
&= \int_0^{2\pi} r\sqrt{\sin^2 t + \cos^2 t}\, dt \\
&= \int_0^{2\pi} r\, dt \\
&= r\, \Big|_0^{2\pi} t \\
&= 2\pi\, r.
\end{aligned}
$$

EXERCISE 1.5. Suppose a circle of radius a sits on the x-axis making contact at $(0,0)$. Let the circle roll along the positive x-axis. Show that the path α followed by the point originally in contact with the x-axis is given by

$$\alpha(t) = (a\,(t - \sin t),\ a\,(1 - \cos t))$$

where t is the angle formed by the (new) point of contact with the axis, the center and the original point of contact. This curve is called a *cycloid* and we will meet it again in Chapter 8. Hints: Recall that $s = a\,t$ where s is arclength. Draw a picture after the circle has rolled s units. Where is the old point of contact?

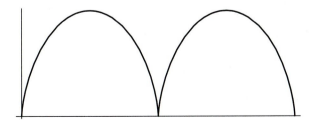

FIGURE 1.7.   A cycloid

EXERCISE 1.6. Consider a cycloid of the form

$$(x(t), y(t)) = (A + a\,(t - \sin t),\ B - a\,(1 - \cos t)).$$

Graph this cycloid to see that it is an inverted form of the one found in the previous exercise. Suppose a unit mass particle starts at rest at a point $(\bar{x}, \bar{y})$ on the cycloid corresponding to an angle $\bar{t}$ in the parametrization above. Under the influence of gravity (and assuming no friction), show that, no matter what initial $\bar{t}$ is chosen, it always takes a time of

$$T = \sqrt{\frac{a}{g}}\,\pi \qquad \text{where } g \text{ is the gravitational constant}$$

for the particle to slide down to the bottom of the cycloid (i.e. $t = \pi$). This property was used by C. Huygens to make clocks without pendula (which he hoped would allow for accurate timekeeping on ships at sea and, thereby, improve navigation). He named the cycloid the *tautochrone* since the latter means "same time." We will revisit the cycloid in Chapter 8 where we will see it has yet another intriguing property and name. Hints: potential energy is turned into kinetic energy by, $\nu^2/2 = g(\bar{y} - y)$. Time equals distance divided by speed, so

$$T = \frac{1}{\sqrt{2g}} \int_{\bar{x}}^{x_{bot}} \frac{\sqrt{1 + y'^2}}{\sqrt{\bar{y} - y}}\, dx$$

$$= \sqrt{\frac{a}{g}} \int_{\bar{t}}^{\pi} \sqrt{\frac{1 - \cos t}{\cos \bar{t} - \cos t}}\, dt$$

$$= \sqrt{\frac{a}{g}} \bigg|_{\bar{t}}^{\pi} - 2 \arcsin \left[\frac{\sqrt{2}\cos\frac{t}{2}}{\sqrt{1 + \cos \bar{t}}} \right]$$

using $y' = (dy/dt)/(dx/dt) = -\sin t/(1 - \cos t)$, $\bar{y} = B - a(1 - \cos \bar{t})$, $y = B - a(1 - \cos t)$ and simplifying. Verify the final step by differentiation and compute T.

Example 1.3: The Astroid. $\alpha(t) = (a\cos^3 t,\ a\sin^3 t)$ for $0 \le t \le 2\pi$. The definition of the astroid (which was discovered by people searching for the best form of gear teeth) is very similar to that of the cycloid. For the astroid, however, a circle is rolled, not on a line, but inside another circle. More precisely, let a circle of radius $a/4$ roll inside a large circle of radius a (centered at $(0,0)$ say). For concreteness, suppose we start the little circle at $(a, 0)$ and follow the path of the point originally in contact with $(a, 0)$ as the circle rolls up. Roll the circle a little bit and notice that a piece of arclength s is used up. The key to understanding this situation is the fact that, for t, the angle from the center of the large circle to the new contact point and θ, the angle through which the small circle has rolled (measured from the point of contact), we have $s = at$ and $s = a\theta/4$. Hence, $\theta = 4t$.

Now we can do two things to parametrize the path of the rolling point. First, we can parametrize the path *with respect to the center of the little*

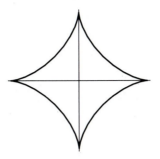

FIGURE 1.8.   An astroid

circle. This is easy because we are just moving an angle θ around a circle. There is a small problem because we measure θ from the point of contact, but this is handled easily as in Figure 1.9. Note that we make real use of the relation $\theta = 4t$ here. The coordinates with respect to the center then become $x = a \cos 3t/4$ and $y = a \sin 3t/4$. Secondly, the center of the little circle always remains on the circle of radius $3a/4$ centered at $(0,0)$, so that the center has moved to $(3a \cos t/4, 3a \sin t/4)$. Hence, with respect to the origin $(0,0)$, the rolling point has moved to

$$\alpha(t) = \left(\frac{3a}{4} \cos t + \frac{a}{4} \cos 3t, \ \frac{3a}{4} \sin t - \frac{a}{4} \sin 3t \right).$$

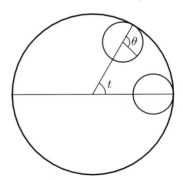

FIGURE 1.9

EXERCISE 1.7. Verify the statements above and show that the formula for the astroid may be reduced to

$$\alpha(t) = (a \cos^3 t, \ a \sin^3 t)$$

with implicit form $x^{2/3} + y^{2/3} = a^{2/3}$.

EXERCISE 1.8. From the origin draw a line through any point P of the circle of radius a centered at $(0, a)$. Find the intersection Q of this line with the horizontal line $y = 2a$. Drop a vertical line from Q and intersect it with a horizontal line passing through P. This intersection is a point on the curve known as the *witch of Agnesi**. Let t denote the between the vertical axis and the line through $(0, 0)$, P and Q. Show that the witch $W(t)$ is given in terms of t by

$$W(t) = \left(2a \tan t, \ 2a \cos^2 t \right).$$

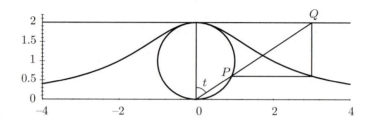

FIGURE 1.10.   The Witch of Agnesi

Example 1.4: The Helix. $\alpha(t) = (a \cos t, \ a \sin t, \ bt)$, for $0 \le t < \infty$. Notice that the the first two coordinates provide circular motion while the third coordinate lifts the curve out of the plane. We calculate $\alpha'(t) = (-a \sin t, \ a \cos t, \ b)$ and $a''(t) = (-a \cos t, \ -a \sin t, \ 0)$. Note that $a''(t)$ points toward the z-axis.

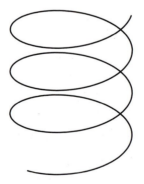

FIGURE 1.11.   A helix

---

*The "witch" was named by Maria Agnesi in the middle 1700's — apparently as the result of a mistranslation! The curve had been studied previously by Grandi who had used the Italian word "versorio," meaning "free to move in any direction," in connection with it. Agnesi mistakenly thought Grandi had used the Italian "versiera", which means "Devil's wife" or "witch" and so Agnesi's witch got "her" name [Ya].

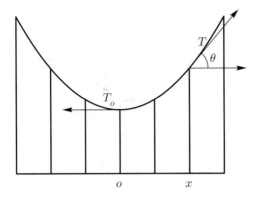

FIGURE 1.12.   Diagram of a suspension bridge

Example 1.5: The Suspension Bridge.

Consider the diagram of a suspension bridge (Figure 1.12) where the cable supports a uniformly distributed load. This means that, over any interval $[a, b]$, the weight supported is given by $W = c\,(b - a)$, where c is constant. Newton's Law $F = ma$ applies to the vertical and horizontal components of tension in the cable T to give force equations

$$T \sin \theta = c\,x \qquad \text{and} \qquad T \cos \theta = T_0$$

where x and θ are as depicted and T_0 is the horizontal force of tension pulling on the length of cable from 0 to x.

EXERCISE 1.9. Using $dy/dx = \tan \theta$, solve these equations to get

$$y = \frac{c}{2\,T_0}\, x^2 + d.$$

Thus, the cable of a suspension bridge hangs as a parabola.

Example 1.6: The Catenary.

Now suppose a cable hangs freely only supporting its own weight (Figure 1.13). What is the curve it follows? The key difference between the preceding example and this is that the weight is not uniformly distributed horizontally, but, rather, is uniformly distributed *along the length of the cable*. Therefore, we must take arclength into account as opposed to the simpler distance along the axis. Let W denote the (weight) density of the cable and let $s = s(x)$ denote the arclength of the cable from $x = 0$.

EXERCISE 1.10. From the diagram, Newton's Law $F = ma$, $dy/dx = \tan \theta$ and the arclength derivative $ds/dx = \sqrt{1 + (y')^2}$, derive the following equations

$$\frac{dy}{dx} = \frac{W}{T_0}\, s(x) \qquad \text{and} \qquad \frac{d^2 y}{dx^2} = \frac{W}{T_0} \sqrt{1 + \frac{dy}{dx}^2}.$$

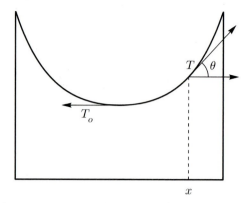

FIGURE 1.13.   Diagram of a hanging cable

Now solve the second equation to get

$$y = C \cosh\left(\frac{x}{C}\right) + D.$$

Hints: to solve the differential equation, let $z = dy/dx$. The equation becomes separable

$$\frac{dz}{\sqrt{1+z^2}} = \frac{W}{T_0}\,dx$$

and the lefthand side may be integrated by substituting in $z = \sinh u$ and recalling $\cosh^2 u - \sinh^2 u = 1$.

A curve of the form $y = c\cosh(x/c)$ is called a *catenary* from the Latin for "chain." We will meet this curve again and again in the rest of the book.

Example 1.7: The Pursuit Curve.

Suppose an enemy plane begins at $(0,0)$ and travels up the y-axis at constant speed ν_p. A missile is fired at $(a,0)$ with speed ν_m and the missile has a heat sensor which always directs it toward the plane. Show that the pursuit curve which the missile follows is given implicitly by the differential-integral equation

$$y = x\,y' + \frac{\nu_p}{\nu_m}\int \sqrt{1 + y'^2}\,dx.$$

Differentiate this expression to get a separable differential equation. Integrate to get the closed form expression for the pursuit curve

$$y = \frac{a^{\frac{\nu_p}{\nu_m}}}{2(1-\frac{\nu_p}{\nu_m})}\left[x^{1-\frac{\nu_p}{\nu_m}} - \frac{\nu_p}{\nu_m}a^{1-\frac{\nu_p}{\nu_m}}\right] - \frac{a^{-\frac{\nu_p}{\nu_m}}}{2(1+\frac{\nu_p}{\nu_m})}\left[x^{1+\frac{\nu_p}{\nu_m}} + \frac{\nu_p}{\nu_m}a^{1+\frac{\nu_p}{\nu_m}}\right].$$

Example 1.8: The Mystery Curve. $\alpha(t) = \left(\frac{1}{\sqrt{2}}\cos t,\ \sin t,\ \frac{1}{\sqrt{2}}\cos t\right)$ for $0 \leq t < 2\pi$.

EXERCISE 1.11. Identify this curve. Find $\alpha'(t)$, $\alpha''(t)$, and $L(\alpha)$. Hints: what kind of curve is this? Where does it lie in $\mathbb{R}^3$?

EXERCISE 1.12. Parametrize a circle which is centered at the point (a, b).

EXERCISE 1.13. Parametrize the ellipse $\frac{x^2}{a^2} + \frac{y^2}{b^2} = 1$.

1.2 ARCLENGTH PARAMETRIZATION

So far, we have thought of the parameter t as being time. More often it is convenient to let the parameter express how far along the curve we are. That is, when we now write $\alpha(s)$, the parameter s is exactly the distance we have travelled along the curve. This is called *parametrization by arclength*. As we will see, arclength parametrization simplifies the differential geometry of curves considerably. Therefore, it would be nice to know that every curve may be parametrized in this way. This is, in fact, the case, although it is only "in principle" that this process works in general. By this we mean to say that, while any curve may be parametrized by arclength, it is quite rare to obtain an explicit formula for the arclength parametrization. With this in mind, we will now first consider the notion of parametrization a little more carefully, then prove that every curve has an arclength parametrization. In the next section we will use the arclength parametrization to derive the Frenet Formulas, the formulas which describe the geometry of curves.

Given a curve $\alpha\colon I \to \mathbb{R}^3$ with parameter t, we may reparametrize the curve by mapping another interval onto I and using the composition as a "new" curve (which, of course, has the same point-set image in $\mathbb{R}^3$). Precisely, let $h\colon J \to I$ be a map of the interval J onto I. Then a *reparametrization* of α is given by

$$\beta = \alpha \circ h\colon J \to \mathbb{R}^3, \qquad \beta(s) = \alpha(h(s)).$$

The curves β and α pass through the same points in $\mathbb{R}^3$, but they reach any one of these points in different "times" (s and t). The velocity vectors of α and β are altered by h.

Lemma 2.1. $\beta'(s) = \alpha'(h(s)) \cdot \frac{dh}{ds}(s)$.

Proof. By definition, $\beta(s) = (\alpha^1(h(s)),\ \alpha^2(h(s)),\ \alpha^3(h(s)))$. The derivative of the coordinate function β^i is,

$$\frac{d\beta^i}{ds}(s) = \frac{d\alpha^i(h(s))}{ds} = \frac{d\alpha^i}{dh}\frac{dh}{ds}$$

by the chain rule. But $d\alpha^i/dh$ is exactly $d\alpha^i/dt$ where $h(s) = t \in I$, so we obtain for $t = h(s)$,

$$\beta'(s) = \left(\frac{d\alpha^1}{dt} \frac{dh}{ds}, \ \frac{d\alpha^2}{dt} \frac{dh}{ds}, \ \frac{d\alpha^3}{dt} \frac{dh}{ds} \right) = \alpha'(t) \frac{dh}{ds}.$$

□

EXERCISE 2.1. Recall that the arclength of a curve $\alpha \colon [a,b] \to \mathbb{R}^3$ is given by $L(\alpha) = \int |\alpha'(t)| \, dt$. Let $\beta(s) \colon [c,d] \to \mathbb{R}^3$ be a reparametrization of α defined by taking a map $h \colon [c,d] \to [a,b]$. Show that the arclength does not change under reparametrization.

A curve $\alpha(t)$ is *regular* if $\alpha'(t) \neq 0$ for all $t \in I$. In general, if a curve is not regular, then we split it into regular pieces and consider each of these separately.

Example 2.1: The Cusp $\alpha(t) = (t^2, t^3)$.
Note that $x = t^2$, $y = t^3$, so $y = \pm x^{3/2}$ with $\alpha'(t) = (2t, 3t^2)$ and $\alpha'(0) = (0,0)$. Thus α is *not* regular, but, the pieces above and below the x-axis *are* regular.

Theorem 2.2. *If α is a regular curve, then α may be reparametrized to have unit speed.*

Proof. Define the "arclength function" to be

$$s(t) = \int_a^t |\alpha'(u)| \, du$$

(where u is just a "dummy" variable). Since α is regular, the Fundamental Theorem of Calculus implies

$$\frac{ds}{dt} = |\alpha'(t)| > 0$$

By The Mean Value Theorem, s is strictly increasing on I and so is one-to-one. Therefore, s has an inverse function which we denote by $t(s)$ and the respective derivatives are inversely related,

$$\frac{dt}{ds}(s) = \frac{1}{\frac{ds}{dt}(t(s))} > 0.$$

Let $\beta(s) = \alpha(t(s))$. Then, using the Lemma, $\beta'(s) = \alpha'(t(s))\frac{dt}{ds}(s)$. But now we have,

$$|\beta'(s)| = |\alpha'(t(s))| \, |\frac{dt}{ds}(s)|$$
$$= \frac{ds}{dt}(t(s)) \frac{dt}{ds}(s)$$
$$= \frac{ds}{dt}(t(s)) \frac{1}{\frac{ds}{dt}(t(s))}$$
$$= 1.$$

$\square$

Without loss of generality, suppose β is defined on the interval $[0, 1]$. Consider the arclength of the reparametrization β out to a certain parameter value s_0,

$$L(s_0) = \int_0^{s_0} |\beta'(s)| \, ds = \int_0^{s_0} 1 \, ds = s_0.$$

This is, of course, exactly what we mean by "arclength parametrization." Hence, a curve is parametrized by arclength exactly when it has unit speed.

Example 2.2: The Helix.
Consider the helix $\alpha(t) = (a \cos t, a \sin t, bt)$ with $\alpha'(t) = (-a \sin t, a \cos t, b)$. We have $|\alpha'(t)| = \sqrt{a^2 + b^2} = c$. Then $s(t) = \int_0^t c \, dt = ct$ and the inverse function is $t(s) = \frac{1}{c}s$. So an arclength reparametrization is given by

$$\beta(s) = \alpha(s/c) = (a \cos \frac{s}{c}, a \sin \frac{s}{c}, \frac{bs}{c}).$$

Note that $|\beta'(s)| = 1$. This example generalizes to the case of any α with constant speed. Such a curve may be parametrized by arclength explicitly. For general curves, however, the integral defining s may be impossible to compute in closed form.

EXERCISE 2.2. Reparametrize the circle of radius r to have unit speed.

EXERCISE 2.3. Can you explicitly reparametrize the ellipse to have unit speed?

EXERCISE 2.4. Let $\alpha(t)$ be a curve not necessarily parametrized by arclength. Suppose a string of length C is wound around the curve α and is attached to the curve at $s(t_C) = C$, where s is the arclength function along α. If the string is kept taut and unwound from the curve, the free end always lies a distance $s(t)$ away from the point $\alpha(t)$ in the direction $-\alpha'(t)$. Thus, show that the curve swept out by the free end is given by

$$\mathcal{I}(t) = \alpha(t) - s(t) \frac{\alpha'(t)}{|\alpha'(t)|}.$$

This curve is an *involute* of α. Calculate and graph the involute of the unit circle $\alpha(t) = (\cos t, \sin t, 0)$. Note that the involute of a unit speed curve $\alpha(s)$ is given by $\mathcal{I}(s) = \alpha(s) - s\,\alpha'(s)$. For an application of circle involutes to gear teeth design see [SU].

EXERCISE 2.5. Show that the involute of a helix is a plane curve.

1.3 FRENET FORMULAS

In this section we shall assume our curves are parametrized by arclength. Consider this problem. Suppose a robot arm with three fingers at its end is to move and grasp an object. How should a path be programmed for the "hand" so that the fingers end up in the proper "grasping" position? Somehow, a curve (which may not be a line, of course, since various obstructions have to be avoided) must be parametrized to align the fingers correctly. We will see that the geometry of a curve (i.e. its twisting and turning) may be completely described by understanding how three "fingers" attached to the curve vary as we move along the curve. The variation is described by what are now called the Frenet Formulas. What are these "fingers" which are attached to the curve?

The first finger is the tangent vector to the curve. Let $\beta\colon I \to \mathbb{R}^3$ be a unit speed curve (i.e. $|\beta'| = 1$). Denote *the unit tangent vector* along β by $T = \beta'$. Because $|T| = 1$ everywhere along β, the derivative T' measures only the rate of change of T's direction. Hence, T' is a good choice to detect some of β's geometry.

Trick to Remember 3.1. $|T| = \sqrt{T \cdot T} = 1$, *so* $T \cdot T = 1$ *as well. The product rule then gives*

$$0 = (T \cdot T)' = T' \cdot T + T \cdot T' = 2T \cdot T'.$$

Hence $T \cdot T' = 0$ *and, therefore,* T' *is perpendicular to* T.

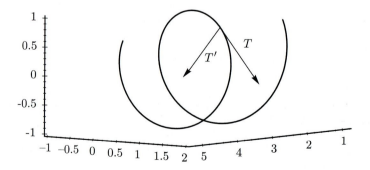

FIGURE 1.14.   Curve with T and T'

We say that T' is *normal* to β. Because there is no reason why the length of T' should be 1, however, we define the *curvature* function of β to be

$$\kappa(s) = |T'(s)|.$$

Of course $\kappa \geq 0$ and κ increases as β turns more sharply. If $\kappa = 0$, then, as we will see in Theorem 3.2 below, we know everything about the curve β already. Therefore, for the discussion below, assume $\kappa > 0$.

EXERCISE 3.1. Suppose β is a curve in the xy-plane. Show that, up to sign, the curvature of β is given by $\frac{d\theta}{ds}$, where θ is the angle between T and $\mathbf{e}_1 = (1, 0)$.

Since we would like our fundamental quantities to be unit vectors, we define the *principal normal vector* along β to be $N = \frac{1}{\kappa}T'$. Note that for this to make sense we require $\kappa > 0$. Then, by the definition of curvature, $|N| = 1$. We need to have a third vector along β as part of our 3-fingered orientation in 3-dimensional space and this third vector should be perpendicular to both T and N (just as T and N are to each other). Before we make the definition, we recall some facts about another vector operation, the *cross product*. Unlike the dot product of two vectors (which produces a real number), the cross product of two vectors produces another vector. Let $\mathbf{v} = (v^1, v^2, v^3)$ and $\mathbf{w} = (w^1, w^2, w^3)$. Then we define the *cross product* of $\mathbf{v}$ and $\mathbf{w}$ to be (in coordinates),

$$\mathbf{v} \times \mathbf{w} = (v^2 w^3 - v^3 w^2, \; v^3 w^1 - v^1 w^3, \; v^1 w^2 - v^2 w^1).$$

An easy way to remember this formula is to compute the "determinant"

$$\mathbf{v} \times \mathbf{w} = \begin{vmatrix} \mathbf{e}_1 & \mathbf{e}_2 & \mathbf{e}_3 \\ v^1 & v^2 & v^3 \\ w^1 & w^2 & w^3 \end{vmatrix}$$

where $\mathbf{e}_1$, $\mathbf{e}_2$, $\mathbf{e}_3$ denote the unit vectors $(1, 0, 0)$, $(0, 1, 0)$ and $(0, 0, 1)$ respectively. Note that the traditional notation of vector calculus denotes these vectors by i, j and k.

EXERCISE 3.2. Compute the determinant and verify that it gives the formula for $\mathbf{v} \times \mathbf{w}$.

EXERCISE 3.3. What happens if the 2nd and 3rd rows of the determinant are interchanged?

EXERCISE 3.4. Verify (by tedious computation) the following formula known as Lagrange's Identity.
$$|\mathbf{v} \times \mathbf{w}|^2 = |\mathbf{v}|^2 |\mathbf{w}|^2 - (\mathbf{v} \cdot \mathbf{w})^2.$$

EXERCISE 3.5. Use the previous exercise to show that

$$|\mathbf{v} \times \mathbf{w}| = |\mathbf{v}|\,|\mathbf{w}|\sin\theta$$

where θ is the angle between $\mathbf{v}$ and $\mathbf{w}$. Note then that $\mathbf{v} \times \mathbf{w} = 0$ if and only if $\mathbf{v}$ and $\mathbf{w}$ are parallel (i.e. $\mathbf{w} = a\mathbf{v}$ for $a \in \mathbb{R}$.

EXERCISE 3.6. Show that $\mathbf{v} \cdot (\mathbf{v} \times \mathbf{w}) = 0 = \mathbf{w} \cdot (\mathbf{v} \times \mathbf{w})$. Therefore, the cross product of two vectors is a vector perpendicular to both of them.

EXERCISE 3.7. Show that $|\mathbf{v} \times \mathbf{w}|$ is the area of the parallelogram spanned by $\mathbf{v}$ and $\mathbf{w}$ (as shown in Figure 1.15).

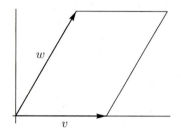

FIGURE 1.15

Now, knowing what we do about the cross product (and always assuming $\kappa > 0$), we may define the *binormal* along β to be $B = T \times N$. Note that $|B| = |T|\,|N|\sin 90° = 1 \cdot 1 \cdot 1 = 1$, so B is already a unit vector. Also, as we have seen above, $B \cdot T = 0 = B \cdot N$. The set $\{T, N, B\}$ is called the *Frenet Frame* along β. The measurement of how T, N, and B vary as we move along β will tell us how β itself twists and turns through space.

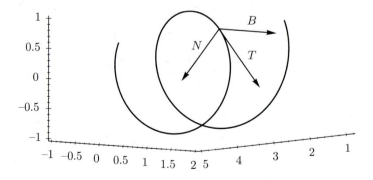

FIGURE 1.16.   Frenet frame along a curve

The variation of T, N, and B will be determined by calculating the derivatives T', N', B'. We already know $T' = \kappa N$ by definition of N, so the curvature κ describes T's variation in direction. We must still find B' and N'. Now, T, N and B are three mutually perpendicular unit (i.e. *orthonormal*) vectors in the 3-dimensional space $\mathbb{R}^3$, so any vector in $\mathbb{R}^3$ is some linear combination of them. In particular, $B' = a\,T + b\,N + c\,B$. If we can identify a, b and c, then we will know B'. To do this we use what we know about T, N and B. First,

$$\begin{aligned}
T \cdot B' &= a\,T \cdot T + b\,T \cdot N + c\,T \cdot B \\
&= a \cdot 1 + b \cdot 0 + c \cdot 0 \\
&= a.
\end{aligned}$$

Similarly, $N \cdot B' = b$ and $B \cdot B' = c$. Therefore,

$$B' = (T \cdot B')\,T + (N \cdot B')\,N + (B \cdot B')\,B.$$

Now let's identify $T \cdot B'$. We know $T \cdot B = 0$, so that $0 = (T \cdot B)' = T' \cdot B + T \cdot B'$ by the product rule. Then, using $N \cdot B = 0$, we obtain

$$\begin{aligned}
T \cdot B' &= -T' \cdot B \\
&= -\kappa N \cdot B \\
&= 0.
\end{aligned}$$

We can also identify $B \cdot B'$. We know $B \cdot B = 1$, so $0 = (B \cdot B)' = B' \cdot B + B \cdot B' = 2B \cdot B'$. Thus, $B \cdot B' = 0$. Now we are left with a single possibly nonzero coefficient in the expression for B'. Since we can't immediately identify it in terms of known quantities, we give it a name. *Define $\tau = -N \cdot B'$ to be the torsion* of the curve β. By what we have said above,

$$B' = -\tau\,N.$$

Now we find N'. Just as for B', we have

$$N' = (T \cdot N')\,T + (N \cdot N')\,N + (B \cdot N')\,B.$$

The same types of calculations give $T \cdot N = 0$, so $0 = T' \cdot N + T \cdot N'$ and $T' = \kappa N$ so $T \cdot N' = -\kappa N \cdot N = -\kappa$. Also, $N \cdot N = 1$, so $N \cdot N' = 0$ and $B \cdot N = 0$, so $B' \cdot N + B \cdot N' = 0$. Hence, $B \cdot N' = -B' \cdot N = -N \cdot B' = \tau$ by definition. Thus,

$$N' = -\kappa\,T + \tau\,B.$$

Theorem 3.1 (The Frenet Formulas). *For a unit speed curve with $\kappa > 0$, the derivatives of the Frenet frame are given by*

$$\begin{aligned}
T' &= \kappa N \\
N' &= -\kappa T + \tau B \\
B' &= -\tau N.
\end{aligned}$$

Example 3.2: The Circle.
Consider a circle $\alpha(t) = (r \cos \frac{t}{r}, r \sin \frac{t}{r}, 0)$. The derivative of α is $\alpha'(t) = 1/r(-r \sin t/r, r \cos t/r, 0) = (-\sin t/r, \cos t/r, 0)$ with $|\alpha'(t)| = 1$, so α has unit speed and $T = \alpha'$. Now $T' = \alpha'' = (-1/r \cos t/r, -1/r \sin t/r, 0) = -1/r(\cos t/r, \sin t/r, 0)$ where $(-\cos t/r, -\sin t/r, 0)$ is a unit vector and thus equal to N. Now, $T' = +1/r \, N$, so $\kappa = 1/r$. That is, for a circle of radius r, the curvature is *constant* and is equal to $1/r$. This makes sense intuitively since, as r increases, the circle becomes less curved. The limit $\kappa = 1/r \to 0$ reflects this. Now, we wish to understand the torsion of a circle.

$$N' = (\frac{1}{r} \sin \frac{t}{r}, \ -\frac{1}{r} \cos \frac{t}{r}, \ 0)$$
$$= -\frac{1}{r} T$$
$$= -\kappa \, T.$$

But the Frenet Formulas say $N' = -\kappa T + \tau B$, so we must have $\tau = 0$. Therefore, the circle has zero torsion. We shall see a general reason for this fact shortly.

EXERCISE 3.8. The Frenet Formulas help us to identify curves even when they are given in rather complicated parametrized forms. The curve below will also appear in Exercise 5.3. Let $\beta(s) = \left(\frac{(1+s)^{3/2}}{3}, \frac{(1-s)^{3/2}}{3}, \frac{s}{\sqrt{2}} \right)$ for $-1 < s < 1$.

(1) Show that β has unit speed.
(2) Show that $\kappa = 1/\sqrt{8(1 - s^2)}$.
(3) Show that $N = (\frac{\sqrt{2(1-s)}}{2}, \frac{\sqrt{2(1+s)}}{2}, 0)$ and
$B = T \times N = (\frac{-\sqrt{1+s}}{2}, \frac{\sqrt{1-s}}{2}, \frac{\sqrt{2}}{2})$.
(4) Show that $\tau = \kappa$.

EXERCISE 3.9. If a rigid body moves along a curve $\alpha(s)$ (which we suppose is unit speed), then the motion of the body consists of translation along α and rotation about α. The rotation is determined by an angular velocity vector ω which satisfies $T' = \omega \times T$, $N' = \omega \times N$ and $B' = \omega \times B$. The vector ω is called the *Darboux vector*. Show that ω, in terms of T, N and B, is given by $\omega = \tau T + \kappa B$. Hint: write $\omega = a \, T + b \, N + c \, B$ and take cross products with T, N and B to determine a, b and c.

EXERCISE 3.10. Show that $T' \times T'' = \kappa^2 \omega$ where ω is the Darboux vector.

EXERCISE 3.11. Instead of taking the Frenet frame along a (unit speed) curve $\alpha: [a, b] \to \mathbb{R}^3$, we can define a frame $\{T, U, V\}$ by taking T to be the tangent vector of α as usual and letting U be any unit vector field along α with $T \cdot U = 0$. That is, $U: [a, b] \to \mathbb{R}^3$ associates a unit vector $U(t)$ to each $t \in [a, b]$ which is perpendicular $T(t)$. Define $V = T \times U$. Show that the natural equations (i.e. "Frenet Formulas") for this frame are

$$T' = \omega_3\, U - \omega_2\, V$$
$$U' = -\omega_3\, T + \omega_1\, V$$
$$V' = \omega_2\, T - \omega_1\, U$$

where ω_1, ω_2 and ω_3 are coefficient functions. Furthermore, show that the Darboux vector ω satisfying $T' = \omega \times T$, $U' = \omega \times U$ and $V' = \omega \times V$ is given by $\omega = \omega_1\, T + \omega_2\, U + \omega_3\, V$.

EXERCISE 3.12 [Sco]. A unit speed *curve of constant precession* is defined by the property that its (Frenet) Darboux vector $\omega = \tau\, T + \kappa\, B$ revolves about a fixed line in space with constant angle and constant speed. Show, by following the steps below, that a curve of constant precession is characterized by having

$$\kappa(s) = a\sin(bs) \qquad \text{and} \qquad \tau(s) = a\cos(bs)$$

where $a > 0$ and b are constants and $c = \sqrt{a^2 + b^2}$. First, show that the following five properties are equivalent for a vector $\mathbf{A} = \omega + b\, N$ and a line ℓ parallel to $\mathbf{A}(0)$:
(i) $|\omega| = a$ (ii) $\cos\theta = a/c$, where θ is the angle between ω and $\mathbf{A}$ (iii) $|N'| = a$ (iv) $\cos(\pi/2 - \theta) = b/c$, where $\pi/2 - \theta$ is the angle between $\mathbf{A}$ and N (v) $|\mathbf{A}| = c$.
Second, given any of the properties just listed, show that $\mathbf{A}$ is always parallel to ℓ (i.e. $\mathbf{A}(0)$) if and only if $\omega' = -b\, N'$. Finally, show that $\mathbf{A}' = 0$ if and only if

$$\tau' = b\kappa \qquad \text{and} \qquad \kappa' = -b\tau.$$

Solve this system of differential equations to get the result. MAPLE's "dsolve" command may be useful. An explicit parametrization for curves of constant precession is given in the MAPLE section at the end of this chapter.

EXERCISE 3.13. Let α and β be two closed curves (i.e. $\alpha(a) = \alpha(b)$ and similarly for β) which do not intersect. The *linking number* of α and β, $Lk(\alpha, \beta)$, is defined by projecting $\mathbb{R}^3$ onto a plane so that at most two points of α and β are mapped to a single image. Then, assigning ± 1 to every α undercrossing according to the orientations shown below and summing the ± 1's gives $Lk(\alpha, \beta)$.

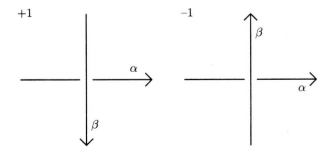

FIGURE 1.17. Linking number orientations

Let $\{T, U, V\}$ be a frame as in Exercise 3.11. The *twist* of U about α is defined to be

$$Tw(\alpha, U) = \frac{1}{2\pi} \int_a^b \omega_1 \, dt.$$

Let $\beta = \alpha + \epsilon U$, where ϵ is small enough so that α and β do not intersect. Show that, if α is a plane curve, then $Lk(\alpha, \beta) = Tw(\alpha, U)$. This is a special case of *White's formula* which has proved important to modern DNA research (see [Poh], [PoR] for instance).

EXERCISE 3.14. For the Frenet frame $\{T, N, B\}$, show that the twist of the principal normal N about α is (up to a multiple of $1/2\pi$) simply the total torsion of α. That is, show

$$Tw(\alpha, N) = \frac{1}{2\pi} \int_a^b \tau \, dt.$$

Example 3.3: The Helix.
Consider the helix $\beta(s) = (a \cos \frac{s}{c}, \ a \sin \frac{s}{c}, \ \frac{bs}{c})$ with $c = (a^2 + b^2)^{1/2}$. Now $\beta'(s) = (-a/c \sin s/c, \ a/c \cos s/c, \ b/c)$ with $|\beta'(s)| = a^2/c^2 + b^2/c^2 = c^2/c^2 = 1$, so β has unit speed. Thus $T = \beta'(s)$ and $T' = \beta''(s) = (-a/c^2 \cos s/c, \ -a/c^2 \sin s/c, \ 0) = a/c^2(-\cos s/c, \ -\sin s/c, \ 0) = a/c^2 N$. Thus

$$\kappa = \frac{a}{c^2} = \frac{a}{(a^2 + b^2)},$$

a constant.

EXERCISE 3.15. Show that $\tau = b/c^2 = b/(a^2 + b^2)$, a constant.

Constraints on curvature and torsion produce concomitant constraints on the geometry of a curve. The simplest constraints are contained in the following

Theorem 3.2. *Let $\beta(s)$ be a unit speed curve. Then,*

(1) $\kappa = 0$ *if and only if β is a line.*
(2) *for $\kappa > 0$, $\tau = 0$ if and only if β is a plane curve.*

Proof.
(1) Suppose β is a line. Then β may be parametrized by $\beta(s) = p + s\mathbf{v}$ with $|\mathbf{v}| = 1$ (so β has unit speed). Then $T = \beta'(s) = \mathbf{v}$ constant, so $T' = 0 = \kappa N$ and, consequently, $\kappa = 0$.

Suppose $\kappa = 0$. Then $T' = 0$ by the Frenet Formulas, so $T = \mathbf{v}$ is a constant (with $|\mathbf{v}| = 1$ since β has unit speed). But $\beta'(s) = T = \mathbf{v}$ implies $\beta(s) = p + s\mathbf{v}$ with p a constant of integration. Hence, β is a line.
(2) Suppose $\tau = 0$. Then, by the Frenet Formulas, $B' = 0$, so B is a constant. But this means $\beta(s)$ should always lie in the plane perpendicular to B. We

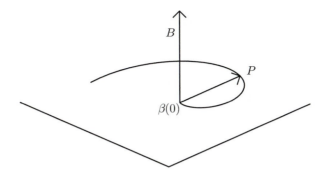

FIGURE 1.18.   Plane through $\beta(0)$ normal to B

show this. Take the plane determined by the point $\beta(0)$ and the normal vector B. Recall that a point P is in this plane if $(P - \beta(0)) \cdot B = 0$.

We will show that, for all s, $(\beta(s) - \beta(0)) \cdot B = 0$. Consider, $((\beta(s) - \beta(0)) \cdot B)' = (\beta(s) - \beta(0))' \cdot B + (\beta(s) - \beta(0)) \cdot B' = \beta'(s) \cdot B = T \cdot B = 0$. Hence, $(\beta(s) - \beta(0)) \cdot B = $ constant. To identify this constant, evaluate the expression at $s = 0$. We get $(\beta(0) - \beta(0)) \cdot B = 0$. Then for all s, $(\beta(s) - \beta(0)) \cdot B = 0$ and, hence, $\beta(s)$ is in the plane determined by $\beta(0)$ and the constant vector B.

Suppose β lies in a plane. Then (as we noted above) the plane is determined by a point P and a normal vector $n \neq 0$. Since β lies in the plane,

$$(\beta(s) - P) \cdot n = 0 \qquad \text{for all } s.$$

By differentiating, we obtain two equations: $\beta'(s) \cdot n = 0$ and $\beta''(s) \cdot n = 0$. That is, $T \cdot n = 0$ and $\kappa N \cdot n = 0$. These equations say that n is perpendicular to both T and N. Thus n is a multiple of B and

$$\pm\, n/|n| = B.$$

Hence B is constant and $B' = 0$. The Frenet Formulas then give $\tau = 0$. $\square$

So now we see that curvature measures the deviation of a curve from being a line and torsion the deviation of a curve from being contained in a plane. We know that the standard circle of radius r in the xy plane has $\tau = 0$ and $\kappa = 1/r$. To see that a circle located anywhere in $\mathbb{R}^3$ has these properties we have two choices. We could give a parametrization for an arbitrary circle in $\mathbb{R}^3$ or we could use the direct definition of a circle as the collection of points *in a plane* which are a fixed distance from a given one. In order to emphasize geometry over analysis (for once), we take the latter approach.

Theorem 3.3. *A curve $\beta(s)$ is part of a circle if and only if $\kappa > 0$ is constant and $\tau = 0$.*

Proof. Suppose β is part of a circle. By definition, β is a plane curve, so $\tau = 0$. Also by definition, for all s, $|\beta(s) - p| = r$. Squaring both sides gives

$(\beta(s) - p) \cdot (\beta(s) - p) = r^2$. If we differentiate this expression, we get (for $T = \beta'$)

$$2T \cdot (\beta(s) - p) = 0 \quad \text{or} \quad T \cdot (\beta(s) - p) = 0.$$

If we differentiate again, then we obtain

$$T' \cdot (\beta(s) - p) + T \cdot T = 0$$

(∗)
$$\kappa\, N \cdot (\beta(s) - p) + 1 = 0$$

$$\kappa\, N \cdot (\beta(s) - p) = -1.$$

This means, in particular, that $\kappa > 0$ and $N \cdot (\beta(s) - p) \neq 0$. Now differentiating (∗) produces

$$\frac{d\kappa}{ds} N \cdot (\beta(s) - p) + \kappa\, N' \cdot (\beta(s) - p) + \kappa\, N \cdot T = 0$$

$$\frac{d\kappa}{ds} N \cdot (\beta(s) - p) + \kappa(-\kappa T + \tau B) \cdot (\beta(s) - p) + 0 = 0.$$

Since $\tau = 0$ and $T \cdot (\beta(s) - p) = 0$ by above, we have

$$\frac{d\kappa}{ds} N \cdot (\beta(s) - p) = 0.$$

Also, $N \cdot (\beta(s) - p) \neq 0$ by above, so $\frac{d\kappa}{ds} = 0$. This means, of course, that $\kappa > 0$ is constant.

Suppose now that $\tau = 0$ and $\kappa > 0$ is constant. To show $\beta(s)$ is part of a circle we must show that each $\beta(s)$ is a fixed distance from a fixed point. For the standard circle, from any point on the circle to the center we proceed in the normal direction a distance equal to the radius. That is, we go $rN = 1/\kappa\, N$. We do the same here. Let γ denote the curve

$$\gamma(s) = \beta(s) + \frac{1}{\kappa} N.$$

Since we want γ to be a single point, the center of the desired circle, we must have $\gamma'(s) = 0$. Computing, we obtain

$$\gamma'(s) = \beta'(s) + \frac{1}{\kappa} N' \qquad \text{since } \kappa \text{ is a constant}$$

$$= T + \frac{1}{\kappa}(-\kappa T + \tau B)$$

$$= T - T \qquad \text{since } \tau = 0$$

$$= 0.$$

Hence $\gamma(s)$ is a constant p. Then we have

$$|\beta(s) - p| = \left| -\frac{1}{\kappa} N \right| = \frac{1}{\kappa},$$

so p is the center of a circle $\beta(s)$ of radius $1/\kappa$. □

EXERCISE 3.16. Show that, if every tangent line of a curve β passes through a fixed point p, then β is a line. Hint: $p = \beta(s) + r(s)\beta'(s)$ for some function $r(s)$.

EXERCISE 3.17. Let α be a unit-speed curve which lies on a sphere of center p and radius R. Show that, if $\tau \neq 0$, then $\alpha(s) - p = -\frac{1}{\kappa}N - (\frac{1}{\kappa})'\frac{1}{\tau}B$ and $R^2 = (\frac{1}{\kappa})^2 + ((\frac{1}{\kappa})'\frac{1}{\tau})^2$.

EXERCISE 3.18. Show that, if $(\frac{1}{\kappa})' \neq 0$ and $(\frac{1}{\kappa})^2 + ((\frac{1}{\kappa})'\frac{1}{\tau})^2$ is a constant, then a (unit speed) curve α lies on a sphere.

EXERCISE 3.19. Compute the curvature κ and torsion τ of the curve

$$\alpha(t) = (a\,\cos^2(t),\, a\,\sin(t)\cos(t),\, a\,\sin(t)).$$

Then, for $a = 1$, graph $\alpha(t)$ and the sphere of radius one centered at the origin on the same set of axes. Also, show that the curve $\alpha(t)$ is a spherical curve by computing

$$\left(\frac{1}{\kappa}\right)^2 + \left(\frac{1}{\tau}\left(\frac{1}{\kappa}\right)'\frac{1}{|\alpha'(t)|}\right)^2$$

to be a constant (which must be the squared radius of the sphere on which α lies). Note that the extra $|\alpha'(t)| = ds/dt$ is necessary since α is not unit speed and the derivative of $1/\kappa$ here is with respect to t whereas the derivative of $1/\kappa$ above was with respect to arclength s. Hint: use MAPLE.

EXERCISE 3.20. Find a parametrization for the curve obtained by intersecting the following sphere and cylinder:

$$x^2 + y^2 + z^2 = 4a^2 \qquad\qquad (x-a)^2 + y^2 = a^2.$$

Hint: $x = a\cos t + a$, $y = a\sin t$ work in the cylinder. Plug these x and y into the sphere's defining equation and solve for z. This curve is known as *Viviani's curve*. Compute the curvature and torsion of Viviani's curve *and* (since we know the curve is spherical!) verify that the formula

$$\left(\frac{1}{\kappa}\right)^2 + \left(\frac{1}{\tau}\left(\frac{1}{\kappa}\right)'\frac{1}{|\alpha'(t)|}\right)^2 = 4a^2$$

holds. Finally, graph Viviani's curve *on the sphere*. Compare Viviani's curve with the curve of Exercise 3.19 and make a conjecture. Can you prove your conjecture?

EXERCISE 3.21. Find κ and τ for the curve $\beta(s) = (\frac{1}{\sqrt{2}}\cos(s),\ \sin(s),\ \frac{1}{\sqrt{2}}\cos(s))$. Identify the curve.

EXERCISE 3.22. Let α be a unit-speed curve. Show that there is a unique circle β with

$$\beta(0) = \alpha(0) \qquad \beta'(0) = \alpha'(0) \qquad \beta''(0) = \alpha''(0).$$

Hint. Start with $\beta(s) = p + R\cos(\frac{s}{R})\mathbf{v}_1 + R\sin(\frac{s}{R})\mathbf{v}_2$ where $\mathbf{v}_1$ and $\mathbf{v}_2$ are orthonormal vectors. Your proof will show that β lies in the plane spanned by T and N. This is the *osculating plane* of α and β is called the *osculating circle* of α. Your proof will also show that the radius of β is exactly $1/\kappa_\alpha(0)$. In this sense β is the circle of closest fit to α at $\alpha(0)$. Here are two things to think about. Does the osculating circle always lie on one side of the curve? How could you use the idea of the osculating circle and a computer to calculate curvature at points of a curve?

1.4 NONUNIT SPEED CURVES

Although all regular curves have theoretical unit speed parametrizations, some (such as the ellipse) do not admit explicit ones. Therefore, in order to understand the geometry of these curves we must obtain modifications of the Frenet Formulas which account for nonconstant speed.

Let $\alpha(t)$ be an arbitrary curve with speed $\nu = |\alpha'(t)| = ds/dt$. Theoretically we may reparametrize α to get a unit speed curve $\overline{\alpha}(s(t))$, so, we *define* curvature and torsion of α in terms of its arclength reparametrization $\overline{\alpha}(s(t))$. Moreover, the tangent vector $\alpha'(t)$ is in the direction of the unit tangent $T(s)$ of the reparametrization, so $T(s(t)) = \alpha'(t)/|\alpha'(t)|$. This says that we should *define* a nonunit speed curve's invariants in terms of its unit speed reparametrization's invariants.

Definition 4.1.

(1) The unit tangent of $\alpha(t)$ is defined to be $T(t) \overset{\text{def}}{=} \overline{T}(s(t))$.

(2) The curvature of $\alpha(t)$ is defined to be $\kappa(t) \overset{\text{def}}{=} \overline{\kappa}(s(t))$.

(3) If $\kappa > 0$, then the principal normal of $\alpha(t)$ is defined to be $N(t) \overset{\text{def}}{=} \overline{N}(s(t))$.

(4) If $\kappa > 0$, then the binormal of $\alpha(t)$ is defined to be $B(t) \overset{\text{def}}{=} \overline{B}(s(t))$.

(5) If $\kappa > 0$, then the torsion of $\alpha(t)$ is defined to $\tau(t) \overset{\text{def}}{=} \overline{\tau}(s(t))$.

Theorem 4.1 (The Frenet Formulas for Nonunit Speed Curves).
For a regular curve α with speed $\nu = ds/dt$ and curvature $\kappa > 0$,

$$
\begin{aligned}
T' &= \kappa\nu\, N \\
N' &= -\kappa\nu\, T + \tau\nu\, B \\
B' &= -\tau\nu\, N
\end{aligned}
$$

Proof. The unit tangent $T(t)$ is $\overline{T}(s)$ by definition. Now $T'(t)$ denotes differentiation with respect to t, so we must use the chain rule on the righthand side to determine κ and τ.

$$T'(t) = \frac{d\overline{T}(s)}{ds}\frac{ds}{dt}$$
$$= \overline{\kappa}(s)\,\overline{N}(s)\nu$$
$$= \kappa(t)\,N(t)\nu(t) \qquad \text{by definition}$$
$$T' = \kappa\nu\,N,$$

so the first formula is proved. For the second and third,

$$N'(t) = \frac{d\overline{N}(s)}{ds}\frac{ds}{dt}$$
$$= (-\overline{\kappa}\,\overline{T} + \overline{\tau}\,\overline{B})\nu$$

by the unit speed Frenet formulas,

$$= -\kappa\nu\,T + \tau\nu\,B.$$

and

$$B'(t) = \frac{d\overline{B}(s)}{ds}\frac{ds}{dt}$$
$$= -\overline{\tau}\,\overline{N}\nu$$
$$= -\tau\nu\,N.$$

$\square$

Lemma 4.2. $\alpha' = \nu T$ and $\alpha'' = \frac{d\nu}{dt}T + \kappa\nu^2\,N$.

Proof. Since $\alpha(t) = \overline{\alpha}(s(t))$, the first calculation is,

$$\alpha'(t) = \overline{\alpha}'(s)\frac{ds}{dt}$$
$$= \nu\,\overline{T}(s)$$
$$= \nu(t)\,T(t)$$

while the second is

$$\alpha''(t) = \frac{d\nu}{dt}\,T(t) + \nu(t)\,T'(t)$$
$$= \frac{d\nu}{dt}\,T(t) + \nu(t)\kappa(t)\nu(t)\,N(t)$$
$$= \frac{d\nu}{dt}\,T(t) + \kappa(t)\nu^2(t)\,N(t).$$

EXERCISE 4.1. Work out the following precisely. The formula for α'' allows an analysis of the following physical situation. Suppose a car is going around a level highway curve at constant speed (only for convenience). Then the centripetal acceleration of the car is $\alpha'' = \kappa \nu^2 N$. The only force acting on the car is the frictional force required to keep the car on the road. As long as the tires roll and do not skid, this force is given by a coefficient of friction μ (typically $\approx .60$) multiplied by the weight (mass time gravitational constant) mg of the car. By Newton's Law, to keep the car on the road we require

$$\mu m g \geq m \kappa \nu^2 \qquad \text{or} \qquad \mu g \geq \kappa \nu^2.$$

For a road of given curvature κ, to stay on the road, speed must satisfy

$$\nu \leq \sqrt{\frac{\mu g}{\kappa}}.$$

What changes if the road is banked?

From the definitions above and what we know about unit speed curves, for a nonunit speed curve α, we have $T = \alpha'/|\alpha'|$, $B = T \times N$ and $N = B \times T$. We also have the

Theorem 4.3. *For any regular curve α, the following formulas hold.*

(1) $B = \dfrac{\alpha' \times \alpha''}{|\alpha' \times \alpha''|}$,

(2) $\kappa = \dfrac{|\alpha' \times \alpha''|}{|\alpha'|^3}$,

(3) $\tau = \dfrac{(\alpha' \times \alpha'') \cdot \alpha'''}{|\alpha' \times \alpha''|^2}$.

Proof. For (1), we use the formulas of the Lemma to get

$$\alpha' \times \alpha'' = (\nu T) \times (\frac{d\nu}{dt} T + \kappa \nu^2 N)$$

$$= \nu \frac{d\nu}{dt} T \times T + \kappa \nu^3 T \times N$$

$$= 0 + \kappa \nu^3 B$$

Hence, $|\alpha' \times \alpha''| = \kappa \nu^3$, so $B = \dfrac{\alpha' \times \alpha''}{|\alpha' \times \alpha''|}$.

For (2), we use the expression for α'' in the Lemma, take cross product with $T = \alpha'/|\alpha'|$ and note that $T \times T = 0$ to isolate the curvature,

$$T \times \alpha'' = 0 + \kappa \nu^2 T \times N$$

$$\frac{\alpha' \times \alpha''}{|\alpha'|} = \kappa \nu^2 T \times N,$$

so taking lengths and using $|B| = 1$ produces

$$\frac{|\alpha' \times \alpha''|}{\nu} = \kappa \nu^2 |B|$$

$$\frac{|\alpha' \times \alpha''|}{\nu^3} = \kappa.$$

For (3), we take the third derivative

$$\alpha''' = \left(\frac{d\nu}{dt} T + \kappa \nu^2 N\right)'$$

$$= \frac{d^2\nu}{dt^2} T + \frac{d\nu}{dt} T' + \frac{d\kappa}{dt} \nu^2 N + \kappa 2\nu \frac{d\nu}{dt} N + \kappa \nu^2 N'.$$

Therefore, since $T' = \kappa N$ and B is perpendicular to T and N, we get

$$B \cdot \alpha''' = \kappa \nu^2 B \cdot N'$$

$$= \kappa \nu^3 B \cdot (-\kappa \nu T + \tau \nu B)$$

$$= \kappa \tau \nu^3$$

since $B \cdot T = 0$ and $B \cdot B = 1$. Now $\alpha' \times \alpha'' = \kappa \nu^3 B$, so

$$(\alpha' \times \alpha'') \cdot \alpha''' = \kappa \nu^3 B \cdot \alpha'''$$

$$= \kappa \nu^3 (\kappa \tau \nu^3)$$

$$= \kappa^2 \nu^6 \tau.$$

Of course, $|\alpha' \times \alpha''|^2 = \kappa^2 \nu^6$, so we have

$$\tau = \frac{(\alpha' \times \alpha'') \cdot \alpha'''}{\kappa^2 \nu^6} = \frac{(\alpha' \times \alpha'') \cdot \alpha'''}{|\alpha' \times \alpha''|^2}.$$

$\square$

EXERCISE 4.2. For the ellipse $\alpha(t) = (a \cos t, \ b \sin t)$, show that $\kappa = ab/(a^2 \sin^2 t + b^2 \cos^2 t)^{3/2}$. Note the case of the circle $a = b$. Show that a formula for the curvature of any plane curve $\alpha(t) = (x(t), y(t))$ is given by

$$\kappa_\alpha = \frac{|x'y'' - x''y'|}{(x'^2 + y'^2)^{3/2}}.$$

EXERCISE 4.3. Compute T, N, B and the curvature and torsion for the curve $\beta(t) = (e^t \cos t, e^t \sin t, e^t)$.

The curve $\alpha(t) = (\cosh t, \sinh t, t)$ is known as a *hyperbolic helix*. Recall that the hyperbolic trigonometric functions are defined by the formulas,

$$\cosh t = \frac{e^t + e^{-t}}{2}, \quad \sinh t = \frac{e^t - e^{-t}}{2}, \quad \tanh t = \frac{e^t - e^{-t}}{e^t + e^{-t}}.$$

We have the fundamental identity $\cosh^2 t - \sinh^2 t = 1$ as well as the derivative formulas

$$\frac{d}{dt}(\cosh t) = \sinh t, \quad \frac{d}{dt}(\sinh t) = \cosh t.$$

EXERCISE 4.4. For the hyperbolic helix, show that

(1) $T = \frac{1}{\sqrt{2}}(\tanh t, \ 1, \ \operatorname{sech} t)$ where $\operatorname{sech} t = \frac{1}{\cosh t}$,

(2) $B = \frac{1}{\sqrt{2}}(-\tanh t, \ 1, \ -\operatorname{sech} t)$,

(3) $N = (\operatorname{sech} t, \ 0, \ -\tanh t)$,

(4) $\kappa = 1/2 \cosh^2 t$,

(5) $\tau = 1/2 \cosh^2 t$.

EXERCISE 4.5. Compute $s(t) = \int_0^t |\alpha'| dt$ for the hyperbolic helix α.

EXERCISE 4.6. Let $\alpha(t)$ be a not necessarily unit speed curve with involute $\mathcal{I}(t)$. Show that the curvature of the involute $\mathcal{I}(t)$ is given by

$$\kappa_{\mathcal{I}}(t) = \frac{\sqrt{\kappa_\alpha(t)^2 + \tau_\alpha(t)^2}}{s(t)\, \kappa_\alpha(t)}$$

where $s(t)$ is α's arclength function. What does the formula become if α is a plane curve? Compute the curvature of the involute of the unit circle directly and from the formula above.

Example 4.2: Plane Evolutes.
The *plane evolute* of a plane curve $\alpha(t)$ is the plane curve whose involute is α. A formula for the evolute is given by

$$\mathcal{E}(t) = \alpha(t) + \frac{1}{\kappa(t)} N(t)$$

where N is the normal of α. Note that the evolute is the curve consisting of the centers of the osculating circles associated to α. To see that the involute of $\mathcal{E}$ is α, first compute

$$\mathcal{E}' = \alpha' + \left(\frac{1}{\kappa}\right)' N + \frac{1}{\kappa} N'$$

$$= \alpha' + \left(\frac{1}{\kappa}\right)' N + \frac{1}{\kappa}(-\kappa\nu T + \tau\nu B)$$

$$= \alpha' + \left(\frac{1}{\kappa}\right)' N - \nu T \qquad \text{since } \tau = 0$$

$$= \left(\frac{1}{\kappa}\right)' N \qquad \text{since } \alpha' = \nu T.$$

The length of $\mathcal{E}'$ is then $|\mathcal{E}'| = |\left(\frac{1}{\kappa}\right)'|$. Note then that the unit tangent vector $T_{\mathcal{E}}$ of the evolute is the normal N of α. The involute of the evolute is then given by

$$\mathcal{I}_{\mathcal{E}}(t) = \mathcal{E}(t) - s(t)\, T_{\mathcal{E}}(t)$$

$$= \mathcal{E}(t) - s(t)\, N(t)$$

$$= \alpha(t) + \frac{1}{\kappa} N - \frac{1}{\kappa} N$$

$$= \alpha(t).$$

Here we have used the fact that the arclength function for the evolute is $s(t) = \frac{1}{\kappa(t)}$. This follows since $s(t) = \int_0^t |\mathcal{E}'|\, du = \int_0^t |\left(\frac{1}{\kappa}\right)'|\, du = \frac{1}{\kappa}$. We note here that evolutes may be defined for 3-space curves also. The formula for a space evolute is $\mathcal{E} = \alpha + \frac{1}{\kappa} N + \frac{c}{\kappa} B$ where c is a constant.

Example 4.3: The Plane Evolute of a Parabola.

Suppose a parabola $y^2 = 2ax$ is given with parametrization $\alpha(t) = (t^2/2a, t)$. Then $\alpha'(t) = (t/a, 1)$, $\nu = |\alpha'| = \sqrt{a^2 + t^2}/a$ and, therefore, $T = (t/\sqrt{a^2 + t^2}, a/\sqrt{a^2 + t^2})$. From this we may calculate curvature,

$$T' = \left(\frac{a^2}{(a^2 + t^2)^{3/2}}, \frac{-at}{(a^2 + t^2)^{3/2}}\right)$$

$$= \kappa\nu N$$

$$= \kappa \frac{\sqrt{a^2 + t^2}}{a} N$$

and taking lengths gives

$$|T'| = \frac{a}{a^2 + t^2}$$

$$= \kappa \frac{\sqrt{a^2 + t^2}}{a}.$$

Hence, $\kappa = a^2/(a^2 + t^2)^{3/2}$ and, putting this in the expression for T', $N = (a/\sqrt{a^2 + t^2},\ -t/\sqrt{a^2 + t^2})$. Finally, the evolute is then given by

$$\mathcal{E}(t) = \alpha(t) + \frac{1}{\kappa(t)}\, N(t)$$

$$= \left(\frac{t^2}{a}, t\right) + \frac{(a^2 + t^2)^{3/2}}{a^2}\left(\frac{a}{\sqrt{a^2 + t^2}},\ \frac{-t}{\sqrt{a^2 + t^2}}\right)$$

$$= \left(\frac{2a^2 + 3t^2}{2a},\ \frac{t^3}{a^2}\right).$$

EXERCISE 4.7. Show that the x and y coordinates of $\mathcal{E}(t)$ satisfy the relation

$$27\,a\,y^2 = 8\,(x - a)^3.$$

This curve is known as *Neil's parabola* and we will meet it again when we discuss minimal surfaces.

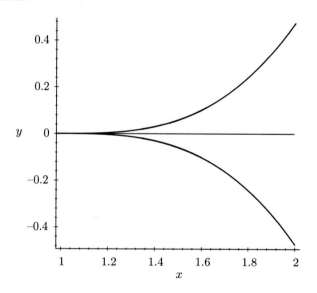

FIGURE 1.19.   Neil's parabola

EXERCISE 4.8. Show that the evolute of the ellipse $\alpha(t) = (a\,\cos t,\ b\,\sin t)$ is the astroid

$$\left(\frac{a^2 - b^2}{a}\,\cos^3 t,\ \frac{b^2 - a^2}{b}\,\sin^3 t,\right).$$

EXERCISE 4.9. Show that an evolute of the catenary $\alpha(t) = (t,\ \cosh t)$ is given by $(t - \cosh t \sinh t,\ 2\cosh t)$.

EXERCISE 4.10. Show that an evolute of the astroid $\alpha(t) = (a \cos^3 t, a \sin^3 t)$ is an expanded, rotated out-of-phase astroid. Hints: After computing the evolute, take an astroid $(2a \cos^3 t, 2a \sin^3 t)$, rotate it by an amount guessed from the graph and let $t = \bar{t} - \pi/4$.

1.5 SOME IMPLICATIONS OF CURVATURE AND TORSION

We have already seen some of the ways in which the Frenet Formulas detect geometric information about the curve. As a more complicated example we will now consider a class of curves called cylindrical helices. In order to understand the definition of these curves, let us consider the standard helix, $\alpha(s) = (a \cos \frac{s}{c}, \ a \sin \frac{s}{c}, \ \frac{bs}{c})$ where $c = (a^2 + b^2)^{1/2}$. Consider $T = \alpha'(s) = (-\frac{a}{c} \sin \frac{s}{c}, \ \frac{a}{c} \cos \frac{s}{c}, \ \frac{b}{c})$. Notice that the third coordinate is constant, so $T \cdot \mathbf{e_3} = \frac{b}{c}$ (where $\mathbf{e_3} = (0, 0, 1)$ as usual). Hence, $\cos \theta = \frac{b}{c}$, so the angle θ between T and $\mathbf{e_3}$ never changes.

We may generalize this property by saying that a curve α, which we assume has unit speed, is a *cylindrical helix* if there is some constant unit vector $\mathbf{u}$ such that $T \cdot \mathbf{u} = \cos \theta$ is constant along the curve. The direction of the vector $\mathbf{u}$ is the *axis* of the helix. We can identify cylindrical helices by a simple condition on torsion and curvature.

Theorem 5.1. *Suppose α has $\kappa > 0$ Then α is a cylindrical helix if and only if τ/κ is constant.*

Proof. Without loss of generality, assume α has unit speed. Suppose $T \cdot \mathbf{u} = \cos \theta$ is constant for some $\mathbf{u}$. Then $0 = (T \cdot \mathbf{u})' = T' \cdot \mathbf{u} = \kappa N \cdot \mathbf{u}$, so $N \cdot \mathbf{u} = 0$ since $\kappa > 0$. Now, $\mathbf{u}$ is perpendicular to N, so

$$\mathbf{u} = (\mathbf{u} \cdot T) T + (\mathbf{u} \cdot B) B$$
$$= \cos \theta \, T + \sin \theta \, B.$$

The diagram in Figure 1.20 explains the coefficient of B.

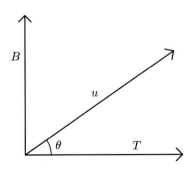

FIGURE 1.20

Because $\mathbf{u}$ is constant, $\mathbf{u}' = 0$. Hence,

$$0 = \cos\theta\, T' + \sin\theta\, B'$$
$$= \kappa\cos\theta\, N - \tau\sin\theta\, N$$
$$= (\kappa\cos\theta - \tau\sin\theta)\, N.$$

Thus, $\kappa\cos\theta - \tau\sin\theta = 0$ which gives $\cot\theta = \tau/\kappa$ constant since θ is.

Now suppose τ/κ is constant. The cotangent function takes on all values, so we can choose θ with $\cot\theta = \tau/\kappa$. Define $\mathbf{u} = \cos\theta\, T + \sin\theta\, B$ to get

$$\mathbf{u}' = (\kappa\cos\theta - \tau\sin\theta)\, N = 0.$$

Hence, $\mathbf{u}$ is a constant vector and clearly $T \cdot \mathbf{u} = \cos\theta$ is constant. Thus α is a cylindrical helix. $\qquad\qquad\square$

In order to get a good geometric picture of a cylindrical helix we might want to project the helix back onto the plane determined by a starting point $(\beta(0))$ and the fixed (normal) vector $\mathbf{u}$. The picture is given in Figure 1.21.

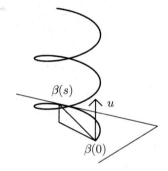

FIGURE 1.21

To do this, we subtract the $\mathbf{u}$-component of $\beta(s) - \beta(0)$ from $\beta(s)$. Now, this $\mathbf{u}$-component has length

$$h(s) = (\beta(s) - \beta(0)) \cdot \mathbf{u}$$

and to see what this is in terms of the definition of β as a cylindrical helix, we differentiate,

$$\frac{dh}{ds} = \beta' \cdot \mathbf{u} = T \cdot \mathbf{u} = \cos\theta,$$

a constant. Hence, $h(s) = s\cos\theta + C$. Now $h(0) = 0 = 0 + C$, so $C = 0$ also. Finally, $h(s) = s\cos\theta$. Therefore the projection of the cylindrical helix $\beta(s)$ onto the plane through $\beta(0)$ and perpendicular to $\mathbf{u}$ is given by:

$$\gamma(s) = \beta(s) - h(s)\, \mathbf{u},$$

where $h(s) = s\cos\theta$. Note that this parametrization of γ is *not* necessarily a unit speed parametrization.

EXERCISE 5.1. Show that $\kappa_\gamma = \kappa_\beta / \sin^2 \theta$, where κ_β is the curvature of β.

EXERCISE 5.2. A cylindrical helix β is said to be a *circular helix* if the projection γ is a circle. Show that β is a circular helix if and only if τ and κ are constant.

EXERCISE 5.3. What can you now say about the curve considered in Exercise 3.8,

$$\beta(s) = \left(\frac{(1+s)^{3/2}}{3}, \frac{(1-s)^{3/2}}{3}, \frac{s}{\sqrt{2}} \right).$$

EXERCISE 5.4. What do you know about the curve $\beta(t) = (t + \sqrt{3}\sin(t), 2\cos(t), \sqrt{3}\,t - \sin(t))$?

EXERCISE 5.5. Show that $\beta(t) = (at,\ bt^2,\ t^3)$ is a cylindrical helix if and only if $4b^4 = 9a^2$. In this case, find the vector $\mathbf{u}$ such that $T = \beta'/|\beta'|$ has $T \cdot \mathbf{u} = $ constant.

EXERCISE 5.6. Let $\beta(s)$ be a curve and suppose that, for every s, $N(s)$ is perpendicular to a fixed vector $\mathbf{u}$. Show that β is a cylindrical helix. Conversely, show that, for any helix, the normal N is perpendicular to the axis $\mathbf{u}$ of the helix.

EXERCISE 5.7. Show that, for any helix, the Darboux vector ω is parallel to the axis $\mathbf{u}$ of the helix.

EXERCISE 5.8. Define a curve $\beta(s)$ by $\beta(s) = \int_0^s B_\alpha(t)\, dt$, where $\alpha(t)$ is a unit speed curve. Show that $\kappa_\beta = \tau_\alpha$, $\tau_\beta = \kappa_\alpha$, $T_\beta = B_\alpha$, $N_\beta = -N_\alpha$ and $B_\beta = T_\alpha$. Now suppose α is a circular helix and show that β is as well.

EXERCISE 5.9. Show that the space evolute $\mathcal{E}(s)$ of a plane curve $\beta(s)$ is a cylindrical helix. Hint: Take $\mathcal{E}(s)$ to be unit speed and note, by definition, that the involute of $\mathcal{E}$ is β. Compute torsion of β and set it to zero since β is a plane curve. This gives conditions on derivatives of $\mathcal{E}$. Use the Frenet formulas applied to $\mathcal{E}$.

We have seen that curvature and torsion, individually and in combination, tell us a great deal about the geometry of a curve. In fact, in a very real sense, they tell us everything. Precisely, if two unit-speed curves have the same curvature and torsion functions, then there is a rigid motion of $\mathbb{R}^3$ taking one curve onto the other. Furthermore, given specified curvature and torsion functions, there is a curve which realizes them as its own curvature and torsion. These results are, essentially, theorems about existence and uniqueness of solutions of systems of differential equations. The Frenet Formulas provide the system and the unique solution provides the curve. We shall not prove these statements, but instead indicate portions of the proofs by the following exercises.

EXERCISE 5.10. Suppose we want a plane curve $\beta(s)$ with curvature a specified curvature function $\kappa(s)$. Show that such a β may be defined by

$$\beta(s) = \left(\int_0^s \cos\theta(u)\, du, \ \int_0^s \sin\theta(u)\, du \right)$$

where $\theta(u) = \int_0^u \kappa(t)\, dt$. Hints: use the Fundamental Theorem of Calculus to show β has unit speed and then compute T_β'.

EXERCISE 5.11. Find a unit speed plane curve $\alpha(s)$ with $\kappa_\alpha(s) = 2s$ and $\alpha(0) = (0,0)$. The resulting curve is called *Euler's spiral* or *the spiral of Cornu*. We will see in Chapter 8 that this curve arises as the solution to a variational problem.

EXERCISE 5.12. Find a unit speed plane curve $\alpha(s)$ with $\kappa_\alpha(s) = 1/(1 + s^2)$ and $\alpha(0) = (0,0)$.

EXERCISE 5.13. Find a unit speed plane curve $\alpha(s)$ with $\kappa_\alpha(s) = 1/s$ and $\alpha(1) = (3,-2)$.

EXERCISE 5.14. Show that, if two unit speed curves $\alpha(s)$ and $\beta(s)$ have $\kappa_\alpha(s) = \kappa_\beta(s)$ and $\tau_\alpha(s) = \tau_\beta(s)$ for all s, then there is a rigid motion (i.e. combination of translations, rotations and reflections) of 3-space which takes α onto β. Hints:

(1) Move $\beta(0)$ to $\alpha(0)$ by a translation.
(2) Rotate and, if necessary, reflect the Frenet frame of β at $\beta(0) = \alpha(0)$ into the Frenet frame of α.
(3) To show that the Frenet frames now coincide for all s, consider the function

$$\mathcal{D}(s) = T_\alpha(s) \cdot T_\beta(s) + N_\alpha(s) \cdot N_\beta(s) + B_\alpha(s) \cdot B_\beta(s)$$

which attempts to measure the deviation of the frames from being identical. Show that $\mathcal{D}(0) = 3$, $\mathcal{D}(s) \leq 3$ for all s and $\mathcal{D}(s_0) = 3$ only when the frames coincide at s_0.
(4) Show that $\mathcal{D}$ is constant by calculating $\mathcal{D}' = 0$. Use the Frenet equations. Hence, $\mathcal{D}(s) = 3$ for all s.
(5) Then, $\alpha'(s) = \beta'(s)$. Integrate and use initial conditions to show $\alpha(s) = \beta(s)$, hence showing that the original translation, rotation and reflection carried β into α.

1.6 THE GEOMETRY OF CURVES AND MAPLE

Computer algebra systems provide powerful tools for computation and visualization in differential geometry. Applications range from simple computations such as numerical evaluation of an arclength integral to more complicated graphics such as drawing geodesics on surfaces. No matter the application, the interaction of human geometer with computer is a very personal thing. Therefore, in this section and all similar ones dealing with computers, I shall give (in the first person) my personal version of *computers and geometry*. I make no claims that the ways I have of doing things are optimal. In fact, I'm certain that many readers will have much better ways of approaching various problems. Nevertheless, I believe it is worthwhile nowadays to expose any prospective geometer to the world of computer algebra systems. In this text, I have chosen to use MAPLEV3 for WINDOWS (hereafter known as MAPLE) because this is the system I was exposed to first (in the post DOS age) and this is the system that I believe is the simplest for students to learn. MAPLE behaves similarly in its MAC (or Power PC) version although there are certain annoying features such as "Scratchpad" which must be dealt with. There are now many introductions to MAPLE on the market. For a straightforward list of commands and package descriptions, however, take a look at [Red]. For a MATHEMATICA approach to geometry, the reader is referred to [Gr]. Let's begin.

Much of what we do in geometry deals with vectors, of course, and the easiest way to enter them in MAPLE is to write the following input.

> A := [a,b,c];

This gives the following output.

$$A: = [a, b, c]$$

Components of the vector A may be isolated as follows:

>A[1];

$$a$$

Notice the semicolon at the end of an input. Without the semicolon, MAPLE does not know that the input has ended. Also, as we shall see below, there are times when a colon is preferred over the semicolon. A colon still causes MAPLE to execute the input, but suppresses the output. To try a simple example, input both of the following.

>with(plots);
>with(plots):

The command with(plots) loads a special plotting package into a MAPLE worksheet (which is just the name of the WINDOWS screen where you do and save your work). This plotting package will be necessary for us later.

Of course, we want computer algebra systems to carry out tasks which are tiresome to do by hand — such as the calculation of curvature κ and torsion τ. In order to see how this may be accomplished, let's first look at some simpler calculations. The first set of *procedures* below are for dot product, the length (or norm) of a vector and for cross product. MAPLE has built-in procedures for these operations (e.g. dotprod), but in the old DOS days, there were some problems with using them in the geometric procedures below. Also, the very

simplicity of the procedures themselves is a painless way to learn the art of
MAPLE procedure writing. To learn more about the built-in dot product, for
example, use MAPLE's most important command — ?dotprod. The ? is a
signal for MAPLE to give you help, in this case on the dot product. Try ?plot.
Now, type in the following. (To go from one line to the next in WINDOWS,
use Shift-Enter or Shift-Return, where the dash means to hold down Shift
while you press Enter. On the Power PC version, the Return key alone takes
you to a new line while the Enter key on the far right enters input.)

```
> dp := proc(X,Y)
>       X[1]*Y[1]+X[2]*Y[2]+X[3]*Y[3];
>       end:
```

Hit the ENTER key on the keyboard (or double-click with the mouse). The
dot product procedure is now entered into MAPLE *for this particular worksheet*. Every time you begin MAPLE and open this worksheet, however, this
procedure must be re-entered — this goes for everything! The dot product procedure above does exactly what we would do as human calculators.
Namely, it takes as input two vectors, X and Y, identifies their components
by the bracket notation $X[1]$ etc. and carries out the dot product multiplication and addition. Notice that there is no colon or semicolon after proc(X,Y)
and there is a colon after the "end" to end the input (i.e. the procedure). For
example, the following input produces a dot product output.

```
> B := [k,l,m]:
> dp(A,B);
```
$$ak + bl + cm$$

We could give this calculation a name as follows.

```
> try := dp(A,B);
```
$$try: \; = ak + bl + cm.$$

The next procedures compute the length of a vector and the cross product
of two vectors. By the way, note that MAPLE uses the usual computerese
versions of multiplication (*), exponentiation (^) and square root (sqrt).

```
> nrm := proc(X)
>       sqrt(dp(X,X));
>       end:
```

```
> xp := proc(X,Y)
>       local a,b,c;
>       a := X[2]*Y[3]-X[3]*Y[2];
>       b := X[3]*Y[1]-X[1]*Y[3];
>       c := X[1]*Y[2]-X[2]*Y[1];
>       [a,b,c];
>       end:
```

EXERCISE 6.1. Compute nrm(A,B) and xp(A,B) for the vectors A and B above.

Once you have written one procedure, you can write a thousand. The following procedures compute curvature and torsion from the nonunit speed formulas of section 1.4. These procedures assume that the curve is given parametrically in the variable t in three coordinates. If the curve is a plane curve in the xy-plane, say, then make the third coordinate zero. Note that, in the line right after proc( ), the word "local" appears. This tells MAPLE that the variables following "local" are to be considered only within the procedure (usually just for intermediate calculational purposes), so that, later on in the worksheet, invoking one of the local variables produces no response.

```
> curv:=proc(alpha)
>       local alphap,alphapp,num;
>       alphap:=diff(alpha,t);
>       alphapp:=diff(alphap,t);
>       num:= nrm(xp(alphap,alphapp));
>       RETURN(kappa=simplify(num/nrm(alphap)^3));
>       end:
```

```
> tor:= proc(alpha)
>       local alphap,alphapp,alphappp;
>       alphap:=diff(alpha,t);
>       alphapp:=diff(alphap,t);
>       alphappp:=diff(alphapp,t);
>       num:=dp(xp(alphap,alphapp),alphappp);
>       RETURN(tau=simplify(num/nrm(xp(alphap,alphapp))^2));
>       end:
```

In both procedures above, MAPLE differentiates each coordinate function in a parametrized curve "alpha" by the simple command diff(alpha,t);. The tangent and acceleration vectors "alphap" and "alphapp" are defined as local variables because they are not the objects of the calculation. They are only intermediate quantities necessary to calculate curvature. In fact, in MAPLEV3, all variables are implicitly assumed to be local, but it is still a good idea to declare them as such — not only to avoid annoying "warnings" telling you the variables are "local," but to make your procedure clear to anyone who reads or uses it. If it should ever be necessary to have a variable in a procedure pass out of the procedure to the rest of the worksheet, then you should declare the variable to be "global" (see ?global). Also, the command (all in capital letters) "RETURN" tells MAPLE to explicitly output whatever is contained within the parentheses. While this looks nice as output (i.e. $\kappa = \dots$), if the output of the procedure is required in a later procedure, the output will be read as an equation rather than a quantity. To get around this, we can pick off the quantity itself by taking its righthandside (see below).

Remark 6.1. I like to keep "separator lines" between individual inputs or between a procedure, which has no separating lines within itself, and another input. This is a particularly good idea on the Power PC since it seems to try to execute every unseparated thing in sight. I will not include separator lines here, though, since it makes for difficult reading. Also, Power PC users should *not* have prompts > before every line in a procedure. PPC MAPLE will think each line is a separate input instead of a procedure. Secondary prompts >> are fine.

Now, finally, here's an example. Let's take a helix.

> hel:=[a*cos(t),a*sin(t),b*t];

$$hel: \; = [a\cos(t), a\sin(t), bt]$$

> curv(hel);

$$\kappa = \frac{a}{a^2 + b^2}$$

> tor(hel);

$$\tau = \frac{b}{a^2 + b^2}$$

To take the righthand sides of these equations, we input

> rhs(curv(hel)); rhs(tor(hel));

$$\frac{a}{a^2 + b^2} \qquad \frac{b}{a^2 + b^2}$$

In order to evaluate the curvature and torsion of a particular helix, it is not necessary to redefine the whole curve. The command "subs" allows substitutions to be made into a general definition as in the following.

EXERCISE 6.2. Input the following to compute curvature and torsion for the helix $(4\cos(t), 4\sin(t), 2t)$.

> hel1:=subs({a=4,b=2},hel);
> curv(hel1);
> tor(hel1);

As I have mentioned above, before special types of plots can be made, the following command must be issued.

> with(plots):

With with(plots) we can now graph a red helix in three dimensions. The plot command is as follows.

> spacecurve(hel1,t=0..10*Pi,scaling=constrained,thickness=2,color=red);

Here, *hel1* is plotted as t varies from zero to 10π. The entries following the domain of t are plotting options which make the plot appear a certain way. For example, "scaling=constrained" makes distance along the axes the same in each direction and "thickness" makes the curve thicker or thinner as required.

EXERCISE 6.3. Plot the helix with options "scaling=unconstrained," "thickness=1," and "color=blue." For other options, type ?plot3d[options]. The command "space-curve" is a curve version of MAPLE's "plot3d" command which we shall use extensively later.

To plot plane curves, we can use the "plot" command. For example, let's take the witch of Agnesi. There will always be some amount of trouble going from space curves to plane curves. For example, our dot product is calculated using three coordinates, so defining a plane curve with two coordinates will cause problems with the curvature procedure. On the other hand, plotting a plane curve with MAPLE's "plot" command requires exactly two coordinates. It's fairly simple to get around this trouble by first defining the witch with a fake third coordinate as

> witch:=[2*tan(t),2*cos(t)^2,0];

and then taking its curvature

> curv(witch);.

However, you cannot say "plot(witch, ...)" as we did for "spacecurve" for two reasons. First, since we've used three coordinates, MAPLE's "plot" command will not work. Secondly, even if we use two coordinates to define the witch, MAPLE's "plot" command uses an annoying syntax which will not allow an input such as "plot(witch,t= -1.3..1.3);." You must say (*in this exact form*)

> plot([2*tan(t),2*cos(t)^2,t=-1.3..1.3],scaling=constrained);

or use the first two coordinates of the witch as follows (with an optional "axes" parameter)

> plot([witch[1],witch[2],t=-1.3..1.3],scaling=constrained,axes= framed,
        title="The Witch,"titlefont=[TIMES,ITALIC,20]);

Notice that, unlike "spacecurve," the square brackets must go around the $t = \ldots$ as well as the parametrization. Also notice that we have included "title options" to give the plot a title with a particular font. For details, see MAPLE'S help page for title (i.e. type in ?title). Now try using options such as "constrained" or "color" or "axes" or "thickness" for the "plot" command and see what you get.

We can use MAPLE procedures to graph more complicated curves. For example, the second procedure below takes input a curve and gives output the involute of the curve. The first procedures gives the arclength of a curve from 0 to t.

```
> arc:=proc(alpha)
>      local alphap ;
>      alphap:=diff(subs(t=u,alpha),u);
>      int(nrm(alphap),u=0..t);
>      end:
```

Notice that this procedure expects the curve "alpha" to be parametrized by
"t" and that, at the very first stage, "t" is replaced by a dummy variable "u."
Finally, the length of the tangent vector in terms of "u" (i.e. the speed) is
integrated to produce an arclength in "t." Using this arclength procedure, we
can write a simple involute procedure.

```
> involute:=proc(alpha)
>     local aa,bb,cc;
>     aa:=simplify(alpha[1]-arc(alpha)*diff(alpha[1],t)/nrm(diff(alpha,t)));
>     bb:=simplify(alpha[2]-arc(alpha)*diff(alpha[2],t)/nrm(diff(alpha,t)));
>     cc:=simplify(alpha[3]-arc(alpha)*diff(alpha[3],t)/nrm(diff(alpha,t)));
>     [aa,bb,cc];
>     end:
```

For example, we have the following input and output.

```
> circ:=[cos(t),sin(t),0];
```

$$circ := [\cos(t), \sin(t), 0]$$

```
> inv:=involute(circ);
```

$$inv := [\cos(t) + t\sin(t), \sin(t) - t\cos(t), 0]$$

If you wish to graph the circle and its involute together, you can do this by
creating a *plot structure* for each and then *displaying* both at once. Also,
while it is possible, as we have seen before, to use MAPLE's "plot" command,
we will use "spacecurve" here to avoid the bracket problem mentioned above.

```
> circplot:=spacecurve(circ,t=0..2*Pi,color=black,thickness=1):
```

```
> invplot:=spacecurve(inv,t=0..2*Pi,color=red,thickness=2):
```

```
> display({circplot,invplot},scaling=constrained,axes=framed);
```

EXERCISE 6.4. Show that the involute of the helix $(\cos t, \sin t, t)$ is a plane curve.
What plane curve is it?

Similarly, we can write a procedure to find the evolute of a given curve.
Recall that, given $\alpha(t)$, its evolute is defined by $\mathcal{E}(t) = \alpha(t) + 1/\kappa(t) N(t)$
where N is the normal of α.

EXERCISE 6.5. Show that

$$\frac{1}{\kappa} N = \frac{|\alpha'|^2}{|\alpha' \times \alpha''|^2} (\alpha' \times \alpha'') \times \alpha'.$$

Hint: use the nonunit speed Frenet formulas.

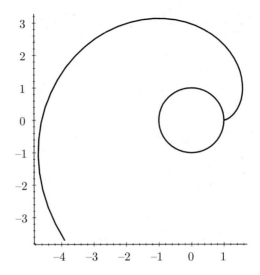

FIGURE 1.22.   Involute of a circle

The exercise leads to the procedure

```
> evolute:=proc(alpha)
>       local aa,bb,alphap,alphapp,A,C;
>       alphap:=diff(alpha,t);
>       alphapp:=diff(alphap,t);
>       A:=nrm(alphap)^2/nrm(xp(alphap,alphapp))^2;
>       C:=xp(xp(alphap,alphapp),alphap);
>       aa:=simplify(alpha[1]+A*C[1]);
>       bb:=simplify(alpha[2]+A*C[2]);
>       RETURN(Evolute=[aa,bb]);
>       end:
```

The following input and output gives the evolute of an ellipse.

```
> ellipse:=[a*cos(t),b*sin(t),0];
```

$$ellipse := [a\cos(t), b\sin(t), 0]$$

```
> evolute(ellipse);
```

$$Evolute = \left[\frac{\cos(t)^3(a^2 - b^2)}{a}, \frac{\sin(t)(-a^2 + a^2\cos(t)^2 + b^2 - b^2\cos(t)^2)}{b}\right]$$

Notice that MAPLE simplifies the first coordinate, but not the second. In fact, although there does not seem to be a way of inducing MAPLE to simplify both coordinates simultaneously, the second coordinate simplifies by hand to

$$\frac{\sin(t)^3(b^2 - a^2)}{b},$$

so that the evolute of an ellipse is an astroid. By the way, to plot the astroid, it is necessary to first peel off the name "Evolute." This may be accomplished by taking

>ev:=rhs(evolute(ellipse));

$$ev := \left[\frac{\cos(t)^3(a^2 - b^2)}{a}, \ \frac{\sin(t)(-a^2 + a^2 \cos(t)^2 + b^2 - b^2 \cos(t)^2}{b} \right]$$

and plotting ev as we have done before.

EXERCISE 6.6. Show that the evolute of the catenary $(t, \cosh(t))$ is given by $(t - \cosh(t)\sinh(t), 2\cosh(t))$. Plot this evolute.

EXERCISE 6.7. Do Exercise 1.4.10 by finding the evolute of the astroid $(a\cos^3(t), a\sin^3(t))$ and then following the instructions below.

(1) Take the astroid $(2a\cos^3(t), 2a\sin^3(t))$ (thought of as a column vector) and multiply it on the left by the rotation matrix

$$\begin{bmatrix} \cos(\pi/4) & -\sin(\pi/4) \\ \sin(\pi/4) & \cos(\pi/4) \end{bmatrix}$$

using MAPLE's "matrix" and "matrix evaluation" commands. Before you can define a matrix, you must load MAPLE's linear algebra package by typing

>with(linalg):

Now the rotation may be accomplished as follows.

> astr3:=evalm(matrix(2,2,[sqrt(2)/2,-sqrt(2)/2,sqrt(2)/2, sqrt(2)/2])*
        [2*a*cos(t)^3,2*a*sin(t)^3]);

(2) Now shift the phase of the rotated astroid by $\pi/4$ by taking

> expand(subs(t=u-Pi/4,astr3[1]));

> expand(subs(t=u-Pi/4,astr3[2]));

Finally, compare your result to the expression for the evolute of the astroid above.

(3) Plot both astroids on the same set of axes using the "display" command. You should get

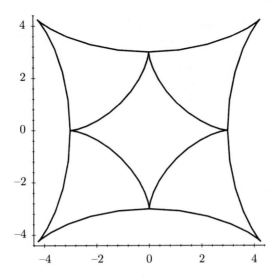

FIGURE 1.23.   Astroid and its evolute

Now let's use Exercise 1.5.10 to try to recreate a plane curve from its curvature alone. While the formula involves integrals which can only rarely be solved, we can still plot the resulting curve. The exercise shows that a plane curve β is determined by its curvature as follows:

$$\beta(s) = \left(\int_0^s \cos\theta(u)\,du, \ \int_0^s \sin\theta(u)\,du \right)$$

where $\theta(u) = \int_0^u \kappa(t)\,dt$. By the Fundamental Theorem of Calculus, we can transform these integrals into a system of differential equations to be solved numerically and plotted. The system is

$$\frac{d\theta}{ds} = \kappa(s) \qquad \frac{d\beta_1}{ds} = \cos(\theta(s)) \qquad \frac{d\beta_2}{ds} = \sin(\theta(s)).$$

The result will be the unit speed curve with the specified curvature. As a first example, let's take $\kappa = s$ itself. In MAPLE, we define the system and then use MAPLE's differential equations solver to get a numeric solution. For details on its operation, type ?dsolve.

```
> sys:=diff(theta(s),s)=s, diff(b1(s),s)=cos(theta(s)), diff(b2(s),s)
       = sin(theta(s));

> p:=dsolve({sys,theta(0)=0,b1(0)=0,b2(0)=0},{theta(s),b1(s),b2(s)},
       type=numeric):
```

Notice that the numeric option in "dsolve" requires initial conditions as well as the system of equations. In fact, for certain equations, "dsolve" will return a solution in closed form from the system alone. The numeric solution may now be plotted using MAPLE's "odeplot" command.

> scurve:=odeplot(p,[b1(s),b2(s)],-5..5,numpoints=200):

> display(scurve,view=[-2..2,-2..2]);

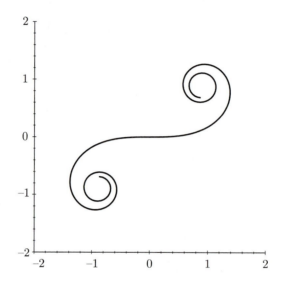

FIGURE 1.24.   Curve with $\kappa(s) = s$

EXERCISE 6.8. Create the unit speed curve with $\kappa(s) = 1/(1 + s^2)$. What is this curve? Can you prove it?

Now let's write a procedure to take an input curvature and output the curve. Additional inputs give bounds on the curve's parameter and viewing region.

```
> recreate:=proc(kap,a,b,c,d,f,g)
>       local sys,b1,b2,p,theta,pl;
>       sys:=diff(theta(s),s)=kap(s), diff(b1(s),s)=cos(theta(s)), diff(b2(s),s)=
>             sin(theta(s));
>       p:=dsolve({sys,theta(0)=0,b1(0)=0,b2(0)=0},{theta(s), b1(s),b2(s)},
>             type=numeric):
>       pl:=odeplot(p,[b1(s),b2(s)],a..b,numpoints=200,thickness=1,
>             axes=framed,color=red):
>       display(pl,view=[c..d,f..g]);
>       end:
```

To specify a curvature function, give the parameter (t say) and use the hyphen and greater than sign to construct an arrow -> to the effect on t. Some examples of curvature functions are

```
> kap1:=t->t;
> kap2:=t->t^2;
> kap3:=t->exp(t);
> kap4:=t->sin(t);
> kap5:=t->sin(t)*t;
> kap6:=t->sin(t)*t^2;
> kap7:=t->1/(1+t^2); .
```

```
> recreate(kap5,-8,8,-2,2,0,3); (See Figure 1.24)
```

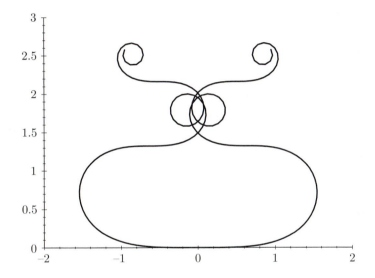

FIGURE 1.25.   Curve with $\kappa(t) = t \sin(t)$

EXERCISE 6.9. Recreate curves with the curvature functions listed above.

We can also recreate curves in space from their curvature and torsion via the Frenet (differential) equations. The following procedure solves the nine Frenet equations together with the three equations defining the unit tangent vector of the curve and allows you to zoom in by defining a viewing region. The inputs kap and ta are the desired curvature and torsion and a,b bound the parameter of the solution curve while c,d,e,f,g,h give ranges for x,y and z respectively.

```
> recreate3dview:=proc(kap,ta,a,b,c,d,e,f,g,h)
>       local sys,p,alph1,alph2,alph3,T1,T2,T3,N1,N2,N3,B1,B2,B3,pl;
```

```
>       sys:=diff(alph1(s),s)=T1(s),diff(alph2(s),s)=T2(s),
>          diff(alph3(s),s)=T3(s),
>          diff(T1(s),s)=kap(s)*N1(s), diff(T2(s),s)=kap(s)*N2(s),
>          diff(T3(s),s)=kap(s)*N3(s),
>          diff(N1(s),s)=-kap(s)*T1(s)+ta(s)*B1(s),
>          diff(N2(s),s)=-kap(s)*T2(s)+ta(s)*B2(s),
>          diff(N3(s),s)=-kap(s)*T3(s)+ta(s)*B3(s),
>          diff(B1(s),s)=-ta(s)*N1(s), diff(B2(s),s)=-ta(s)*N2(s),
>          diff(B3(s),s)=-ta(s)*N3(s); print(sys);
>       p:=dsolve({sys, alph1(0)=0,alph2(0)=0,alph3(0)=0,T1(0)=1,
>          T2(0)=0,T3(0)=0,N1(0)=0,N2(0)=1,N3(0)=0,B1(0)=0,B2(0)=0,
              B3(0)=1}, {alph1(s),alph2(s),alph3(s),T1(s),T2(s),T3(s),N1(s),
              N2(s),N3(s),B1(s),B2(s),B3(s)},type=numeric):
>       pl:=odeplot(p,[alph1(s),alph2(s),alph3(s)],a..b,numpoints=200,
              thickness=1, axes=framed, color=red):
>       display(pl,scaling=constrained,view=[c..d,e..f,g..h]);
>       end:
```

We can recreate a circular helix by taking κ and τ to be constants. For example, take $\kappa = 1/5$ and $\tau = 1/10$ with the solution curve parameter ranging from 0 to 100, all viewed in a $20 \times 20 \times 20$-cube, to get a standard helix.

```
> recreate3dview(1/5,1/10,0,100,-10,10,-10,10,-10,10);
```

EXERCISE 6.10. Use recreate3dview to recreate curves as follows.

```
> recreate3dview(1/5,1/10,-20,100,0,18,0,8,0,30);
```

```
> kap3d1:=t->t;
```

```
> recreate3dview(kap3d1,1/10,0,10,0,2,0,2,0,1);
```

```
> tau3d1:=t->t/10;
```

```
> recreate3dview(kap3d1,tau3d1,0,10,0,2,0,2,0,1);
```

Here is Viviani's curve from Exercise 1.3.19.

```
> viv:=[a*(1+cos(t)),a*sin(t),a*2*sin(t/2)];
```

```
> vi:=spacecurve(subs(a=1,viv),t=-Pi..Pi,color=black,thickness=2):
```

```
> sphere:=plot3d([2*cos(u)*cos(v),2*sin(u)*cos(v), 2*sin(v)],u=0..2*Pi,
              v=-Pi/2..Pi/2, style=wireframe, color=grey):
```

```
> cyl:=plot3d([cos(u)+1,sin(u),v],u=0..2*Pi,v=-2..2,style=wireframe,
              color=red):
```

```
> display({vi,sphere,cyl},scaling=constrained,orientation=[0,78]);
```

Now let's use MAPLE to verify that Viviani's curve is a spherical curve. We use the criterion for spherical curves given in Exercises 1.3.17-1.3.19.

```
> kk:=simplify(rhs(curv(viv)));
```

```
> tt:=simplify(rhs(tor(viv)));
```

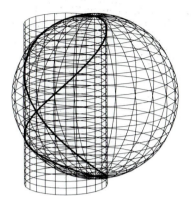

FIGURE 1.26.   Viviani's curve

> zz:=simplify((1/kk)^2+(1/tt)^2*(diff(1/kk,t)/ nrm(diff(viv,t)))^2);

The curvature and torsion above therefore satisfy the equation

$$\left(\frac{1}{\kappa}\right)^2 + \left(\frac{1}{\tau}\left(\frac{1}{\kappa}\right)'\frac{1}{|\alpha'(t)|}\right)^2 = 4a^2$$

so that Viviani's curve is truly spherical.

EXERCISE 6.11. Show that the following curve is spherical.

$$(a\cos^2(t), a\sin(t)\cos(t), a\sin(t))$$

In Exercise 3.12, curves of constant precession were discussed. The recreate procedures above may be used to create curves of constant procession by choosing $\kappa(s) = a\sin(bs)$ and $\tau(s) = a\cos(bs)$ for constants $a > 0$ and b. However, in [Sco], an explicit parametrization for such curves is derived. Furthermore, it is shown that any such curve lies on a hyperboloid of one sheet (which we shall consider several times in this book) with the formula

$$x^2 + y^2 - \frac{b^2}{a^2}z^2 = \frac{4b^2}{a^4}.$$

The explicit parametrization is given by

$$x(s) = \frac{c+b}{2c(c-b)}\sin([c-b]s) - \frac{c-b}{2c(c+b)}\sin([c+b]s)$$

$$y(s) = -\frac{c+b}{2c(c-b)}\cos([c-b]s) + \frac{c-b}{2c(c+b)}\cos([c+b]s)$$

$$z(s) = \frac{a}{bc}\sin(bs)$$

where $c = \sqrt{a^2 + b^2}$. The following procedure graphs curves of constant precession on the associated hyperboloids.

```
> constprec:=proc(a,b)
>        local c,alpha,a1,a2,hyp,pl1,pl2 ;
>        c:=sqrt(a^2+b^2);
>        a1:=c+b;
>        a2:=c-b;
>        alpha:=[a1/(2*c*a2)*sin(a2*s)-a2/(2*c*a1) *sin(a1*s),
>                -a1/(2*c*a2)*cos(a2*s)+ a2/(2*c*a1) *cos(a1*s),
>                a/(b*c)*sin(b*s)];
>        hyp:=[2*b/a^2*cosh(r)*cos(q),2*b/a^2 *cosh(r)*sin(q),
>                2/a*sinh(r)];
>        pl1:=spacecurve(alpha,s=0..2*Pi,color=black,thickness=2,
>                numpoints=200):
>        pl2:=plot3d(hyp,r=-1..1,q=0..2*Pi):
>        display({pl1,pl2},scaling=constrained,style=wireframe,
>                shading=XY,axes=framed);
>        end:
```

EXERCISE 6.12. Graph the curves of constant precession with $a = 15$, $b = 8$ and with $a = \sqrt{6}$, $b = \sqrt{2}$. Can you make a conjecture about when curves of constant precession are closed curves? See [Sco].

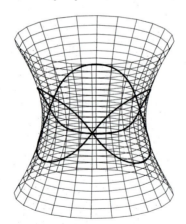

FIGURE 1.27.   Constant precession curve with $a = 4$ and $b = 3$

EXERCISE 6.13. Graph the unit tangent vector $T(s)$ (i.e. the tangent indicatrix) for curves of constant precession with $a = 15$, $b = 8$ and with $a = \sqrt{6}$, $b = \sqrt{2}$. Use "display" to graph $T(s)$ on the unit sphere. Can you make a conjecture about the number of cusps of such curves? For a discussion of tangent indicatrices of spherical curves, see [Sol].

Chapter 2

SURFACES

2.1 INTRODUCTION

We see examples of surfaces in everyday life. Balloons, innertubes, cans and soap films provide physical models of surfaces. In order to study the geometry of these objects, what is required? Somehow, we must have coordinates to make calculations. But just because these surfaces reside in $\mathbb{R}^3$ we shouldn't let ourselves think of them as 3-dimensional. For instance, if we cut a cylinder lengthwise, then we can unroll it to lie flat on a tabletop. This indicates that surfaces are inherently 2-dimensional objects and, therefore, should be describable by two coordinates. This gives us our first clue as to an approach for describing the geometry of surfaces. Namely, we should try to spread part of the plane around a surface and, in terms of the required twisting and stretching, understand how the surface curves in space. The spreading provides coordinates to do calculus on a surface and various calculus-based quantities associated with the spreading describe the shape of the surface. Just as the calculus of the Frenet Formulas enabled us to describe the geometry of a curve, so also will the calculus of a surface enable us to describe the geometry of a surface.

Let D denote an open set in the plane $\mathbb{R}^2$. The open set D will typically be an open disk or an open rectangle. Let

$$\mathbf{x}\colon D \to \mathbb{R}^3 \qquad (u, v) \mapsto (x^1(u, v),\ x^2(u, v),\ x^3(u, v))$$

denote a mapping of D into 3-space. The $x^i(u, v)$ are the component functions of the mapping $\mathbf{x}$. We can do calculus one variable at a time on $\mathbf{x}$ by partial differentiation. Fix $v = v_0$ and let u vary. Then $\mathbf{x}(u, v_0)$ depends on one parameter and is, therefore, a curve. It is called a *u-parameter curve*. Similarly, if we fix $u = u_0$, then the curve $\mathbf{x}(u_0, v)$ is a *v-parameter curve*. Note that both curves pass through $\mathbf{x}(u_0, v_0)$ in $\mathbb{R}^3$. Tangent vectors for the u-parameter and v-parameter curves are given by differentiating the component functions of $\mathbf{x}$ with respect to u and v respectively. We write,

$$\mathbf{x}_u = \left(\frac{\partial x^1}{\partial u},\ \frac{\partial x^2}{\partial u},\ \frac{\partial x^3}{\partial u} \right) \qquad \text{and} \qquad \mathbf{x}_v = \left(\frac{\partial x^1}{\partial v},\ \frac{\partial x^2}{\partial v},\ \frac{\partial x^3}{\partial v} \right).$$

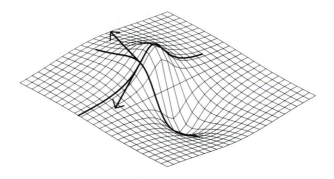

FIGURE 2.1.　Parameter curves with tangent vectors

We can evaluate these partial derivatives at (u_0, v_0) to obtain the tangent, or velocity, vectors of the parameter curves at that point, $\mathbf{x}_u(u_0, v_0)$ and $\mathbf{x}_v(u_0, v_0)$.

Of course, to obtain true coordinates on a surface, we need two properties. The first is that $\mathbf{x}$ must be one-to-one. In fact, as we will see later, we can relax this condition slightly to allow for certain self-intersections of a surface. Namely, we will allow those intersections which still give unambiguous normal vectors. Secondly, $\mathbf{x}$ must not "crinkle" parameter curves together so that $\mathbf{x}_u$ and $\mathbf{x}_v$ have the same direction. For, then, we lose 2-dimensionality — the very attribute of a surface we are trying to describe. The first property is verified by simple algebra or *imposed by decreasing the size of D*. But how can we show that the vectors $\mathbf{x}_u$ and $\mathbf{x}_v$ are never linearly dependent so that the second property holds? The following exercise provides a simple test.

EXERCISE 1.1. Show that $\mathbf{x}_u$ and $\mathbf{x}_v$ are linearly dependent if and only if $\mathbf{x}_u \times \mathbf{x}_v = 0$. Recall that two vectors are linearly dependent if one is a scalar multiple of the other.

A mapping $\mathbf{x} \colon D \to \mathbb{R}^3$ from an open set $D \subseteq \mathbb{R}^2$ is *regular* if $\mathbf{x}_u \times \mathbf{x}_v \neq 0$ for all points in D. (Of course, $\mathbf{x}_u \times \mathbf{x}_v$ must exist as well!) A *coordinate patch* (or *parametrization*) is a one-to-one regular mapping $\mathbf{x} \colon D \to \mathbb{R}^3$ of an open set $D \subseteq \mathbb{R}^2$ to $\mathbb{R}^3$. A *surface in $\mathbb{R}^3$* is a subset $M \subseteq \mathbb{R}^3$ such that each point of M has a neighborhood (in M) contained in the image of some coordinate patch $\mathbf{x} \colon D \to M \subseteq \mathbb{R}^3$.

Because each patch (e.g. $\mathbf{x}$, $\mathbf{y}$, $\mathbf{z}$) on a surface is one-to-one (at least in a smaller open portion of the original domain), then the respective inverse functions exist (and, in fact, are continuous). We may, therefore, consider the composition of any pair of patches $\mathbf{x}^{-1}\mathbf{y} \colon D \to \mathbb{R}^2$. Such a map is *differentiable* (or *smooth*) if all partial derivatives $\frac{\partial}{\partial u}$, $\frac{\partial^2}{\partial u^2}$, $\frac{\partial^3}{\partial u^3}$, $\cdots$ of each component function exist and are continuous. A surface is *differentiable* (or *smooth*) if $\mathbf{x}^{-1}\mathbf{y}$ is differentiable for any pair of patches on the surface. The surfaces we

FIGURE 2.2.   A patch

deal with will generally be smooth or close to it. That is, there may be a finite number of points which must be discarded to achieve smoothness. In virtually all of our examples, a surface M will be defined by a single patch with the exception of only a finite number of points. Indeed, we mention the notion of overlapping patches only because we use the idea later in our discussion of isometries and Gauss curvature.

A function $f\colon M \to \mathbb{R}$ from a surface M is *differentiable* (or *smooth*) if the composition $f \circ \mathbf{x}\colon D \to \mathbb{R}$ is smooth for each patch $\mathbf{x}$ on M. A curve on a surface is simply a mapping from an interval of real numbers $I = [a, b] \subseteq \mathbb{R}$ to the surface M, $\alpha\colon I \to M$. The surface M is said to be *path connected* if, for any two points p, $q \in M$, there is a curve $\alpha\colon [0, 1] \to M$ with $\alpha(0) = p$ and $\alpha(1) = q$. In this book *all surfaces, unless explicitly mentioned otherwise, will be assumed to be path connected.* A curve is *differentiable* (or *smooth*) if $\mathbf{x}^{-1} \circ \alpha\colon I \to \mathbb{R}^2$ is smooth for all patches $\mathbf{x}$ on M. The following lemma shows one reason why smoothness of surfaces is important for us. Namely, in order to understand curves on surfaces, it will be sufficient to understand parametric curves back in the open domains of $\mathbb{R}^2$ rather than in the surfaces themselves.

Important Lemma 1.1. *Let M be a surface. If $\alpha\colon I \to \mathbf{x}(D) \subseteq M$ is a curve in $\mathbb{R}^3$ which is contained in the image of a patch $\mathbf{x}$ on M, then for unique smooth functions $u(t)$, $v(t)\colon I \to \mathbb{R}$,*

$$\alpha(t) = \mathbf{x}(u(t), v(t)).$$

Proof. Because α is smooth, the composition $\mathbf{x}^{-1} \circ \alpha\colon I \to D$ is smooth by definition. Now, $D \subseteq \mathbb{R}^2$, so $\mathbf{x}^{-1}\alpha(t) = (u(t), v(t))$. Hence

$$\alpha(t) = \mathbf{x}(\mathbf{x}^{-1}\alpha(t)) = \mathbf{x}(u(t), v(t)).$$

In order to see that $u(t)$ and $v(t)$ are unique, suppose $\alpha = \mathbf{x}(\bar{u}(t), \bar{v}(t))$ for two other functions $\bar{u}$ and $\bar{v}$. Note that we can assume that $\bar{u}$ and $\bar{v}$ are defined on I as well by reparametrizing. Then

$$(u(t), v(t)) = \mathbf{x}^{-1}\alpha(t) = \mathbf{x}^{-1}\mathbf{x}(\bar{u}(t), \bar{v}(t)) = (\bar{u}(t), \bar{v}(t)).$$

$\square$

To study a curve on a smooth surface now, we can look at the functions of one variable $u(t)$ and $v(t)$. Also, note that, if a curve α on M is not contained in a single patch, then we may study it by considering the parts of it which lie in separate patches and then piece the information together. Again we note that our examples shall deal only with curves contained in a single patch. Finally, suppose $M \colon \mathbf{x}(u, v)$ and $N \colon \mathbf{y}(r, s)$ are two surfaces defined by single patches. A mapping $F \colon M \to N$ simply associates a point of N, unambiguously, to a point of M. We write $F(p) = q$ for $p \in M$ and $q \in N$. A mapping F is *differentiable* (or *smooth*) if the composition $\mathbf{y}^{-1} \circ F \circ \mathbf{x} \colon D_\mathbf{x} \to D_\mathbf{y}$ is smooth as a map of open subsets of $\mathbb{R}^2$. That is, the usual partial derivatives must all exist and be continuous. Of course, we may make the same definition for surfaces covered by many patches. We simply require that all such compositions are smooth. Mappings of surfaces may be quite complicated as we shall see later. We shall also see, however, that linear algebra allows us to get a reasonable linear approximation of the mapping which is quite understandable.

Examples of Patches on Surfaces

Example 1.1: The Monge Patch.
The graph of a real-valued function of two variables $z = f(x, y)$ is a surface in $\mathbb{R}^3$. To see this, define a patch by

$$\mathbf{x}(u, v) = (u, v, f(u, v))$$

where u and v range over the domain of f. Then, $\mathbf{x}_u = (1, 0, \frac{\partial f}{\partial u})$ and $\mathbf{x}_v = (0, 1, \frac{\partial f}{\partial v})$. From now on, when convenient, we shall denote partial derivatives of functions by

$$\frac{\partial f}{\partial u} = f_u \qquad\qquad \frac{\partial f}{\partial v} = f_v.$$

The patch is regular, as can be seen from

$$\mathbf{x}_u \times \mathbf{x}_v = \begin{vmatrix} i & j & k \\ 1 & 0 & f_u \\ 0 & 1 & f_v \end{vmatrix} = (-f_u, -f_v, 1) \neq 0.$$

As an example, consider the *paraboloid* $z = x^2 + y^2$.

A Monge patch is given by

$$\mathbf{x}(u, v) = (u, v, u^2 + v^2).$$

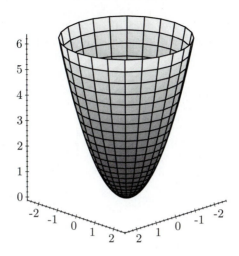

FIGURE 2.3.   A paraboloid

EXERCISE 1.2. What are the parameter curves for the paraboloid when $u_0 = 0$, $v_0 = 0$?

EXERCISE 1.3. Find a patch for the cone $z = \sqrt{x^2 + y^2}$. Does something go wrong here?

EXERCISE 1.4. Find a Monge patch on (part of) the unit sphere $x^2 + y^2 + z^2 = 1$. Why can't a Monge patch be defined on all of the sphere?

Example 1.2: Spherical Coordinates.
Let M be a sphere of radius R (centered at $(0,0,0)$ for convenience). Recall that spherical coordinates make use of two angles and a radius from the center (which in our case is fixed at R).

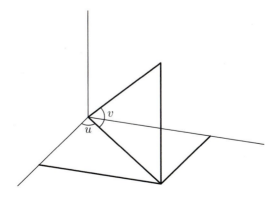

FIGURE 2.4.   Spherical coordinates

For a point a distance R away from the origin, draw a line segment from $(0,0,0)$ to the point. The parameter v, which varies as $-\pi/2 < v < \pi/2$, measures the angle (in radians) from this line down to the xy-plane. If the line is projected onto the xy-plane, then u (with $0 \le u < 2\pi$) measures the angle of the projection from the positive x-axis. Consider the following triangles.

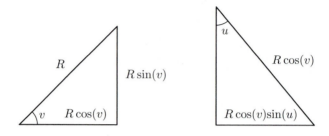

FIGURE 2.5

Hence, the xyz-coordinates of the point are

$$(R\cos u \cos v, \ R\sin u \cos v, \ R\sin v) = \mathbf{x}(u,v)$$

with $\mathbf{x}_u = (-R\sin u \cos v, \ R\cos u \cos v, \ 0)$ and $\mathbf{x}_v = (-R\cos u \sin v, \ -R\sin u \sin v, \ R\cos v)$. Further, we compute the cross product to be $\mathbf{x}_u \times \mathbf{x}_v = (R^2 \cos u \cos^2 v, \ R^2 \sin u \cos^2 v, \ R^2 \sin v \cos v)$, where, in the 3rd coordinate, we use $\sin^2 u + \cos^2 u = 1$. Note that $|\mathbf{x}_u \times \mathbf{x}_v| = R^2 \cos v$. From now on, we shall use the notation $S^2(R)$ for the sphere of radius R centered at the origin and simply S^2 for the sphere of radius 1 centered at the origin.

Example 1.3: Surfaces of Revolution.
Suppose C is a curve in the xy-plane and is parametrized by $\alpha(u) = (g(u), h(u), 0)$. Revolve C about the x-axis.

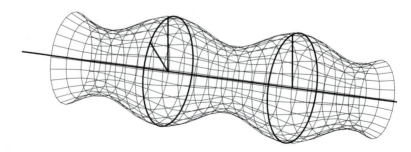

FIGURE 2.6.   A surface of revolution

The coordinates of a typical point P may be found as follows. The x-coordinate is that of the curve itself since we rotate about the x-axis. If v denotes the angle of rotation from the xy-plane, then the y-coordinate is shortened to $y \cos v = h(u) \cos v$ and the z-coordinate increases to $h(u) \sin v$ by

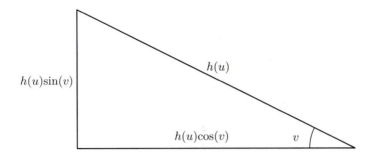

FIGURE 2.7

Hence, a patch may be defined by:

$$\mathbf{x}(u, v) = (g(u), \ h(u) \cos v, \ h(u) \sin v).$$

If we revolve about a different axis, then the coordinates permute. For example, a parametrized curve $(h(u), 0, g(u))$ in the xz- plane rotated about the z-axis gives a surface of revolution $(h(u) \cos v, h(u) \sin v, g(u))$.

EXERCISE 1.5. Check that

$$\mathbf{x}_u \times \mathbf{x}_v = h \left(\frac{dh}{du}, \ -\frac{dg}{du} \cos v, \ -\frac{dg}{du} \sin v \right).$$

Why is $\mathbf{x}_u \times \mathbf{x}_v \neq 0$ for all u, v?

In general, $g(u)$ measures the distance along the axis of revolution and $h(u)$ measures the distance from the axis of revolution. The nicest situation is, of course, when $g(u) = u$, for then it is easy to see where you are on the surface in terms of the parameter u.

EXERCISE 1.6. Find a patch for the *catenoid* obtained by revolving the catenary $y = \cosh(x)$ about the x-axis.

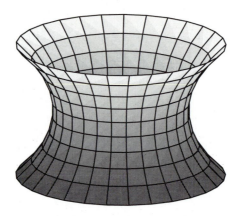

FIGURE 2.8.   A catenoid

EXERCISE 1.7: The Torus.
Consider the diagram

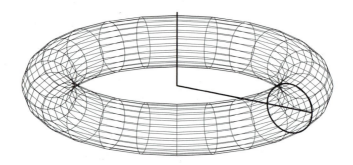

FIGURE 2.9.   A torus of revolution

and revolve the circle of radius r about the z-axis. Show that a patch is given by

$$\mathbf{x}(u,v) = ((R + r\cos u)\cos v, \ (R + r\cos u)\sin v, \ r\sin u).$$

How do u and v vary?

Example 1.4: The Helicoid.
Take a helix $\alpha(u) = (a\cos u, \ a\sin u, \ bu)$ and draw a line through $(0,0,bu)$ and $(a\cos u, \ a\sin u, \ bu)$. The surface swept out by this rising, rotating line is a helicoid.

The line required is given by $(0,0,bu) + v(a\cos u, a\sin u, 0)$, so a patch for the helicoid is given by

$$\mathbf{x}(u,v) = (av\cos u, av\sin u, bu).$$

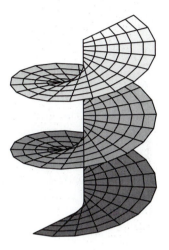

FIGURE 2.10. A helicoid

EXERCISE 1.8. Show that $\mathbf{x}(u, v)$ is a patch by verifying that $|\mathbf{x}_u \times \mathbf{x}_v| \neq 0$.

(5) **EXERCISE 1.9: Enneper's Surface.**
Define a surface by,

$$\mathbf{x}(u, v) = (u - \frac{u^3}{3} + uv^2, \ v - \frac{v^3}{3} + vu^2, \ u^2 - v^2).$$

Show that this definition gives a regular patch. Then show that, for $u^2 + v^2 < 3$, Enneper's surface has no self-intersections. Hint: use polar coordinates $u = r \cos \theta$, $v = r \sin \theta$, show that the equality $x^2 + y^2 + \frac{4}{3} z^2 = \frac{1}{9} r^2 (3 + r^2)^2$ holds, where x, y and z are the coordinate functions of Enneper's surface, and then show that the equality implies that points in the (u, v)-plane on different circles about $(0, 0)$ cannot be mapped to the same point. Finally, find two points on the circle $u^2 + v^2 = 3$ which *do* map to the same point of the surface.

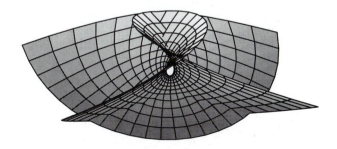

FIGURE 2.11. Enneper's surface

Ruled Surfaces

A surface is *ruled* if it has a *parametrization*

$$\mathbf{x}(u,v) = \beta(u) + v\delta(u)$$

where β and δ are curves. That is, the entire surface is covered by this one patch which consists of lines emanating from a curve $\beta(u)$ going in the direction $\delta(u)$. The curve $\beta(u)$ is called the *directrix* of the surface and and a line having $\delta(u)$ as direction vector is called a *ruling*. A surface is *doubly ruled* if it has two different ruled patches $\mathbf{x}(u,v) = \beta(u) + v\delta(u)$ and $y(u,v) = \alpha(u) + v\gamma(u)$. Ruled surfaces may have points with $\mathbf{x}_u \times \mathbf{x}_v = \beta'(u) \times \delta(u) + v\delta'(u) \times \delta(u) = 0$. These points are well-controlled, as Exercise 1.15 below shows. We have two very well known examples of ruled surfaces.

Example 1.5: Cones. $\mathbf{x}(u,v) = p + v\delta(u)$ where p is a fixed point.

Example 1.6: Cylinders. $\mathbf{x}(u,v) = \beta(u) + vq$ where q is a fixed direction vector.

EXERCISE 1.10. Find patches for the standard cone $z = \sqrt{x^2 + y^2}$ and standard cylinder $x^2 + y^2 = 1$ which are ruling patches in the sense of (1) and (2) above. This explains why the names cone and cylinder are used for the general patches of types (1) and (2) above.

EXERCISE 1.11. Show that the *saddle surface* $z = xy$ is doubly ruled. See Figure 2.12.

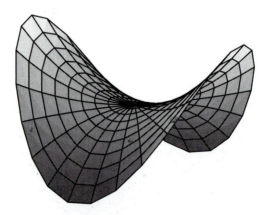

FIGURE 2.12.   A saddle

EXERCISE 1.12. Show that the helicoid is a ruled surface.

EXERCISE 1.13. The hyperboloid of one sheet

$$\frac{x^2}{a^2} + \frac{y^2}{b^2} - \frac{z^2}{c^2} = 1$$

has a useful parametrization

$$\mathbf{x}(u, v) = (a \cosh u \cos v, \ b \cosh u \sin v, \ c \sinh u).$$

FIGURE 2.13. A hyperboloid of one sheet

However, this is not a ruling patch. Find two (related) ruling patches for this surface. That is, show the surface is doubly ruled. Hint: A directrix is $\beta(u) = (a \cos u, \ b \sin u, \ 0)$. Let $\delta(u) = \beta'(u) + (0, 0, c)$. Because they are ruled surfaces, hyperboloids of one sheet are useful in engineering. For example, one reason nuclear cooling towers are built in the shape of hyperboloids of one sheet is that they may be built with straight beams along the rulings (see [Whi]). Therefore, stress on the tower is small (i.e. there is no internal bending) and the tower is easy to build. What do you think another reason is for using such shapes? Also, look at the weave in the base of a wicker chair with a sea-shell back. What do you notice and why?

EXERCISE 1.14. Another patch for the hyperboloid of one sheet is given by

$$\mathbf{x}(u, v) = (a \frac{1 - uv}{1 + uv}, b \frac{u + v}{1 + uv}, c \frac{u - v}{1 + uv}).$$

For simplicity, take $a = b = c = 1$ corresponding to $x^2 + y^2 - z^2 = 1$. Show that the u and v-parameter curves of this patch are, in fact, the rulings of the two patches in the previous exercise. Hence, the hyperboloid has an orthogonal patch whose parameter curves are straight lines. Hint: (1) in $\mathbf{x}(u, v)$, let $w = \frac{u}{1 + uv}$ (2) for fixed v, find the intersection of the ruling with the unit circle in the xy-plane (3) show that $\mathbf{x}(u, v) = (\frac{1 - v^2}{1 + v^2}, \frac{2v}{1 + v^2}, 0) + w(-2v, 1 - v^2, 1 + v^2)$ has direction vector, through a point $(a, b, 0)$ with $a^2 + b^2 = 1$, a multiple of $(-\sin r, \cos r, 1)$, where the first ruled patch of the exercise above is $\mathbf{y}(r, s) = (\cos r, \sin r, 0) + s(-\sin r, \cos r, 1)$.

EXERCISE 1.15. Suppose $M : \mathbf{x}(u, v) = \beta(u) + v\delta(u)$ is a ruled surface with $|\beta'| = 1$ and $|\delta| = 1$. Also suppose $\delta' \neq 0$, so that M is noncylindrical. Show that M may be reparametrized by $\mathbf{y}(u, w) = \gamma(u) + w\delta(u)$, where $\gamma' \cdot \delta' = 0$. Note that γ may not be unit speed. A curve such as γ is called a *line of striction* for M. Show that any point on M where $\mathbf{x}_u \times \mathbf{x}_v = 0$ must lie on the line of striction. Hint: write $\gamma(u) = \beta(u) + r(u)\delta(u)$, use $\gamma' \cdot \delta = 0$ and solve for $r(u)$. Let $w = v - r(u)$.

EXERCISE 1.16. Find the lines of striction for the helicoid and the hyperboloid of one sheet.

EXERCISE 1.17. Suppose $M : \mathbf{x}(u, v) = \beta(u) + v\delta(u)$ is a ruled surface. Show that M may be reparametrized by $\mathbf{y}(u, w) = \epsilon(u) + w\delta(u)$, where $\epsilon' \cdot \delta = 0$. Hint: write $\epsilon(u) = \beta(u) + s(u)\delta(u)$, use $\epsilon' \cdot \delta = 0$ and solve for $s(u)$. Your answer may be an integral. We will use this result in proving Catalan's Theorem later. Find such a parametrization for a cylinder $\mathbf{x}(u, v) = \beta(u) + vq$.

The patches defined in this section provide a variety of examples of surfaces on which we will later test our notion of curvature (and its effect on geometry). We must note several important points, however, before we go on.

(1) Occasionally our patches miss one or several points on a surface. This defect in a particular defining patch usually means that a single patch is not really sufficient to completely define the surface. In general, many patches may be required to cover a surface. Ask why, for example, in our patch on the sphere, the North and South poles are missing.

(2) Other easily checked conditions also give us surfaces. For example, if we have a level set $g(x, y, z) = c$, then the Implicit (or Inverse) Function Theorem (IFT) may be invoked to show that the nonvanishing of the gradient of g, $\nabla g = \left(\frac{\partial g}{\partial x}, \frac{\partial g}{\partial y}, \frac{\partial g}{\partial z} \right)$, at all points of the level set, ensures that the level set is a surface. Essentially the IFT constructs a patch, so, in some sense, what we have done is more general. Furthermore, the "patch approach" allows much cleaner calculations later on. Thus, we have emphasized patches over other conditions for surfaces.

FORESHADOWING EXERCISE 1.18. In the plane our standard coordinate system uses the vectors $\mathbf{e}_1 = (1, 0)$ and $\mathbf{e}_2 = (0, 1)$ as $\mathbf{x}_u$ and $\mathbf{x}_v$ since the parameter curves are horizontal and vertical lines respectively. We have $\mathbf{e}_1 \cdot \mathbf{e}_1 = 1$, $\mathbf{e}_1 \cdot \mathbf{e}_2 = 0$ and $\mathbf{e}_2 \cdot \mathbf{e}_2 = 1$. Use the notation $E = \mathbf{x}_u \cdot \mathbf{x}_u$, $F = \mathbf{x}_u \cdot \mathbf{x}_v$, $G = \mathbf{x}_v \cdot \mathbf{x}_v$ and compute E, F, and G for all the patches given in this section. What is the same and what is different when a specific patch is compared to the plane itself? In particular, what does F tell you?

2.2 THE GEOMETRY OF SURFACES

Now that we have an idea of what a surface is, how do we detect its geometry? One of the most important techniques in mathematics and, indeed, all of the natural sciences, is that of linear approximation. By this we mean the following. We recognize that the nonlinear or curved object at hand is too complicated to study directly, so we approximate it by something linear: a line, a plane, a Euclidean space. We then study the linear object and, from it, infer results about the original curved object. Of course this is exactly what we do when we attach the linear Frenet frame to a curve. This process also is useful in subjects ranging from differential equations to algebraic topology.

The question is, of course, what type of linear space may be used to approximate a surface? Just as, in one-variable calculus, we use a tangent line to approximate a curve near a point, so we can use a *tangent plane* $T_p(M)$ to approximate M near $p \in M$. A plane consists of vectors and the natural idea for the make-up of $T_p(M)$ is to use the vectors which arise as velocity vectors of curves on M.

Formally, say that a vector $\mathbf{v}_p \in T_p(M)$ is *tangent to* M *at* p if $\mathbf{v}_p$ is the velocity vector of some curve on M. That is, there is some $\alpha: I \to M$ with $\alpha(0) = p$ and $\alpha'(0) = \mathbf{v}_p$. Usually we write $\mathbf{v}$ instead of $\mathbf{v}_p$ when no confusion will arise. Then, the *tangent plane of* M *at* p is defined to be

$$T_p(M) = \{\mathbf{v} \mid \mathbf{v} \text{ is tangent to } M \text{ at } p\}.$$

Immediately we know two curves which pass through $p = \mathbf{x}(u_0, v_0)$: the u and v-parameter curves with velocity vectors $\mathbf{x}_u$ and $\mathbf{x}_v$. The following result says that *every* tangent vector is made up of a (unique) linear combination of $\mathbf{x}_u$ and $\mathbf{x}_v$. Hence, $\{\mathbf{x}_u, \mathbf{x}_v\}$ form a *basis* for the vector space $T_p(M)$. (Recall that this is what the term 'basis' means.)

Lemma 2.1. $\mathbf{v} \in T_p(M)$ *if and only if* $\mathbf{v} = \lambda_1 \mathbf{x}_u + \lambda_2 \mathbf{x}_v$, *where* $\mathbf{x}_u$, $\mathbf{x}_v$ *are evaluated at* $(u_0,\ v_0)$.

Proof. First suppose that α is a curve with $\alpha(0) = p$ and $\alpha'(0) = \mathbf{v}$. We saw before that $\alpha(t) = \mathbf{x}(u(t), v(t))$, so the chain rule gives $\alpha' = \mathbf{x}_u(du/dt) + \mathbf{x}_v(dv/dt)$. Now, $\alpha(0) = p = \mathbf{x}(u(0), v(0))$, so $u(0) = u_0$ and $v(0) = v_0$ (since $\mathbf{x}$ is one-to-one) and

$$\mathbf{v} = \alpha'(0) = \mathbf{x}_u(u_0, v_0)\frac{du}{dt}(0) + \mathbf{x}_v(u_0, v_0)\frac{dv}{dt}(0).$$

Hence, $\lambda_1 = (du/dt)(0)$ and $\lambda_2 = (dv/dt)(0)$.

Now suppose $\mathbf{v} = \lambda_1 \mathbf{x}_u + \lambda_2 \mathbf{x}_v$ (where x_u and x_v are evaluated at (u_0, v_0)). We must find a curve on M with $\alpha(0) = p$ and $\alpha'(0) = \mathbf{v}$. Define this curve using the patch $\mathbf{x}$ by

$$\alpha(t) = \mathbf{x}(u_0 + t\lambda_1,\ v_0 + t\lambda_2).$$

Then $\alpha(0) = \mathbf{x}(u_0, v_0) = p$ and

$$\alpha'(t) = \mathbf{x}_u(u_0 + t\lambda_1, v_0 + t\lambda_2)\frac{d(u_0 + t\lambda_1)}{dt}$$
$$+ \ \mathbf{x}_v(u_0 + t\lambda_1, v_0 + t\lambda_2)\frac{d(v_0 + t\lambda_2)}{dt}$$
$$= \mathbf{x}_u(u_0 + t\lambda_1, v_0 + t\lambda_2)\lambda_1 + \mathbf{x}_v(u_0 + t\lambda_1, v_0 + t\lambda_2)\lambda_2.$$

Thus, $\alpha'(0) = \mathbf{x}_u\lambda_1 + \mathbf{x}_v\lambda_2$. $\qquad\qquad\qquad\qquad\qquad\qquad\qquad\square$

How do we identify the tangent plane via an equation? The same idea works for any plane in $\mathbb{R}^3$. A plane is determined by a point p and a normal vector (at p) to the plane N. Any q is in the plane exactly when $q - p$ is perpendicular to N. We then have, $N \cdot (q - p) = 0$ or, for $q = (x, y, z)$, $p = (x_0, y_0, z_0)$ and $N = (a, b, c)$,

$$ax + by + cz - [ax_0 + by_0 + cz_0] = 0$$
$$ax + by + cz + d = 0$$

where $d = -[ax_0 + by_0 + cz_0]$. This is, of course, the usual equation of a plane.

In fact, we already know how to get a normal vector for $T_p(M)$. Since $\{\mathbf{x}_u, \mathbf{x}_v\}$ forms a basis for $T_p(M)$, we need only find a vector perpendicular to both $\mathbf{x}_u$ and $\mathbf{x}_v$ (and hence to any $\mathbf{v} = \lambda_1\mathbf{x}_u + \lambda_2\mathbf{x}_v$). Such a vector is given by the cross product $\mathbf{x}_u \times \mathbf{x}_v$. Therefore, if $\mathbf{x}$ is a patch for a surface with $\mathbf{x}(u_0, v_0) = p$, then a normal $\mathcal{N}$ for $T_p(M)$ is given by,

$$\mathcal{N} = x_u(u_0, v_0) \times \mathbf{x}_v(u_0, v_0).$$

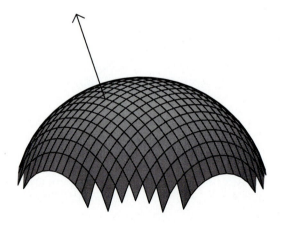

FIGURE 2.14. Surface with normal

Example 2.1: A Function of Two Variables. $M: z = f(x, y)$. We have a Monge patch $\mathbf{x}(u, v) = (u, v, f(u, v))$ with $\mathbf{x}_u \times \mathbf{x}_v = (-f_u, -f_v, 1)$. If $p = (u_0, v_0, f(u_0, v_0))$, then the equation describing $T_p(M)$ is,

$$-f_u(u_0, v_0)(x - u_0) - f_v(u_0, v_0)(y - v_0) + (z - f(u_0, v_0)) = 0.$$

For instance, if $z = x^2 + y^2$, then we have, for $p = (x_0, y_0, x_0^2 + y_0^2)$,

$$-2x_0(x - x_0) - 2y_0(y - y_0) + (z - (x_0^2 + y_0^2)) = 0$$

which reduces to

$$2x_0 x + 2y_0 y + z = x_0^2 + y_0^2.$$

Specifically, for $p = (1, 0, 1)$, we have $2x + z = 1$ as the equation of the tangent plane.

Example 2.2: The Sphere of radius R. $x^2 + y^2 + z^2 = R^2$.
Recall that the unit normal may be written,

$$\mathbf{x}_u \times \mathbf{x}_v = (R^2 \cos u \cos^2 v, R^2 \sin u \cos^2 v, R^2 \sin v \cos v).$$

(Note that $\mathbf{x}_u \times \mathbf{x}_v = (R \cos v) p$, a multiple of the point on p the sphere. This verifies that the radial vector from the origin to the sphere meets the sphere at an angle of $\pi/2$.) Suppose $p = (R, 0, 0)$, so that $u = 0$ and $v = 0$. Then $\mathbf{x}_u \times \mathbf{x}_v$ at p is $(R^2, 0, 0)$ and $T_p(M)$ is given by,

$$(R^2, 0, 0) \cdot (x - R, y, z) = 0$$
$$R^2 x = R^3$$
$$x = R.$$

This is, of course, the plane which is tangent to the sphere at $(R, 0, 0)$

In fact, we shall rarely be interested in the equation of the tangent plane — *it is the normal itself which will detect the geometry of the surface.* If we know how the direction of the normal to a surface changes as we move in the direction determined by a tangent vector, then we will know how the surface itself curves in that direction. There is a point which must be made here. When a surface requires more than one patch to cover it, there is the possibility that the normals defined from the various patches are incompatible with each other in a fundamental way. We refer the reader to [DoC1, p. 102] for details. Such a surface is said to be *nonorientable* and the first example is

Example 2.3: The Möbius strip. Consider the regular parametrization

$$\mathbf{x}(u, v) = \left(\left(2 - v \sin\left(\frac{u}{2}\right)\right) \sin u, \left(2 - v \sin\left(\frac{u}{2}\right)\right) \cos u, v \cos\left(\frac{u}{2}\right)\right)$$

for $0 < u < 2\pi$ and $-1 < v < 1$. This patch omits the interval where $u = 0$,
so another patch is needed. This patch may be given by $\mathbf{y}(\bar{u}, \bar{v}) =$

$$\left(\left(2 - \bar{v}\sin\left(\frac{\pi}{4} + \frac{\bar{u}}{2}\right)\right)\cos\bar{u}, \ \left(2 - \bar{v}\sin\left(\frac{\pi}{4} + \frac{\bar{u}}{2}\right)\right)\sin\bar{u}, \bar{v}\cos\left(\frac{\pi}{4} + \frac{\bar{u}}{2}\right)\right)$$

which omits $\bar{u} = 0$ corresponding to $u = \pi/2$. No matter how we try to
arrange it, these patches lead to the conclusion that the unit normal (of $\mathbf{x}$
say) must be equal to its negative! A picture of the Möbius strip indicates
how a normal may reverse itself as it travels around a closed curve.

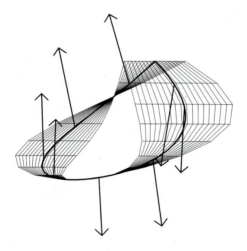

FIGURE 2.15.   A Möbius strip

We have been rather vague about this notion of orientability since it will
rarely come up in the rest of the book. In particular, a single regular patch $\mathbf{x}$
is always *orientable* since an explicit unit normal

$$U = \frac{\mathbf{x}_u \times \mathbf{x}_v}{|\mathbf{x}_u \times \mathbf{x}_v|}$$

exists at each point. (Note that switching the order of $\mathbf{x}_u$ and $\mathbf{x}_v$ gives $-U$ *at
every point of the patch*.) This leads to the general definition that a surface is
orientable if it has a smooth vector field of unit normals defined everywhere
on it. Except for isolated examples such as Henneberg's surface, which has a
Möbius strip contained in it, the surfaces we deal with will be orientable.

Now let's return to our discussion of how the change in unit normal detects
geometry. In order to see how we might calculate this change in a given
direction, first let us see how a tangent vector acts to change a function.

Let $g(x, y, z)$ be a function and let $\alpha(t) = (\alpha^1(t), \alpha^2(t), \alpha^3(t))$ be a curve. Then the chain rule of multivariable calculus gives

$$
\begin{aligned}
\frac{d}{dt}(g(\alpha(t))) &= \frac{\partial g}{\partial x}\frac{dx}{dt} + \frac{\partial g}{\partial y}\frac{dy}{dt} + \frac{\partial g}{\partial z}\frac{dz}{dt} \\
&= \frac{\partial g}{\partial x}\frac{d\alpha^1}{dt} + \frac{\partial g}{\partial y}\frac{d\alpha^2}{dt} + \frac{\partial g}{\partial z}\frac{d\alpha^3}{dt} \\
&= \left(\frac{\partial g}{\partial x}, \frac{\partial g}{\partial y}, \frac{\partial g}{\partial z}\right) \cdot \left(\frac{d\alpha^1}{dt}, \frac{d\alpha^2}{dt}, \frac{d\alpha^3}{dt}\right) \\
&= \nabla g(\alpha(t)) \cdot \alpha'(t) \qquad \text{where} \quad \nabla g \stackrel{\text{def}}{=} \left(\frac{\partial g}{\partial x}, \frac{\partial g}{\partial y}, \frac{\partial g}{\partial z}\right).
\end{aligned}
$$

Now let $\mathbf{v} \in T_p(M)$ and suppose g is restricted to $M \subseteq \mathbb{R}^3$. We might ask how g changes on M *in the $\mathbf{v}$ direction*. But the $\mathbf{v}$-direction on M itself means *in the direction of a curve α with $\alpha' = \mathbf{v}$ at p*. The *directional derivative of g in the $\mathbf{v}$-direction* is

$$
\mathbf{v}[g](p) \stackrel{\text{def}}{=} \frac{d}{dt}(g(\alpha(t)) \mid_{t=0} = \nabla g(p) \cdot \mathbf{v},
$$

where $\alpha(0) = p \in M$ and $\alpha'(0) = \mathbf{v}$. (Note that the last equality shows that the directional derivative does not depend on the curve chosen through p with velocity vector $\mathbf{v}$.) Hence, $\mathbf{v}$ acts on a function $g: M \to \mathbb{R}$ to produce a scalar $\mathbf{v}[g]$. This is why the notation $\mathbf{v}[g]$ is chosen to denote directional derivative. Notice that the usual directional derivative of multivariable calculus has precisely the same definition, but the given function g may not be restricted to a surface M and the direction vector $\mathbf{v}$ may be any vector of $\mathbb{R}^3$. In many cases $\mathbf{v}$ may denote a *vector field* on M — a smoothly varying choice of tangent vector at some collection of points $p \in M$ — and then, $\mathbf{v}[g]$ gives a new function on M defined by $\mathbf{v}[g](p) = \mathbf{v}_p[g]$. (Here we say that a vector field $\mathbf{v}$ *smoothly varies* on M if, for any curve $\alpha: I \to M$, the assignment $t \mapsto \mathbf{v}(\alpha(t))$ is differentiable.)

If f is a function on M, then the composition with a patch $\mathbf{x}$ may be formed, $f(\mathbf{x}(u, v)) = f \circ \mathbf{x}$, so that on the piece of M given by the patch, f is a function of u and v. Hence we may write

$$
\frac{\partial f}{\partial u} = \frac{\partial(f \circ \mathbf{x})}{\partial u}
$$

and similarly for v. Then, since $\frac{d}{du}(\mathbf{x}(u, v_0)) = \mathbf{x}_u$, we have

$$
\mathbf{x}_u[f] = \frac{d}{du}(f(\mathbf{x}(u, v_0)) \mid_{u=u_0} = \frac{\partial f}{\partial u} \mid_{u=u_0}
$$

and this works for all f. That is, a u-parameter velocity vector applied to a function gives the u-partial derivative of the function (composed with the patch).

EXERCISE 2.1. Show the Leibniz Rule (Product Rule) holds. That is, $\mathbf{v}[fg] = \mathbf{v}[f]g + f\mathbf{v}[g]$.

EXERCISE 2.2. Show $\mathbf{v}[x] = v^1$, $\mathbf{v}[y] = v^2$, $\mathbf{v}[z] = v^3$ where $\mathbf{v} = (v^1, v^2, v^3)$.

EXERCISE 2.3. Suppose M is a surface which is the level set of a function g, $g(x, y, z) = c$. Show that $\nabla g(p)$ is a normal vector for M at every $p \in M$. Hint: take $\mathbf{v} \in T_p(M)$ and show $\nabla g(p) \cdot \mathbf{v} = 0$ using the definition of $\mathbf{v}$ and the chain rule. Hence, for level sets we have another way of obtaining a normal. Try this for the sphere $g(x, y, z) = x^2 + y^2 + z^2 = R^2$.

Important Note 2.4. In calculus, the directional derivative is generally defined using a *unit vector* $\mathbf{v}$. The reason for this is simply that we wish to detect the change in a function in the $\mathbf{v}$-direction and we do not want the magnitude of $\mathbf{v}$ artificially affecting our computation. For this very reason we will now take *unit* normals to surfaces. Let

$$U = \frac{\mathbf{x}_u \times \mathbf{x}_v}{|\mathbf{x}_u \times \mathbf{x}_v|}.$$

Now when we ask how U changes in a given direction, only U's change in direction will be measured — and it is this quantity which will detect the shape of the surface. A normal vector U to a surface M takes any point of M, p, and assigns a vector $U(p) \in \mathbb{R}^3$ to it. Now, $U(p) = (u^1(p), u^2(p), u^3(p))$ since it is a vector in $\mathbb{R}^3$, so we may write

$$U = (u^1, u^2, u^3)$$

or, in the basis $\mathbf{e}_1 = (1, 0, 0)$, $\mathbf{e}_2 = (0, 1, 0)$, $\mathbf{e}_3 = (0, 0, 1)$ for $\mathbb{R}^3$

$$U = u^1\mathbf{e}_1 + u^2\mathbf{e}_2 + u^3\mathbf{e}_3$$

$$= \sum_{i=1}^{3} u^i\mathbf{e}_i$$

where u^1, u^2, u^3 are functions from M to $\mathbb{R}$. Because U assigns vectors to each $p \in M$, we say that U is a *vector field* on M. Similarly, any assignment of a tangent vector to each point of M which varies smoothly over M is also called a *vector field* on M.

How should we describe the change in U in a direction $\mathbf{v} \in T_p(M)$? We can simply look at the changes in the functions u^1, u^2, u^3 in the $\mathbf{v}$-direction. That is, we consider the directional derivatives of the u^i in the direction $\mathbf{v}$. Of course, we are only interested in the initial rate of change of U in the $\mathbf{v}$-direction because $\mathbf{v}$ is situated at p, so we evaluate derivatives at 0 (for $\alpha(0) = p$). We give the name *covariant derivative* to this initial rate of change of U in the v-direction. Although the notation is somewhat strange, remember that the covariant derivative is just the usual directional derivative applied to each coordinate function of U. This is denoted by,

$$\nabla_{\mathbf{v}} U \stackrel{\text{def}}{=} (\mathbf{v}[u^1], \mathbf{v}[u^2], \mathbf{v}[u^3])$$

$$= \sum_{i=1}^{3} \mathbf{v}[u^i] \mathbf{e}_i.$$

From the definition of $\mathbf{v}[\cdot]$, we see that

$$\nabla_{\mathbf{v}} U = \sum_{i=1}^{3} \frac{d}{dt}(u^i(\alpha(t)))|_{t=0} \, \mathbf{e}_i$$

where $\alpha(0) = p$ and $\alpha'(0) = \mathbf{v}$. Notice that, if $U(t)$ is defined along a curve $\alpha(t)$, then the chain rule applied to each coordinate allows us to write $U_t = \frac{dU}{dt} = \nabla_{\alpha'(t)} U$. From our discussion above, we know that $\nabla_{\mathbf{v}} U$ tells us how M curves in the $\mathbf{v}$-direction. That is, $\nabla_{(\cdot)} U$ tells us about shape. We, therefore, define,

$$S_p(\mathbf{v}) = -\nabla_{\mathbf{v}} U$$

to be the *Shape Operator (or Weingarten Map) of M at p*. The negative sign in the definition of S_p is simply a convention which will allow us to give "positive" and "negative" curvature appropriate meanings. Also, note that choosing $-U$ as normal reverses the sign of S as well. S is called an operator because of the following

Lemma 2.2. *S_p is a linear transformation from $T_p(M)$ to itself.*

Proof. Recall that $T: V \to V$ is a *linear transformation* of the vector space V to itself if it has two properties: (1) $T(\mathbf{v}_1 + \mathbf{v}_2) = T(\mathbf{v}_1) + T(\mathbf{v}_2)$ and (2) $T(c \cdot \mathbf{v}) = c \cdot T(\mathbf{v})$ for any constant c. To prove the lemma, the first thing we have to show is that $S_p(\mathbf{v})$ is another vector in $T_p(M)$. To show this, it suffices to prove that $S_p(\mathbf{v})$ is perpendicular to U (since being perpendicular to U at p defines $T_p(M)$). Now $U \cdot U = (u^1)^2 + (u^2)^2 + (u^3)^2 = 1$ since U is a unit vector field and the usual Leibniz rule gives $\mathbf{v}[U \cdot U] = 2\nabla_{\mathbf{v}} U \cdot U$ (show this!). Thus, for any $\mathbf{v}$, $0 = \mathbf{v}[1] = \mathbf{v}[U \cdot U] = 2\nabla_{\mathbf{v}} U \cdot U$. Hence, $\nabla_{\mathbf{v}} U \cdot U = 0$ and $\nabla_{\mathbf{v}} U$ is perpendicular to U.

To show S_p linear, we must show that (1) $S_p(a\mathbf{v}) = aS_p(\mathbf{v})$ for $a \in \mathbb{R}$ and (2) $S_p(\mathbf{v} + \mathbf{w}) = S_p(\mathbf{v}) + S_p(\mathbf{w})$. Property (1) follows immediately from the definition of $\nabla_{\mathbf{v}} U$. Property (2) follows once we know that, for any function f, $(\mathbf{v} + \mathbf{w})[f] = \mathbf{v}[f] + \mathbf{w}[f]$. By the following exercise, we are done. □

EXERCISE 2.4. Let $\alpha(t) = \mathbf{x}(u(t), v(t))$, $\beta(t) = \mathbf{x}(\bar{u}(t), \bar{v}(t))$ with $\alpha(0) = p = \beta(0)$ and $\alpha'(0) = \mathbf{v}$, $\beta'(0) = \mathbf{w}$. Show that $\gamma'(0) = \mathbf{v} + \mathbf{w}$ for $\gamma(t) = \mathbf{x}(u(t) + \bar{u}(t), v(t) + \bar{v}(t))$. Then, using the gradient formula, show that $(\mathbf{v} + \mathbf{w})[f] = \mathbf{v}[f] + \mathbf{w}[f]$.

Because $S_p \colon T_p(M) \to T_p(M)$ is a linear transformation, we may bring to bear all the tools of linear algebra to analyze it. In particular, S_p has a matrix form (which we discuss below) and standard invariants such as determinant, trace and eigenvalues have deep geometric significance. In this sense, the shape operator is the most fundamental of all the tools used to study the geometry of surfaces embedded in $\mathbb{R}^3$.

Example 2.5: M is a plane in $\mathbb{R}^3$.
We know $U = \sum u^i e_i$ is *constant*. Because the u^i are constant, $\mathbf{v}[u^i] = 0$ for all $\mathbf{v}$. Hence, $S_p(\mathbf{v}) = -\nabla_{\mathbf{v}} U = -\sum \mathbf{v}[u^i] e_i = 0$ for all $\mathbf{v}$. This makes sense since a plane is flat — that is, the shape operator detects no "shape."

Example 2.6: $S^2(R)$ is the sphere of the radius R.
We shall use the standard parametrization of the R-sphere given by $\mathbf{x}(u, v) = (R \cos u \cos v, \ R \sin u \cos v, \ R \sin v)$ with tangent basis vectors,

$$\mathbf{x}_u = (-R \sin u \cos v, R \cos u \cos v, 0),$$
$$\mathbf{x}_v = (-R \cos u \sin v, -R \sin u \sin v, R \cos v).$$

Also, the unit normal is given by $U = (\cos u \cos v, \ \sin u \cos v, \ \sin v)$. To understand the shape operator S on $S^2(R)$ it suffices to know what S does to the basis $\{\mathbf{x}_u, \mathbf{x}_v\}$. From our discussion above we know that $\mathbf{x}_u[f] = \frac{\partial f}{\partial u} = f_u$ and similarly for v. Therefore,

$$S(\mathbf{x}_u) = -\nabla_{\mathbf{x}_u} U$$
$$= -(\mathbf{x}_u[\cos u \cos v], \mathbf{x}_u[\sin u \cos v], \mathbf{x}_u[\sin v])$$
$$= -(-\sin u \cos v, \cos u \cos v, 0)$$
$$= -\frac{\mathbf{x}_u}{R}.$$

A similar calculation shows $S(\mathbf{x}_v) = -\frac{\mathbf{x}_v}{R}$. Thus, the shape operator multiplies every tangent vector by $-\frac{1}{R}$. The corresponding matrix is $-\frac{1}{R} I$, where I is the 2×2 identity matrix.

EXERCISE 2.5. Let M be the cylinder $x^2 + y^2 = R^2$ parametrized by $\mathbf{x}(u, v) = (R \cos u, R \sin u, v)$. Show that the shape operator on M is described on a basis by $S(\mathbf{x}_u) = -\frac{1}{R} \mathbf{x}_u$ and $S(\mathbf{x}_v) = 0$. Therefore, in the u-direction the cylinder resembles a sphere and in the v-direction a plane. Of course, intuitively, this is exactly right. Why?

EXERCISE 2.6. Show that the shape operator S for the torus
$\mathbf{x}(u, v) = ((R + r \cos u) \cos v, (R + r \cos u) \sin v, \ r \sin u)$, on a basis, is given by $S(\mathbf{x}_u) = -\frac{\mathbf{x}_u}{r}$ and $S(\mathbf{x}_v) = \frac{-\cos u}{R + r \cos u} \mathbf{x}_v$.

EXERCISE 2.7. Show that the shape operator S for the saddle surface $z = xy$ parametrized by $\mathbf{x}(u, v) = (u, v, uv)$ is given on a basis by

$$S(\mathbf{x}_u) = \frac{-uv}{(1 + u^2 + v^2)^{\frac{3}{2}}} \mathbf{x}_u + \frac{1 + v^2}{(1 + u^2 + v^2)^{\frac{3}{2}}} \mathbf{x}_v$$

$$S(\mathbf{x}_v) = \frac{1 + u^2}{(1 + u^2 + v^2)^{\frac{3}{2}}} \mathbf{x}_u - \frac{uv}{(1 + u^2 + v^2)^{\frac{3}{2}}} \mathbf{x}_v.$$

We have already seen that a plane has zero shape operator. Intuitively, since the shape operator detects the change in the unit normal U, a zero shape operator for a surface M should imply that M is a plane. This is verified by the following result. Notice that the algebraic condition on the shape operator must be translated into a geometric restriction on M — a restriction which characterizes a plane. Recall that a plane is described by a point p and a normal vector U which we think of as originating from p. Then the plane is given by the collection of all points q with $(q - p) \cdot U = 0$.

Theorem 2.3. *If $S_p = 0$ at every $p \in M$, then M is contained in a plane.*

Proof. Fix $p \in M$ with unit normal $U(p)$ at p and take an arbitrary point $q \in M$. We will show that q is in the plane determined by p and $U(p)$. That is, we will show $(q - p) \cdot U(p) = 0$. Since q is arbitrary, then all of M will be in this plane. Take a curve α in M with $\alpha(0) = q$ and $\alpha(1) = p$ and define a function

$$f(t) = (q - \alpha(t)) \cdot U(\alpha(t)).$$

The product rule and the fact that $\frac{dU}{dt} = \nabla_{\alpha'(t)} U$ allow us to compute the derivative of this function.

$$f'(t) = -\alpha'(t) \cdot U(\alpha(t)) + (q - \alpha(t)) \cdot \nabla_{\alpha'(t)} U(\alpha(t))$$
$$= 0$$

since $\alpha'(t) \in T_{\alpha(t)}M$ is perpendicular to $U(\alpha(t))$ and $\nabla_{\alpha'(t)} U(\alpha(t)) = -S_{\alpha(t)}(\alpha'(t)) = 0$ by hypothesis. Thus, since the derivative of f is zero, the function $f(t)$ is constant. To see what constant this is, we can evaluate f at 0. But clearly $f(0) = 0$ since $\alpha(0) = q$, so $(q - \alpha(t)) \cdot U(\alpha(t)) = 0$ for all t. In particular, for $t = 1$ we get $(q - p) \cdot U(p) = 0$ and we are done. $\square$

2.3 THE LINEAR ALGEBRA OF SURFACES

Now, it should be clear that even for slightly complicated examples, an exact global representation for S on a surface M may be difficult to achieve. Even if we do obtain an exact representation of the shape operator (on a basis say), the form of the representation may not give much information (e.g. the saddle above). We can still cull the essential geometry of M, however, from computable quantities associated to the shape operator. *Therefore, the usefulness of the shape operator itself is derived not from its computability, but rather from its theoretical description of geometry in terms of linear algebra.* With this in mind, we shall prove one theoretical fact about the shape operator. The computations in the proof will be of importance later. Before we do this, it is worthwhile to recall a few notions of linear algebra.

Suppose $T\colon V \to V$ is a *linear transformation* of the vector space V to itself. If we are given a basis $\mathcal{B} = \{\mathbf{x}_1, \ldots, \mathbf{x}_n\}$ for V, then T may be represented by a matrix A. This correspondence works in the following way. First, $T(\mathbf{x}_i) = \sum_{j=1}^{n} a^{ji}\mathbf{x}_j$ since $\mathcal{B}$ is a basis. Second, for fixed i, the a^{ji} may be assembled into the ith column of a matrix A. Doing this for each i gives $A = (a^{ji})$, an $n \times n$ matrix which takes $\mathbf{e}_i = (0, \ldots, 0, 1^i, 0, \ldots, 0)$ (thought of as a column vector and representing $\mathbf{x}_i$) to $\mathbf{a} = (a^{1i}, a^{2i}, \ldots, a^{ni})$. Here the vectors $\mathbf{x}_i$ and $T(\mathbf{x}_i)$ have been identified with their respective matrices of coefficients $\mathbf{e}_i$ and $\mathbf{a}$ *with respect to the basis* $\mathcal{B}$. Notice that, while the linear transformation T is defined without regard to a chosen basis, its representation as a matrix A depends essentially on which basis is chosen. Choosing a different basis for V changes the coefficients in the expansion of $T(\mathbf{x}_i)$ and so a new matrix arises. For a linear transformation T, if there is some vector $\mathbf{v}$ and real number λ such that

$$T(\mathbf{v}) = \lambda\mathbf{v}$$

then λ is called an *eigenvalue* of T associated to the *eigenvector* $\mathbf{v}$. If V is n-dimensional and there are n linearly independent eigenvectors $\mathbf{v}_1, \cdots, \mathbf{v}_n$ with associated eigenvalues $\lambda_1, \cdots, \lambda_n$, then, in this basis for V, the matrix for T is diagonal (and hence very simple)

$$\begin{bmatrix} \lambda_1 & & & 0 \\ & \lambda_2 & & \\ & & \ddots & \\ 0 & & & \lambda_n \end{bmatrix}.$$

Therefore, $\det(T) = \prod_{i=1}^{n} \lambda_i$ and $\operatorname{tr}(T) = \sum_{i=1}^{n} \lambda_i$. In general, the determinant of a matrix is the product of its eigenvalues and the trace is the sum of the eigenvalues.

Example 3.1. Let $T\colon \mathbb{R}^2 \to \mathbb{R}^2$ be a linear transformation with $T(1,0) = (4,3)$ and $T(0,1) = (-2,-1)$. The matrix for T with respect to the basis $\{(1,0),(0,1)\}$ is

$$\begin{bmatrix} 4 & -2 \\ 3 & -1 \end{bmatrix}.$$

Suppose we have

$$\begin{bmatrix} 4 & -2 \\ 3 & -1 \end{bmatrix}\begin{bmatrix} a \\ b \end{bmatrix} = \begin{bmatrix} \lambda a \\ \lambda b \end{bmatrix} = \begin{bmatrix} \lambda & 0 \\ 0 & \lambda \end{bmatrix}\begin{bmatrix} a \\ b \end{bmatrix}.$$

Subtracting, we get

$$\begin{bmatrix} 4 - \lambda & -2 \\ 3 & -1 - \lambda \end{bmatrix} \begin{bmatrix} a \\ b \end{bmatrix} = \begin{bmatrix} 0 \\ 0 \end{bmatrix}.$$

But a matrix can only take nonzero vectors to zero if $\det(T) = 0$. Hence

$$\det \begin{bmatrix} 4 - \lambda & -2 \\ 3 & -1 - \lambda \end{bmatrix} = 0,$$

so $(4 - \lambda)(-1 - \lambda) + 6 = 0$ or $\lambda^2 - 3\lambda + 2 = 0 = (\lambda - 1)(\lambda - 2)$. Thus $\lambda = 1$ or $\lambda = 2$. These are the eigenvalues of T. To find the eigenvectors, plug λ back into the matrix equation to get,

For $\lambda = 1$: $\begin{bmatrix} 3 & -2 \\ 3 & -2 \end{bmatrix} \begin{bmatrix} a \\ b \end{bmatrix} = \begin{bmatrix} 0 \\ 0 \end{bmatrix}$ or $3a - 2b = 0$ or $\dfrac{3}{2}a = b$.

This means that any vector of the form $\left[a, \frac{3a}{2} \right] = v$ will satisfy $T(\mathbf{v}) = \mathbf{v}$ (since $\lambda = 1$). Indeed, take $[2, 3]$ and note that

$$\begin{bmatrix} 4 & -2 \\ 3 & -1 \end{bmatrix} \begin{bmatrix} 2 \\ 3 \end{bmatrix} = \begin{bmatrix} 2 \\ 3 \end{bmatrix}.$$

EXERCISE 3.1. Show that the other eigenvector is given by $[a, a]$.

EXERCISE 3.2. Show that, if $[2, 3]$ and $[1, 1]$ are chosen as a basis for $\mathbb{R}^2$, then the matrix for T relative to this basis is

$$\begin{bmatrix} 1 & 0 \\ 0 & 2 \end{bmatrix}.$$

Finally, note that $\lambda_1 \lambda_2 = 2$, $\lambda_1 + \lambda_2 = 3$ and, calculating from the original matrix, these are equal to

$$\det \begin{pmatrix} 4 & -2 \\ 3 & -1 \end{pmatrix} = -4 + 6 = 2 \qquad \operatorname{tr} \begin{pmatrix} 4 & -2 \\ 3 & -1 \end{pmatrix} = 4 - 1 = 3.$$

A linear transformation $T \colon \mathbb{R}^2 \to \mathbb{R}^2$ is said to be *symmetric* if $T(\mathbf{v}) \cdot \mathbf{w} = \mathbf{v} \cdot T(\mathbf{w})$ for all vectors $\mathbf{v}$ and $\mathbf{w}$ in $\mathbb{R}^2$.

EXERCISE 3.3.

(1) Show that, with respect to any orthonormal basis, if the 2×2 matrix $\begin{bmatrix} a & b \\ c & d \end{bmatrix}$ represents a symmetric linear transformation, then $b = c$. Such a matrix is called a *symmetric matrix*. Recall that a basis is *orthnormal* if it consists of unit vectors which are perpendicular.

(2) Show that any 2×2 symmetric matrix has real eigenvalues.

Theorem 3.1. *The shape operator is a symmetric linear transformation.*
Further, $S(\mathbf{x}_u) \cdot \mathbf{x}_u = \mathbf{x}_{uu} \cdot U$, $S(\mathbf{x}_u) \cdot \mathbf{x}_v = \mathbf{x}_{uv} \cdot U$ *and* $S(\mathbf{x}_v) \cdot \mathbf{x}_v = \mathbf{x}_{vv} \cdot U$.

Proof. Since $\{\mathbf{x}_u, \mathbf{x}_v\}$ is a basis for $T_p(M)$ we need only show the equation on $\mathbf{x}_u$ and $\mathbf{x}_v$. Note first that $U \cdot \mathbf{x}_u = 0 = U \cdot \mathbf{x}_v$ since U is a normal to $T_p(M)$. Now $U \cdot \mathbf{x}_u$ and $U \cdot \mathbf{x}_v$ are functions, so a vector may be applied to them.

$$0 = \mathbf{x}_u[0] = \mathbf{x}_u[U \cdot \mathbf{x}_v]$$

$$= \mathbf{x}_u \left[\sum_i u^i \frac{\partial x^i}{\partial v} \right]$$

$$= \sum_i (\mathbf{x}_u[u^i] \frac{\partial x^i}{\partial v} + u^i \mathbf{x}_u[\frac{\partial x^i}{\partial v}]) \qquad \text{by the Leibniz Rule}$$

$$= \sum_i \mathbf{x}_u[u^i]\mathbf{e}_i \cdot \mathbf{x}_v + U \cdot \frac{\partial^2 x^i}{\partial u \partial v}$$

$$= \nabla_{\mathbf{x}_u} U \cdot \mathbf{x}_v + U \cdot \mathbf{x}_{vu}.$$

Hence, $S_p(\mathbf{x}_u) \cdot \mathbf{x}_v = -\nabla_{\mathbf{x}_u} U \cdot \mathbf{x}_v = U \cdot \mathbf{x}_{vu}$. A similar computation shows,

$$S_p(\mathbf{x}_v) \cdot \mathbf{x}_u = -\nabla_{\mathbf{x}_v} U \cdot \mathbf{x}_u = U \cdot \mathbf{x}_{uv}.$$

But $\mathbf{x}_{uv} = \mathbf{x}_{vu}$ so the result follows. Finally, the same calculation as above shows $S(\mathbf{x}_u) \cdot \mathbf{x}_u = \mathbf{x}_{uu} \cdot U$ and $S(\mathbf{x}_v) \cdot \mathbf{x}_v = \mathbf{x}_{vv} \cdot U$. $\square$

Corollary 3.2. *The shape operator has real eigenvalues.*

There is yet another way linear algebra makes its presence felt in differential geometry. Given a map of surfaces $F \colon M \to N$, we may define its *derivative* to be a certain linear transformation on all tangent planes $F_{*p} \colon T_p(M) \to T_{F(p)}(N)$. This linear transformation (which we shorten to) F_* is defined by letting a tangent vector $\mathbf{v}$ be represented by a curve $\alpha \colon I \to M$ with $\alpha(0) = p$ and $\alpha'(0) = \mathbf{v}$ and saying

$$F_*(\alpha'(0)) \overset{\text{def}}{=} \frac{d}{dt}(F(\alpha(t)) \mid_{t=0} .$$

Now, $\beta(t) = F(\alpha(t))$ is simply a curve on N and the righthand-side of the definition of F_* is simply the velocity vector of β at $F(p)$. The geometry of the definition may now be seen. In order to understand F, linearly approximate M near p by the tangent plane $T_p(M)$. Each vector in $T_p(M)$ has a curve whose velocity vector is that vector, so use composition with F to map the curves over to N. The curves and their images under F fill out regions near p and $F(p)$ respectively. To see the geometry of even this local mapping, however, we again must linearly approximate by taking derivatives of the image curves in N. Then, by definition, the velocity vector of the image curve is the image of the velocity vector of the original curve in M.

We have made a big deal of this notion of the derivative of a surface mapping not only because it is a difficult idea, but also because it plays an important role in what is to follow. In particular, as we shall see later, surface mappings and their derivatives will give us the right tools to compare surfaces and their geometries. Of course, a linear transformation is determined by its effect on a basis, so, given a parametrization $M \colon \mathbf{x}(u,v)$, F_* may be determined completely by calculating $F_*(\mathbf{x}_u)$ and $F_*(\mathbf{x}_v)$.

Example 3.2. Let $F \colon \mathbb{R}^2 \to \mathbb{R}^2$ be a map of the plane to itself. In coordinates, we may write $F(u,v) = (f(u,v), g(u,v))$. To calculate $F_*(\mathbf{x}_u)$ we take the composition of the u-parameter curve (u, v_0) with F, $F(u, v_0) = (f(u, v_0), g(u, v_0))$ and differentiate with respect to u to get (f_u, g_u). A similar calculation for the v-parameter curve gives

$$F_*(\mathbf{x}_u) = (f_u, g_u) \qquad \text{and} \qquad F_*(\mathbf{x}_v) = (f_v, g_v).$$

By what we have said about the correspondence between linear transformations and matrices, we see that, with respect to the basis $\{\mathbf{x}_u, \mathbf{x}_v\}$, the matrix for F_* is

$$J(F) = \begin{bmatrix} f_u & f_v \\ g_u & g_v \end{bmatrix}.$$

This is the Jacobian matrix of several variable calculus.

The derivative mapping of a surface map is, therefore, just a natural extension of the Jacobian from maps of Euclidean spaces to maps of surfaces. Now, the shape operator is a linear transformation of the tangent plane, so we might well ask if there is a surface mapping whose derivative is $\pm S$? The answer is, in fact, yes. The mapping whose derivative is $\pm S$ is called the *Gauss map* and its behavior provides a geometric alternative to the linear-algebraic approach of the shape operator. The Gauss map is a mapping $G \colon M \to S^2$ from the surface M to the unit sphere S^2 given by $G(p) = U(p)$, where $U(p)$ is the unit normal of M at p. Since $U(p)$ is a unit vector in $\mathbb{R}^3$, we may represent it as a point on S^2, so this definition makes sense. Now, the Gauss map has an induced derivative map $G_* \colon T_p M \to T_{G(p)} S^2$ and, by definition,

$$G_*(\mathbf{v}) = \frac{d}{dt}(G(\alpha(t))|_{t=0}$$

$$= \frac{d}{dt}(U(\alpha(t))|_{t=0}$$

$$= \nabla_{\alpha'(0)} U$$

$$= \nabla_{\mathbf{v}} U$$

$$= -S(\mathbf{v}),$$

where $\alpha(0) = p$ and $\alpha'(0) = \mathbf{v}$. One thing needs to be made clear. Namely, the shape operator has range $T_p(M)$ while the Gauss map has range $T_{G(p)}S^2$, so how can $G_* = -S$? We must look at the geometry of the situation for the answer. The tangent plane, no matter where we visualize it meeting the surface, is a vector space and, therefore, passes through the origin. The plane $T_p(M)$ is the plane through the origin perpendicular to $U(p)$. For any point $q \in S^2$, the tangent plane $T_q(S^2)$ has the beautiful property that it is perpendicular to q itself, thought of as a vector in $\mathbb{R}^3$. Because $G(p) = U(p)$ by definition, the plane $T_{G(p)}(S^2)$ is the plane through the origin perpendicular to $U(p)$. That is, $T_p(M) = T_{G(p)}(S^2)$. This explains how the computation above can make sense. We will see later that the Gauss map has many uses. For now, try

EXERCISE 3.4. Compute the Gauss map and its derivative for the cone $\mathbf{x}(u, v) = (v \cos u, v \sin u, v)$ and estimate the amount of area the image of the Gauss map takes up on the sphere.

EXERCISE 3.5. Compute the Gauss map and its derivative for the cylinder $\mathbf{x}(u, v) = (R \cos u, R \sin u, v)$ and estimate the amount of area the image of the Gauss map takes up on the sphere.

EXERCISE 3.6. Compute the Gauss map and its derivative for the catenoid $\mathbf{x}(u, v) = (u, \cosh u \cos v, \cosh u \sin v)$. Show that the Gauss map is a one-to-one map from the catenoid to the sphere. Hint: focus on the first coordinate first.

EXERCISE 3.7. Compute the Gauss map for Enneper's surface $\mathbf{x}(u, v) = (u - \frac{u^3}{3} + uv^2, \ v - \frac{v^3}{3} + vu^2, \ u^2 - v^2)$ and show that it is a one-to-one map from Enneper's surface to the sphere. Also show that the image of the disk $\{(u, v)|u^2 + v^2 \le 3\}$ under G covers more than a hemisphere of the sphere. Hint: first, write U in polar coordinates by letting $u = r\cos\theta$ and $v = r\sin\theta$ and, then focus on the third coordinate.

2.4 NORMAL CURVATURE

We would now like to use the shape operator to obtain a notion of *curvature* of a surface. After a very geometric approach to this *normal curvature* we will prove a theorem linking the geometry to the linear algebra of S. First we note,

Lemma 4.1. *If α is a curve in M, then $\alpha'' \cdot U = S(\alpha') \cdot \alpha'$.*

Proof. We will give two proofs of this result. The first reinforces the computations above and the second foreshadows the types of calculations we will do when we consider geodesics.

(1.) We know $\alpha' \cdot U = 0$ since U is normal to $T_p(M)$. Then,

$$0 = \alpha'[\alpha' \cdot U]$$

$$= \alpha'\left[\sum_i \alpha^{i'} u^i\right]$$

$$= \sum_i \alpha'[\alpha^{i'}]u^i + \sum_i \alpha^{i'}\alpha'[u^i] \quad \text{by the Leibniz Rule}$$

$$= \sum_i \frac{d^2\alpha^i}{dt^2}u^i + \sum_i \alpha'[u^i]\mathbf{e}_i \cdot \alpha' \quad \text{since} \quad \alpha^{i'}(\alpha(t)) = \frac{d\alpha^i}{dt}(t)$$

$$= \alpha'' \cdot U + \nabla_{\alpha'} U \cdot \alpha'.$$

Hence, $S(\alpha') \cdot \alpha' = -\nabla_{\alpha'} U \cdot \alpha' = \alpha'' \cdot U$.

(2.) By Lemma 1.1, write $\alpha(t) = \mathbf{x}(u(t), v(t))$ and $\alpha' = u'\mathbf{x}_u + v'\mathbf{x}_v$ by the chain rule. Applying the chain rule again gives

$$\alpha'' = u'^2\mathbf{x}_{uu} + 2u'v'\mathbf{x}_{uv} + v'^2\mathbf{x}_{vv} + u''\mathbf{x}_u + v''\mathbf{x}_v.$$

Now, $\mathbf{x}_u \cdot U = 0$ and $\mathbf{x}_v \cdot U = 0$, so

$$\alpha'' \cdot U = (u'^2\mathbf{x}_{uu} + 2u'v'\mathbf{x}_{uv} + v'^2\mathbf{x}_{vv}) \cdot U.$$

Because S is a linear transformation, we have

$$
\begin{aligned}
S(\alpha') \cdot \alpha' &= S(u'\mathbf{x}_u + v'\mathbf{x}_v) \cdot (u'\mathbf{x}_u + v'\mathbf{x}_v) \\
&= (u'S(\mathbf{x}_u) + v'S(\mathbf{x}_v)) \cdot (u'\mathbf{x}_u + v'\mathbf{x}_v) \\
&= u'^2 S(\mathbf{x}_u) \cdot \mathbf{x}_u + u'v'S(\mathbf{x}_u) \cdot \mathbf{x}_v + u'v'S(\mathbf{x}_v) \cdot \mathbf{x}_u + v'^2 S(\mathbf{x}_v) \cdot \mathbf{x}_v \\
&= u'^2\mathbf{x}_{uu} \cdot U + u'v'\mathbf{x}_{uv} \cdot U + u'v'\mathbf{x}_{uv} \cdot U + v'^2\mathbf{x}_{vv} \cdot U \\
&= (u'^2\mathbf{x}_{uu} + 2u'v'\mathbf{x}_{uv} + v'^2\mathbf{x}_{vv}) \cdot U \\
&= \alpha'' \cdot U. \qquad \qquad \square
\end{aligned}
$$

We interpret $\alpha'' \cdot U$ as the component of acceleration due to the bending of M. Of course, we assume that α has unit speed so that the magnitude of α' does not affect our measurement. With this in mind, we make the following definition. For a *unit* vector $\mathbf{u} \in T_p(M)$, the *normal curvature* of M in the $\mathbf{u}$-direction is

$$k(\mathbf{u}) = S_p(\mathbf{u}) \cdot \mathbf{u}.$$

Let α be a curve with $\alpha(0) = p$, $\alpha'(0) = \mathbf{u}$. Then

$$
\begin{aligned}
k(\mathbf{u}) &= S_p(\mathbf{u}) \cdot \mathbf{u} \\
&= S_p(\alpha'(0)) \cdot \alpha'(0) \\
&= \alpha''(0) \cdot U(p) \\
&= \kappa(0)N(0) \cdot U(p) \\
&= \kappa(0)\cos\theta
\end{aligned}
$$

where N is the Frenet normal to the curve α and κ is α's curvature. The angle θ is the angle between $N(0)$ and $U(p)$.

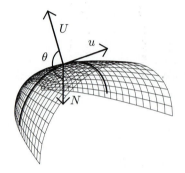

FIGURE 2.16.   Normal curvature

EXERCISE 4.1. The *total torsion* of a curve $\alpha\colon [a, b] \to \mathbb{R}^3$ is $\int_a^b \tau\, dt$. Show that a closed curve on an R-sphere has zero total torsion. Hints: use the formula above, $k(T) = \kappa_\alpha \cos\theta$, where $T = \alpha'$ for a unit speed curve α, the shape operator of the sphere and differentiate $\cos\theta = N \cdot U$. According to Exercise 1.3.12, this means the linking number of α and $\alpha + \epsilon N$ is zero. Why is this true geometrically?

While the normal curvature is a multiple of the curvature of α, we might ask if there is a curve σ whose curvature is exactly equal to the normal curvature $k(\mathbf{u})$. The answer turns out to be yes.

Proposition 4.2. Let P denote the plane determined by $U(p)$ and $\mathbf{u}$ (at $p \in M$) and let σ denote the unit speed curve formed by $P \cap M$ with $\sigma(0) = p$. Then $k(\mathbf{u}) = \pm\kappa_\sigma(0)$.

Proof. First we show that $\sigma'(0) = \mathbf{u}$. But this follows since $\mathbf{u}$ and $\sigma'(0)$ are tangent vectors and $\mathbf{u}$, $\sigma'(0)$ and $U(p)$ lie in a single plane. Namely, the only way for $\mathbf{u}$ and $\sigma'(0)$ to both be perpendicular to $U(p)$ and in P is for $\sigma'(0) = \pm\mathbf{u}$. Take a parametrization with $\sigma'(0) = \mathbf{u}$. Now σ's normal $N_\sigma(0)$ is perpendicular to $T_\sigma(0) = \sigma'(0) = \mathbf{u}$ and $N_\sigma(0)$ is in the plane P since σ ia a plane curve. Thus, $N_\sigma(0) = \pm U(p)$. Then $\cos\theta = \pm 1$ since $\theta = 0$ or π. Hence, $k(\mathbf{u}) = \pm\kappa_\sigma(0)$. $\qquad\square$

What does normal curvature tell us? If $k(\mathbf{u}) > 0$, then the normal $N_\sigma(0)$ is equal to $U(p)$ (i.e. $\theta = 0°$) and σ bends up toward $U(p)$. Hence, so does M along σ. If $k(\mathbf{u}) < 0$, then the normal $N_\sigma(0)$ is equal to $-U(p)$, so, along σ, M bends away from $U(p)$. If $k(\mathbf{u}) = 0$, then $\kappa_\sigma(0) = 0$. This does not mean that σ is a line, of course, since we don't know that $\kappa_\sigma(t) = 0$ for all t. But it does say that *near* p the rate of bending is small.

Therefore, the sign of normal curvature tells us about the bending of M toward or away from its normal in a given direction. Note that by changing the normal to $-U(p)$ the signs on $k(\mathbf{u})$ reverse. Therefore, to avoid this ambiguity, we must fix a convention for normals. For instance, we could say that we will always take outward pointing normals.

Normal curvature is a function from unit vectors in a plane — that is, a circle of radius one — to real numbers. In fact, it can be shown that this normal curvature function is continuous. Just as any continuous function on a closed interval attains its maximum and minimum, so too does any continuous function on a compact (i.e. closed and bounded) set like the circle. Hence, there are unit vectors $\mathbf{u}_1$ and $\mathbf{u}_2$ such that

$$k(\mathbf{u}_1) = k_1 = \max_{\mathbf{u}} k(\mathbf{u}) \qquad k(\mathbf{u}_2) = k_2 = \min_{\mathbf{u}} k(\mathbf{u}).$$

The unit vectors $\mathbf{u}_1$ and $\mathbf{u}_2$ are called *principal vectors* and k_1 and k_2 are *principal curvatures*.

EXERCISE 4.2. A curve $\alpha\colon I \to M$ is a *line of curvature* if $\alpha'(t)$ is an eigenvector of the shape operator S for every $t \in I$. This is equivalent to saying that the unit tangent T_α is always either $\mathbf{u}_1$ or $\mathbf{u}_2$. Show that α is a line of curvature if and only if α' is parallel to $U' = \nabla_{\alpha'} U$ along α. Further, suppose α is a plane curve. That is $\alpha = M \cap P$ for some plane P. Then show that α is a line of curvature if the angle between M and P is constant along α.

Let's do an example before we go on just to see how normal curvature may be computed. We will make use of the Frenet Formulas.

Example 4.1. Let M be a saddle surface $z = x^2 - y^2$. We use the normal U given by the gradient of $g(x, y, z) = z - x^2 + y^2 = 0$,

$$U = \frac{\nabla g}{|\nabla g|} = \frac{(-2x, 2y, 1)}{\sqrt{1 + 4x^2 + 4y^2}}.$$

Let $p = (0, 0, 0)$. Then $U(p) = (0, 0, 1)$. Take $\mathbf{u} = (1, 0, 0)$ for example. The curve σ determined by the intersection of the plane spanned by $(0, 0, 1)$ and $(1, 0, 0)$ with M is the parabola $z = x^2$ bending toward the normal. Therefore, we should find $k(\mathbf{u}) > 0$. Let's see how to compute explicitly.

The plane through $\mathbf{u}$ and $U(p)$ is the xz-plane with normal vector $\mathbf{u} \times U(p) = (0, -1, 0)$. Hence, $y = 0$ and σ is the parabola $z = x^2$ which may be parametrized by $\sigma(t) = (t, 0, t^2)$ say. Then $\sigma'(t) = (1, 0, 2t)$ and $\sigma''(t) = (0, 0, 2)$. Then $k(1, 0, 0) = +\kappa_\sigma(0) = \frac{|\sigma'(0) \times \sigma''(0)|}{|\sigma'(0)|^3} = \frac{2}{1} = 2 > 0$ just as we thought. **Note.** We could find the entire Frenet Frame: $T = \frac{\sigma'}{|\sigma'|} = \left(\frac{1}{\sqrt{1+4t^2}}, 0, \frac{2t}{\sqrt{1+4t^2}}\right)$ with $T(0) = (1, 0, 0)$; $B = \frac{\sigma' \times \sigma''}{|\sigma' \times \sigma''|} = (0, -1, 0)$; $N = B \times T = (0, 0, 1) = U$ *with a $+$ sign. This justifies* $k(1, 0, 0) = +\kappa_\sigma(0)$.

EXERCISE 4.3. Find $k(\mathbf{u})$ at $p = (0, 0, 0)$ where $\mathbf{u} = (0, 1, 0)$. What sign should your answer have?

EXERCISE 4.4. Find $k(\mathbf{u})$ at $p = (0, 0, 0)$ where $\mathbf{u} = \left(\frac{1}{\sqrt{2}}, \frac{1}{\sqrt{2}}, 0\right)$.

Example 4.2. Let $M : x^2 + y^2 = 1$ be a cylinder with $p = (1, 0, 0)$ and $U(p) = (1, 0, 0)$. A unit vector $\mathbf{u} \in T_p(M)$ has the form $\mathbf{u} = (0, u^1, u^2)$ with $(u^1)^2 + (u^2)^2 = 1$. A normal for the plane determined by $\mathbf{u}$ and $U(p)$ is $(0, -u^2, u^1)$, so the plane's equation is $z = (u^2/u^1)y$. The intersection of the plane with M is the set $\left\{\left(\sqrt{1-y^2}, y, (u^2/u^1)y\right)\right\}$ for any y. Parametrize σ by $\sigma(t) = \left(\sqrt{1-t^2}, t, (u^2/u^1)t\right)$ with

$$\sigma'(t) = \left(-t/\sqrt{1-t^2}, 1, u^2/u^1\right) \quad \text{and} \quad \sigma''(t) = \left(-1/(1-t^2)^{3/2}, 0, 0\right).$$

Then we have,

$$T(0) = (0, 1, u^2/u^1)/\sqrt{1 + (u^2/u^1)^2},$$
$$B(0) = \sigma'(0) \times \sigma''(0)/|\sigma'(0) \times \sigma''(0)|$$
$$= (0, -u^2/u^1, 1)/\sqrt{1 + (u^2/u^1)^2}$$
$$N(0) = \left(-(u^2/u^1)^2 - 1/1 + (u^2/u^1)^2, 0, 0\right)$$
$$= (-1, 0, 0)$$

since $(u^1)^2 + (u^2)^2 = 1$. Hence, $N(0) = -U(p)$, so we need a $-$ sign in $k(\mathbf{u}) = -\kappa_\sigma(0)$. Further,

$$\kappa_\sigma(0) = \frac{|\sigma'(0) \times \sigma''(0)|}{|\sigma'(0)|^3} = \frac{((u^2/u^1)^2 + 1)^{1/2}}{(1 + (u^2/u^1)^2)^{3/2}} = (u^1)^2.$$

Hence, $k(\mathbf{u}) = -(u^1)^2$. This is negative or zero. Now, since $\mathbf{u} = (0, u^1, u^2)$ is on the unit circle in the yz-plane, $\max k(\mathbf{u}) = 0$ occurs when $u^1 = 0$ and $\min k(\mathbf{u}) = -1$ occurs when $u^1 = 1$. That is,

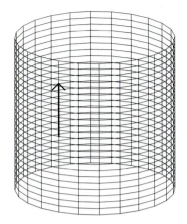

FIGURE 2.17.   max; $k(\mathbf{u}) = 0$; $u^1 = 0$; $\mathbf{u} = (0,0,1)$ in direction of rulings

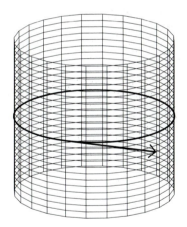

FIGURE 2.18.   min; $k(\mathbf{u}) = -1$; $u^1 = 1$; $\mathbf{u} = (0,1,0)$ in direction of velocity vector of directrix

The corresponding geometry is clear. The cylinder M is flat in the ruling directions and bends away from the normal in directrix directions. Indeed, the bending is what we might call circular.

EXERCISE 4.5. Work through the example above.

Although it was defined in terms of the shape operator originally, we have seen that normal curvature may be described completely geometrically as the curvature of the curve of intersection of the surface with a particular plane. We will now present a fundamental theorem linking the geometry of normal curvature to the linear algebra of the shape operator. Say that a point $p \in M$ is an *umbilic point* if the principal curvatures at p are equal (i.e. $k_1(p) = k_2(p)$). Note that this implies that the normal curvature at p is constant. For instance, every point on a sphere is an umbilic point.

Theorem 4.3.

   (1) If $p \in M$ is umbilic, then $S_p(\mathbf{u}) = k\mathbf{u}$ where $k = k_1 = k_2$.
   (2) If $p \in M$ is not umbilic, then there are exactly two perpendicular unit eigenvectors of S_p with associated eigenvalues the principal curvatures at p.

Proof. We will prove the theorem by going in reverse. We shall consider the eigenvalues of the shape operator and show that these are, in fact, k_1 and k_2. Let $\mathbf{u}_1$ be a unit eigenvector for S with eigenvalue λ_1; $S(\mathbf{u}_1) = \lambda_1 \mathbf{u}_1$. Let $\mathbf{u}_2$ be the unit tangent vector rotated $90°$ counterclockwise from $\mathbf{u}_1$. We may write $S(\mathbf{u}_2) = a\,\mathbf{u}_1 + b\,\mathbf{u}_2$. But $a = S(\mathbf{u}_2) \cdot \mathbf{u}_1 = S(\mathbf{u}_1) \cdot \mathbf{u}_2 = \lambda_1 \mathbf{u}_1 \cdot \mathbf{u}_2 = 0$, so $S(\mathbf{u}_2) = \lambda_2 \mathbf{u}_2$, where $b = \lambda_2$. Thus, $\mathbf{u}_2$ is the other eigenvector of S and is perpendicular to $\mathbf{u}_1$. Also, $k(\mathbf{u}_i) = S(\mathbf{u}_i) \cdot \mathbf{u}_i = \lambda_i \mathbf{u}_i \cdot \mathbf{u}_i = \lambda_i$, so λ_1 and λ_2 are normal curvatures at p as well.

   Note that, if $\lambda_1 = \lambda_2 = \lambda$, then $S(\mathbf{u}) = \lambda\mathbf{u}$ for all $\mathbf{u}$ and p is umbilic. So, suppose without loss of generality that $\lambda_2 < \lambda_1$ and take a unit vector $\mathbf{u}$ which may be written as $\mathbf{u} = \cos\theta\,\mathbf{u}_1 + \sin\theta\,\mathbf{u}_2$, where θ is the angle between $\mathbf{u}$ and $\mathbf{u}_1$. We may now compute the normal curvature at $\mathbf{u}$, denoting the dependence of $\mathbf{u}$ on θ by writing $k(\theta)$.

$$
\begin{aligned}
k(\theta) &= S(\mathbf{u}) \cdot \mathbf{u} \\
&= S(\cos\theta\,\mathbf{u}_1 + \sin\theta\,\mathbf{u}_2) \cdot (\cos\theta\,\mathbf{u}_1 + \sin\theta\,\mathbf{u}_2) \\
&= (\cos\theta\,S(\mathbf{u}_1) + \sin\theta\,S(\mathbf{u}_2)) \cdot (\cos\theta\,\mathbf{u}_1 + \sin\theta\,\mathbf{u}_2) \\
&= \cos^2\theta\,S(\mathbf{u}_1) \cdot \mathbf{u}_1 + \sin\theta\cos\theta\,S(\mathbf{u}_1) \cdot \mathbf{u}_2 \sin\theta\cos\theta\,S(\mathbf{u}_2) \cdot \mathbf{u}_1 \\
&\qquad + \sin^2\theta\,S(\mathbf{u}_2) \cdot \mathbf{u}_2 \\
&= \cos^2\theta\,\lambda_1 + \sin^2\theta\,\lambda_2.
\end{aligned}
$$

Now write $k(\theta) = \lambda_1 + (\lambda_2 - \lambda_1)\sin^2\theta$ using $\sin^2\theta + \cos^2\theta = 1$. Since $\lambda_2 < \lambda_1$, $k(\theta)$ is a maximum when $\theta = 0$; that is, when $\mathbf{u} = \mathbf{u}_1$. Hence, $k_1 = k_{\max}(\theta) = k(\mathbf{u}_1) = \lambda_1$.

   Similarly, we may write $k(\theta) = (\lambda_1 - \lambda_2)\cos^2\theta + \lambda_2$ and, since $\lambda_2 < \lambda_1$, $k(\theta)$ is a minimum when $\theta = \frac{\pi}{2}$. That is, $k(\theta)$ is a minimum when $\mathbf{u} = \mathbf{u}_2$. Hence, $k_2 = k_{\min}(\theta) = k(\mathbf{u}_2) = \lambda_2$. $\qquad\square$

Corollary 4.4. For $\mathbf{u} = \cos\theta\,\mathbf{u}_1 + \sin\theta\,\mathbf{u}_2$ as above, the normal curvature is given by Euler's formula

$$
k(\mathbf{u}) = \cos^2\theta\,k_1 + \sin^2\theta\,k_2.
$$

2.5 PLOTTING SURFACES IN MAPLE

At the end of Chapter 1, rather offhandedly, I wrote the MAPLE commands necessary to plot a sphere and a cylinder together in order to illustrate Viviani's curve. In this short section, I want to describe how MAPLE plots surfaces in general. The basic command is "plot3d." Let's take Enneper's surface

$$\mathbf{x}(u,v) = (u - u^3/3 + uv^2, v - v^3/3 + vu^2, u^2 - v^2).$$

MAPLE input which may be used to plot Enneper's surface is as follows.

> with(plots):

> Ennep:=[u-u^3/3+u*v^2,v-v^3/3+v*u^2, u^2 - v^2];

> plot3d(Ennep,u=-1.5..1.5,v=-1.5..1.5,scaling=constrained, shading=zhue);

The "shading" option tells MAPLE to use a particular color scheme for plotting the surface. Different schemes may be tried by clicking on "color" in the toolbar which appears when an object is graphed. On the Power PC, you should click on the upper left corner of the graphics window to get the menu of options. Note that, just as for "spacecurve," the square brackets do not enclose the domain of the parameters. In the plot command above change the bounds of the parameters u and v to u=-1.9..1.9, v=-1.9..1.9. What is different? Now replace shading=zhue by shading=XY or shading=zgrayscale to change the color scheme at the command line instead of from the toolbar.

EXERCISE 5.1. Graph the hyperboloid of one sheet according to the parametrizations given in Exercise 2.1.13. Experiment with different bounds on the parameters as well as plotting options.

MAPLE may be used to conveniently create surfaces from particular data. For example, a ruled surface $\mathbf{x}(u,v) = \beta(u) + v\,\delta(u)$ may be defined by specifying β and δ.

```
> rulsurf:=proc(beta,delta)
>        local x1,x2,x3;
>        x1:=beta[1]+v*delta[1];
>        x2:=beta[2]+v*delta[2];
>        x3:=beta[3]+v*delta[3];
>        [x1,x2,x3];
>        end:
```

A particular type of ruled surface is a *tangent developable*. Given a curve $\beta(u)$, its tangent developable is defined to be the surface parametrized by $\mathbf{x}(u,v) = \beta(u) + v\beta'(u)$. We may create and graph the tangent developable of the helix (for instance) $\beta(u) = (\cos u, \sin u, u)$. See Figure 3.1 for the result.

> bb:=[cos(u),sin(u),u]; dd:=diff(bb,u);

> helixtandev:=rulsurf(bb,dd);

> plot3d(tandev,u=0..6*Pi,v=0..6,scaling=constrained,grid=[40,6]);

EXERCISE 5.2. Create a ruled surface with $\beta(u) = (0, 0, \sin(ku))$ and $\delta(u) = (\cos u, \sin u, 0)$. Plot this ruled surface for varying k. What do you notice? See Figure 2.19 for the case $k = 1$.

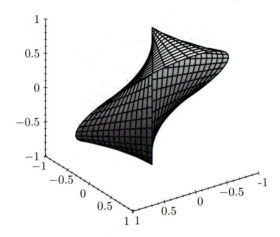

FIGURE 2.19

EXERCISE 5.3. Write a MAPLE procedure to create a parametrization for a surface of revolution given a plane curve $\alpha(u) = (g(u), h(u))$. Use your procedure to plot surfaces of revolution such as the catenoid and the torus.

EXERCISE 5.4. Use the two procedures above to create ruled and revolution surface parametrizations for the hyperboloid of one sheet $x^2 + y^2 - z^2 = 1$.

EXERCISE 5.5. Write a procedure to create a Monge parametrization for any given input function.

Chapter 3

CURVATURE(S)

As we have seen, objects such as the shape operator and the principal curvatures give a great deal of information to us about the geometry of a surface. We have also seen, however, that it may be difficult or impossible to actually compute these objects precisely. In this chapter we will introduce two computable "invariants" of a surface which are associated to the shape operator via linear algebra. In the next chapter we will then consider various results which indicate the kind of information *these* objects provide.

The two most basic linear algebraic invariants associated to a linear transformation are its determinant and its trace. Because the shape operator at a point p is a linear transformation, we may define two *geometric* quantities in terms of the shape operator's determinant and trace.

Definition 1.1.
The *Gaussian curvature* of a surface M at $p \in M$ is defined to be $K(p) = \det(S_p)$.
The *mean curvature* of a surface M at $p \in M$ is defined to be $H(p) = \frac{1}{2} \operatorname{trace}(S_p)$.

These two quantities are fundamental to the study of the geometry of surfaces, as we shall see. Moreover, even though K and H are defined in terms of the shape operator, we will see that they may be calculated by *calculus alone*. Therefore, we are led back to our main premise that linear algebra is the bridge which allows geometry to be studied via calculus. We have seen that the matrix of the shape operator with respect to a basis of principal vectors is given by

$$\begin{bmatrix} k_1(p) & 0 \\ 0 & k_2(p) \end{bmatrix},$$

with determinant and trace equal to $k_1 k_2$ and $k_1 + k_2$ respectively. Hence,

$$K = k_1 k_2, \qquad H = \frac{k_1 + k_2}{2}.$$

Recall that if $-U$ is chosen as a normal instead of U, then $k(u)$ changes sign also. Since Gaussian curvature is the product of two such changes, *it does not change sign*. Note, however, that H *does* change sign under a change of unit normal. These observations are important because the sign of K has meaning.

Suppose $K(p) > 0$. Since $K = k_1 k_2$, k_1 and k_2 must have the same sign. But $k_1 = \max k(u)$ and $k_2 = \min k(u)$, so $k(u)$ has the same sign for all u. If $k(u) > 0$ for all u then M bends toward U in every direction. If $k(u) < 0$ for all u then M bends away from U in every direction.

EXERCISE 1.1. Interpret $K(p) < 0$ and $K(p) = 0$. What are the pictures?

Examples 1.2.
1. For $K(p) > 0$, consider the elliptic paraboloid $z = x^2 + y^2$ at $p = (0,0,0)$. The paraboloid is tangent to the xy-plane at p and, clearly, whichever vertical normal is chosen, either all $k(u)$ are positive or all are negative.
2. For $K(p) < 0$, consider the hyperbolic paraboloid $z = x^2 - y^2$ at $p = (0,0,0)$. Again, the tangent plane at p is the xy-plane. This time, however, the surface both bends away and towards either chosen normal.

EXERCISE 1.2. The case $K(p) = 0$ may arise in two ways exemplified by (1) a plane and (2) a cylinder $x^2 + y^2 = 1$. Discuss these in terms of normal curvature.

Now we will focus on straightforward ways to compute curvature in terms of calculus and linear algebra. Let v and w be linearly independent tangent vectors at $p \in M$. This means simply that one vector is not a multiple of the other. Since they are linearly independent, v and w form a basis for $T_p(M)$ and any vector is a linear combination of them. Hence, we may write the effects of the shape operator as

$$S(\mathbf{v}) = a\mathbf{v} + b\mathbf{w} \qquad \text{and} \qquad S(\mathbf{w}) = c\mathbf{v} + d\mathbf{w}.$$

This says that the matrix of S with respect to the basis $\{\mathbf{v}, \mathbf{w}\}$ is

$$\begin{bmatrix} a & c \\ b & d \end{bmatrix}.$$

Now $\det(S) = ad - bc = K$ and $\mathrm{tr}(S) = a + d = 2H$ by definition, so

$$S(\mathbf{v}) \times S(\mathbf{w}) = (a\mathbf{v} + b\mathbf{w}) \times (c\mathbf{v} + d\mathbf{w})$$
$$= ac(\mathbf{v} \times \mathbf{v}) + ad(\mathbf{v} \times \mathbf{w})$$
$$+ bc(\mathbf{w} \times \mathbf{v}) + bd(\mathbf{w} \times \mathbf{w})$$
$$= 0 + (ad - bc)(\mathbf{v} \times \mathbf{w}) + 0$$
$$= \det(S)\,\mathbf{v} \times \mathbf{w}$$
$$= K\,\mathbf{v} \times \mathbf{w}$$

$$S(\mathbf{v}) \times \mathbf{w} + \mathbf{v} \times S(\mathbf{w}) = (a\mathbf{v} + b\mathbf{w}) \times \mathbf{w} + \mathbf{v} \times (c\mathbf{v} + d\mathbf{w})$$
$$= a(\mathbf{v} \times \mathbf{w}) + d(\mathbf{v} \times \mathbf{w})$$
$$= (a + d)(\mathbf{v} \times \mathbf{w})$$
$$= \operatorname{tr}(S)\,\mathbf{v} \times \mathbf{w}$$
$$= 2H\,\mathbf{v} \times \mathbf{w}.$$

EXERCISE 1.3 (Lagrange Identity). Show that for vectors $\mathbf{v}$, $\mathbf{w}$, $\mathbf{a}$ and $\mathbf{b}$,

$$(\mathbf{v} \times \mathbf{w}) \cdot (\mathbf{a} \times \mathbf{b}) = (\mathbf{v} \cdot \mathbf{a})(\mathbf{w} \cdot \mathbf{b}) - (\mathbf{v} \cdot \mathbf{b})(\mathbf{w} \cdot \mathbf{a}).$$

Hint: Write out the vectors in coordinates $(v^1, v^2, v^3) = \mathbf{v}$ etc.

Now combine Exercise 1.3 with the formulas above to get,

$$(S(\mathbf{v}) \times S(\mathbf{w})) \cdot (\mathbf{v} \times \mathbf{w}) = K\,(\mathbf{v} \times \mathbf{w}) \cdot (\mathbf{v} \times \mathbf{w})$$
$$(S(\mathbf{v}) \cdot \mathbf{v})(S(\mathbf{w}) \cdot \mathbf{w}) - (S(\mathbf{v}) \cdot \mathbf{w})(S(\mathbf{w}) \cdot \mathbf{v})$$
$$= K\,((\mathbf{v} \cdot \mathbf{v})(\mathbf{w} \cdot \mathbf{w}) - (\mathbf{v} \cdot \mathbf{w})(\mathbf{w} \cdot \mathbf{v}))$$

$$K = \frac{(S(\mathbf{v}) \cdot \mathbf{v})(S(\mathbf{w}) \cdot \mathbf{w}) - (S(\mathbf{v}) \cdot \mathbf{w})(S(\mathbf{w}) \cdot \mathbf{v})}{(\mathbf{v} \cdot \mathbf{v})(\mathbf{w} \cdot \mathbf{w}) - (\mathbf{v} \cdot \mathbf{w})(\mathbf{w} \cdot \mathbf{v})}.$$

and

$$(S(\mathbf{v}) \times \mathbf{w} + \mathbf{v} \times S(\mathbf{w})) \cdot (\mathbf{v} \times \mathbf{w}) = 2H\,(\mathbf{v} \times \mathbf{w}) \cdot (\mathbf{v} \times \mathbf{w})$$
$$(S(\mathbf{v}) \cdot \mathbf{v})(\mathbf{w} \cdot \mathbf{w}) - (S(\mathbf{v}) \cdot \mathbf{w})(\mathbf{w} \cdot \mathbf{v})$$
$$+ (\mathbf{v} \cdot \mathbf{v})(S(\mathbf{w}) \cdot \mathbf{w}) - (\mathbf{v} \cdot \mathbf{w})(S(\mathbf{w}) \cdot \mathbf{v})$$
$$= 2H\,((\mathbf{v} \cdot \mathbf{v})(\mathbf{w} \cdot \mathbf{w}) - (\mathbf{v} \cdot \mathbf{w})(\mathbf{w} \cdot \mathbf{v}))$$

Finally, by dividing through, we obtain

$$H =$$

$$\frac{(S(\mathbf{v}) \cdot \mathbf{v})(\mathbf{w} \cdot \mathbf{w}) - (S(\mathbf{v}) \cdot \mathbf{w})(\mathbf{w} \cdot \mathbf{v}) + (\mathbf{v} \cdot \mathbf{v})(S(\mathbf{w}) \cdot \mathbf{w}) - (\mathbf{v} \cdot \mathbf{w})(S(\mathbf{w}) \cdot \mathbf{v})}{2\,((\mathbf{v} \cdot \mathbf{v})(\mathbf{w} \cdot \mathbf{w}) - (\mathbf{v} \cdot \mathbf{w})(\mathbf{w} \cdot \mathbf{v}))}.$$

EXERCISE 1.4. Use Euler's formula (Corollary 2.4.4) to show

(1) the mean curvature H at a point is the average normal curvature

$$H = \frac{1}{2\pi} \int_0^{2\pi} k(\theta) \, d\theta.$$

(2) $H = \frac{k(\mathbf{v}_1) + k(\mathbf{v}_2)}{2}$ for any two unit vectors $\mathbf{v}_1$ and $\mathbf{v}_2$ which are perpendicular. Hint: If the angle from $\mathbf{v}_1$ to $\mathbf{u}_1$ is ϕ, then the angle from $\mathbf{v}_2$ to $\mathbf{u}_1$ is $\phi + \frac{\pi}{2}$.

EXERCISE 1.5. Show that the principal curvatures are given in terms of K and H by

$$k_1 = H + \sqrt{H^2 - K} \quad \text{and} \quad k_2 = H - \sqrt{H^2 - K}.$$

Hint: k_1 and k_2 are the eigenvalues of S. Find these for $S = \begin{bmatrix} a & b \\ b & c \end{bmatrix}$ by setting the characteristic polynomial $\lambda^2 - \mathrm{tr}(S)\lambda + \det(S)$ equal to zero and using the quadratic formula to solve.

There is another more geometric way to view Gaussian curvature in terms of the Gauss map $G\colon M \to S^2$. We have seen in Chapter 2 that the derivative of the Gauss map is the negative of the shape operator, $G_*(\mathbf{v}) = -S(\mathbf{v})$. Take the basis $\{\mathbf{x}_u, \mathbf{x}_v\}$ and consider

$$G_*(\mathbf{x}_u) \times G_*(\mathbf{x}_v) = (-S(\mathbf{x}_u)) \times (-S(\mathbf{x}_v))$$
$$= K \, \mathbf{x}_u \times \mathbf{x}_v.$$

Now, $|G_*(\mathbf{x}_u) \times G_*(\mathbf{x}_v)|$ and $|\mathbf{x}_u \times \mathbf{x}_v|$ may be thought of as infinitesimal pieces of area of the image of the Gauss map on S^2 and M respectively. The formula above then says that the ratio of these infinitesimal areas is precisely $|K|$. Another way to say this is the following. Let $\mathcal{U}$ be a small open neighborhood of $p \in M$ and suppose that $\mathcal{U}$ is contracting down to p. Then

$$|K| = \lim_{\mathcal{U} \to p} \frac{\mathrm{Area}\, G(\mathcal{U})}{\mathrm{Area}\, \mathcal{U}}.$$

This expression means that the magnitude of Gaussian curvature measures the way in which the unit normal expands or contracts area. The Gauss map will be of greater importance to us when we study the complex-analytic approach to minimal surfaces in Chapter 7. With this in mind, we introduce some convenient terminology. A surface M is said to be *flat* if $K(p) = 0$ for every $p \in M$ and it is said to be *minimal* if $H(p) = 0$ for every $p \in M$. We will consider minimal surfaces extensively in two later chapters and we will see exactly where the term "minimal" comes from. The term "flat" is derived from the fact that the prime example of a flat surface is the plane. We have already seen that the shape operator of a plane is identically zero, so it is immediate that its determinant, the Gaussian curvature, is identically zero as well. It may not be so obvious however that other surfaces may be flat.

EXERCISE 1.6. Use your knowledge of the shape operator for the cylinder $x^2 + y^2 = R^2$ to show that a right circular cylinder is flat, but not minimal.

EXERCISE 1.7. Show that if M is minimal, then $K \leq 0$ on M.

EXERCISE 1.8. Show that the R-sphere $S^2(R)$ has $K = 1/R^2$, first by considering the determinant of the R-sphere's shape operator and, secondly by considering the Gauss map of the R-sphere and how area changes.

3.2 CALCULATING CURVATURE

The formulae for K and H for general $\mathbf{v}$ and $\mathbf{w}$ may be particularized to $\mathbf{x}_u$ and $\mathbf{x}_v$ when a patch $\mathbf{x}$ is given for M. With this in mind we introduce the following traditional notation:

$$E = \mathbf{x}_u \cdot \mathbf{x}_u \qquad\qquad l = S(\mathbf{x}_u) \cdot \mathbf{x}_u$$
$$F = \mathbf{x}_u \cdot \mathbf{x}_v \qquad\qquad m = S(\mathbf{x}_u) \cdot \mathbf{x}_v = S(\mathbf{x}_v) \cdot \mathbf{x}_u$$
$$G = \mathbf{x}_v \cdot \mathbf{x}_v \qquad\qquad n = S(\mathbf{x}_v) \cdot \mathbf{x}_v$$

Then, replacing the general $\mathbf{v}$, $\mathbf{w}$ by $\mathbf{x}_u$, $\mathbf{x}_v$, we have the two curvature formulas

$$K = \frac{(S(\mathbf{x}_u) \cdot \mathbf{x}_u)(S(\mathbf{x}_v) \cdot \mathbf{x}_v) - (S(\mathbf{x}_u) \cdot \mathbf{x}_v)(S(\mathbf{x}_v) \cdot \mathbf{x}_u)}{(\mathbf{x}_u \cdot \mathbf{x}_u)(\mathbf{x}_v \cdot \mathbf{x}_v) - (\mathbf{x}_u \cdot \mathbf{x}_v)(\mathbf{x}_v \cdot \mathbf{x}_u)}$$
$$= \frac{ln - m^2}{EG - F^2}.$$

$$H = \frac{(S(\mathbf{x}_u) \cdot \mathbf{x}_u)(\mathbf{x}_v \cdot \mathbf{x}_v) - (S(\mathbf{x}_u) \cdot \mathbf{x}_v)(\mathbf{x}_v \cdot \mathbf{x}_u) + (\mathbf{x}_u \cdot \mathbf{x}_u)(S(\mathbf{x}_v) \cdot \mathbf{x}_v) - (\mathbf{x}_u \cdot \mathbf{x}_v)(S(\mathbf{x}_v) \cdot \mathbf{x}_u)}{2((\mathbf{x}_u \cdot \mathbf{x}_u)(\mathbf{x}_v \cdot \mathbf{x}_v) - (\mathbf{x}_u \cdot \mathbf{x}_v)(\mathbf{x}_v \cdot \mathbf{x}_u))}$$
$$H = \frac{Gl + En - 2Fm}{2(EG - F^2)}.$$

Remark 2.1. The quantity $EG - F^2$ has already made an appearance as $|\mathbf{x}_u \times \mathbf{x}_v|^2$. See Exercise 1.3.4 (Lagrange's identity).

Now we come to the fundamental result which allows for computation of K and H without reference to the shape operator S. The result follows immediately from Lemma 2.4.1, but we write out the proof here explicitly because it is so important.

Lemma 2.1. $l = U \cdot \mathbf{x}_{uu}, \quad m = U \cdot \mathbf{x}_{uv}, \quad n = U \cdot \mathbf{x}_{vv}.$

Proof. We shall prove the formula for m. We know $U \cdot \mathbf{x}_v = 0$ since U is normal and $\mathbf{x}_v \in T_p(M)$. As we have seen previously,

$$0 = \mathbf{x}_u(U \cdot \mathbf{x}_v) = \mathbf{x}_u \left(\sum_i u_i \frac{\partial x_i}{\partial v} \right)$$

$$= \sum_i \mathbf{x}_u[u_i] \frac{\partial x_i}{\partial v} + \sum_i u_i \mathbf{x}_u \left[\frac{\partial x_i}{\partial v} \right]$$

$$= \nabla_{\mathbf{x}_u} U \cdot \mathbf{x}_v + U \cdot \mathbf{x}_{uv} \quad \text{since } \mathbf{x}_u \left[\frac{\partial x_i}{\partial v} \right] = \frac{\partial^2 x_i}{\partial v \partial u}$$

$$= -S(\mathbf{x}_u) \cdot \mathbf{x}_v + U \cdot \mathbf{x}_{uv}.$$

Hence, $S(\mathbf{x}_u) \cdot \mathbf{x}_v = U \cdot \mathbf{x}_{uv}.$ □

EXERCISE 2.1. Prove the formulas for l and n.

EXERCISE 2.2. Suppose $\mathbf{x}(u, v)$ and $\mathbf{y}(r, s)$ are two patches for the same surface M. For $\mathbf{y}^{-1} \circ \mathbf{x}(u, v) = (r(u, v), s(u, v))$, show that

$$E_\mathbf{x} G_\mathbf{x} - F_\mathbf{x}^2 = [r_u s_v - r_v s_u]^2 (E_\mathbf{y} G_\mathbf{y} - F_\mathbf{y}^2)$$

$$l_\mathbf{x} n_\mathbf{x} - m_\mathbf{x}^2 = [r_u s_v - r_v s_u]^2 (l_\mathbf{y} n_\mathbf{y} - m_\mathbf{y}^2)$$

and, consequently, $K_\mathbf{x} = K_\mathbf{y}$.

EXERCISE 2.3. Suppose $\mathbf{x}(u, v)$ is a patch. Define a new patch by $\mathbf{y}(u, v) = c\,\mathbf{x}(u, v)$, where c is a constant. Show that

$$K_\mathbf{y} = \frac{1}{c^2} K_\mathbf{x}.$$

EXERCISE 2.4. From a surface $M \colon \mathbf{x}(u, v)$, construct a *parallel surface* M^t: $\mathbf{x}^t(u, v) = \mathbf{x}(u, v) + t U(u, v)$. Show that the Gaussian and mean curvatures are given by the formulas

$$K^t = \frac{K}{1 - 2Ht + Kt^2} \qquad H^t = \frac{H - Kt}{1 - 2Ht + Kt^2}$$

where K and H are the Gaussian and mean curvatures of M. Then show the following: (1) if M has constant mean curvature $H = c$, then $M^{\frac{1}{2c}}$ has constant Gauss curvature $K^{\frac{1}{2c}} = 4c^2$, (2) if M has constant Gauss curvature $K = c^2$, then $M^{\frac{1}{c}}$ has constant mean curvature $H^{\frac{1}{c}} = \frac{-c}{2}$. Hint: compute $E^t G^t - F^{t2} = |\mathbf{x}_u^t \times \mathbf{x}_v^t|^2 = [1 - 2Ht + Kt^2]^2 (EG - F^2)$ directly from the definition of $\mathbf{x}^t(u, v)$ and compare. We will use parallel surfaces when we discuss minimal and constant mean curvature surfaces in Chapter 4.

EXERCISE 2.5. Let $M: \mathbf{x}(u, v)$ be a surface. Recall that a curve $\alpha: I \to M$ is a line of curvature if the tangent vector $\alpha'(t)$ is an eigenvector of the shape operator for each t. Show that all u and v-parameter curves are lines of curvature if and only if $F = 0$ and $m = 0$.

Example 2.2: Enneper's surface. Let M denote Enneper's surface

$$\mathbf{x}(u, v) = (u - \frac{u^3}{3} + uv^2, \ v - \frac{v^3}{3} + vu^2, \ u^2 - v^2).$$

Then, $\mathbf{x}_u = (1 - u^2 + v^2, 2uv, 2u)$, $\mathbf{x}_v = (2uv, 1 - v^2 + u^2, -2v)$ and we compute E, F and G as

$$E = \mathbf{x}_u \cdot \mathbf{x}_u$$
$$= 1 + 2u^2 + 2v^2 + u^4 + 2u^2v^2 + v^4$$
$$= (1 + u^2 + v^2)^2,$$

$$F = \mathbf{x}_u \cdot \mathbf{x}_v$$
$$= 2uv - 2u^3v + 2uv^3 + 2uv - 2uv^3 + 2u^3v - 4uv$$
$$= 0,$$

$$G = \mathbf{x}_v \cdot \mathbf{x}_v$$
$$= 4u^2v^2 + 1 - v^2 + u^2 - v^2 + v^4 - v^2u^2 + u^2 - u^2v^2 + u^4 + 4v^2$$
$$= 1 + 2u^2 + 2v^2 + u^4 + 2u^2v^2 + v^4$$
$$= (1 + u^2 + v^2)^2.$$

The unit normal U is found by taking

$$\mathbf{x}_u \times \mathbf{x}_v = (-4uv^2 - 2u + 2uv^2 - 2u^3,$$
$$2v - 2u^2v + 2v^3 + 4u^2v,$$
$$1 - (u^2 - v^2)^2 - 4u^2v^2)$$
$$= (-2u(1 + u^2 + v^2), 2v(1 + u^2 + v^2), 1 - (u^2 + v^2)^2)$$

with length squared

$$|\mathbf{x}_u \times \mathbf{x}_v|^2 = 4u^2(1 + u^2 + v^2)^2 + 4v^2(1 + u^2 + v^2)^2$$
$$+ (u^2 + v^2)^4 - 2(u^2 + v^2)^2 + 1$$
$$= 4(u^2 + v^2) + 8(u^2 + v^2)^2 + 4(u^2 + v^2)^3$$
$$+ (u^2 + v^2)^4 - 2(u^2 + v^2)^2 + 1$$
$$= (u^2 + v^2)^4 + 4(u^2 + v^2)^3 + 6(u^2 + v^2)^2 + 4(u^2 + v^2) + 1$$
$$= (u^2 + v^2 + 1)^4$$

which also may be computed by noting the consequence of the Lagrange identity, $|\mathbf{x}_u \times \mathbf{x}_v| = \sqrt{EG - F^2}$. Then,

$$U = \left(\frac{-2u}{u^2 + v^2 + 1}, \; \frac{2v}{u^2 + v^2 + 1}, \; \frac{1 - (u^2 + v^2)^2}{(u^2 + v^2 + 1)^2} \right).$$

The second partials $\mathbf{x}_{uu} = (-2u, 2v, 2)$, $\mathbf{x}_{uv} = (2v, 2u, 0)$ and $\mathbf{x}_{vv} = (2u, -2v, -2)$ then give

$$l = \mathbf{x}_{uu} \cdot U$$

$$= \frac{4u^2}{u^2 + v^2 + 1} + \frac{4v^2}{u^2 + v^2 + 1} + \frac{2 - 2(u^2 + v^2)^2}{(u^2 + v^2 + 1)^2}$$

$$= \frac{2(u^2 + v^2 + 1)^2}{(u^2 + v^2 + 1)^2}$$

$$= 2.$$

Similarly, $n = \mathbf{x}_{vv} \cdot U = -2$ and, clearly, $m = \mathbf{x}_{uv} \cdot U = 0$. Then the Gauss curvature of Enneper's surface is

$$K = \frac{ln - m^2}{EG - F^2}$$

$$= \frac{-4}{(u^2 + v^2 + 1)^4}$$

and the mean curvature is

$$H = \frac{Gl + En - 2Fm}{2(EG - F^2)}$$

$$= \frac{(u^2 + v^2 + 1)^2(2) + (u^2 + v^2 + 1)^2(-2) - 0}{2(u^2 + v^2 + 1)^4}$$

$$= 0.$$

Hence, Enneper's surface is a minimal surface. We will see exactly how its parametrization arises in Chapter 7.

EXERCISE 2.6. Show that the u and v parameter curves of the catenoid and Enneper's surface are lines of curvature. Also, show that these curves are planar (Hint: calculate torsion). Finally, show that the following surface also has parameter curves which are planar lines of curvature:

$$\mathbf{x}(u, v) = \left(\frac{cu \pm \sin u \cosh v}{\sqrt{1 - c^2}}, \; \frac{v \pm c \cos u \sinh v}{\sqrt{1 - c^2}}, \; \pm \cos u \cosh v \right).$$

Hint: use MAPLE. Enneper's surface, the catenoid and this last surface are the only minimal surfaces with planar lines of curvature.

EXERCISE 2.7. Using $K = \frac{ln - m^2}{EG - F^2}$, show that the Gaussian curvature of the R-sphere $S^2(R)$ is $1/R^2$.

Example 2.3: The hyperboloid of two sheets. Let M denote the hyperboloid of two sheets

$$\frac{x^2}{a^2} + \frac{y^2}{b^2} - \frac{z^2}{c^2} = -1$$

parametrized by $\mathbf{x}(u, v) = (a \sinh u \cos v, b \sinh u \sin v, c \cosh u)$. Then,

$$\mathbf{x}_u = (a \cosh u \cos v, b \cosh u \sin v, c \sinh u)$$
$$\mathbf{x}_v = (-a \sinh u \sin v, b \sinh u \cos v, 0)$$

and $\mathbf{x}_u \times \mathbf{x}_v = (-bc \sinh^2 u \cos v, -ac \sinh^2 u \sin v, ab \sinh u \cosh u)$. Dividing by $|\mathbf{x}_u \times \mathbf{x}_v|$ gives

$$U = \frac{\mathbf{x}_u \times \mathbf{x}_v}{W}$$

where $W = \sqrt{b^2 c^2 \sinh^4 u \cos^2 v + a^2 c^2 \sinh^4 u \sin^2 v + a^2 b^2 \sinh^2 u \cosh^2 u}$.

We then have

$$E = a^2 \cosh^2 u \cos^2 v + b^2 \cosh^2 u \sin^2 v + c^2 \sinh^2 u,$$
$$F = -a^2 \sinh u \cosh u \sin v \cos v + b^2 \sinh u \cosh u \sin v \cos v$$
$$G = a^2 \sinh^2 u \sin^2 v + b^2 \sinh^2 u \cos^2 v$$

with

$$EG - F^2 = b^2 c^2 \sinh^4 u \cos^2 v + a^2 c^2 \sinh^4 u \sin^2 v + a^2 b^2 \sinh^2 u \cosh^2 u$$
$$= W^2.$$

The following second partials then give l, m and n.

$$\mathbf{x}_{uu} = (a \sinh u \cos v, b \sinh u \sin v, c \cosh u)$$
$$\mathbf{x}_{uv} = (-a \cosh u \sin v, b \cosh u \cos v, 0)$$
$$\mathbf{x}_{vv} = (-a \sinh u \cos v, -b \sinh u \sin v, 0)$$

$$l = \mathbf{x}_{uu} \cdot U$$

$$= \frac{-abc \sinh^3 u \cos^2 v - abc \sinh^3 u \sin^2 v + abc \sinh u \cosh^2 u}{W}$$

$$= \frac{abc \sinh u}{W} \qquad \text{using} \quad 1 + \sinh^2 u = \cosh^2 u,$$

$$m = \mathbf{x}_{uv} \cdot U$$

$$= \frac{abc \sinh^2 u \cosh u \sin v \cos v - abc \sinh^2 u \cosh u \sin v \cos v}{W}$$

$$= 0,$$

$$n = \mathbf{x}_{vv} \cdot U$$

$$= \frac{abc \sinh^3 u \cos^2 v + abc \sinh^3 u \sin^2 v}{W}$$

$$= \frac{abc \sinh^3 u}{W}.$$

Hence, we obtain the Gauss curvature

$$K = \frac{ln - m^2}{EG - F^2} = \frac{a^2 b^2 c^2 \sinh^4 u}{W^4}$$

which we may write as

$$K = 1 \Big/ \left(\frac{W^2}{abc \sinh^2 u} \right)^2$$

where

$$\frac{W^2}{abc \sinh^2 u} = \frac{bc}{a} \sinh^2 u \cos^2 v + \frac{ac}{b} \sinh^2 u \sin^2 v + \frac{ab}{c} \cosh^2 u.$$

Now, the coordinate functions of the parametrization are $x = a \sinh u \cos v$, $y = b \sinh u \sin v$ and $z = c \cosh u$, so the reader can check that the Gaussian curvature may be written in terms of x, y and z as

$$K = \frac{1}{a^2 b^2 c^2 \left[\frac{x^2}{a^4} + \frac{y^2}{b^4} + \frac{z^2}{c^4} \right]^2}.$$

EXERCISE 2.8. Show that the Gaussian curvature of the hyperboloid of one sheet

$$\mathbf{x}(u, v) = (a \cosh u \cos v, \; b \cosh u \sin v, \; c \sinh u)$$

may be written in Cartesian coordinates as

$$K = -\frac{1}{a^2 b^2 c^2 \left[\frac{x^2}{a^4} + \frac{y^2}{b^4} + \frac{z^2}{c^4} \right]^2}.$$

Hint: MAPLE may be useful here.

EXERCISE 2.9. Show that the Gaussian curvatures of the elliptic and hyperbolic paraboloids (respectively)

$$z = \frac{x^2}{a^2} + \frac{y^2}{b^2} \quad \text{and} \quad z = \frac{x^2}{a^2} - \frac{y^2}{b^2}$$

may be written in cartesian coordinates (respectively) as

$$K = \frac{1}{4a^2 b^2 \left[\frac{x^2}{a^4} + \frac{y^2}{b^4} + \frac{1}{4} \right]^2} \quad \text{and} \quad K = -\frac{1}{4a^2 b^2 \left[\frac{x^2}{a^4} + \frac{y^2}{b^4} + \frac{1}{4} \right]^2}.$$

Hint: MAPLE may be useful here.

EXERCISE 2.10. Find the Gaussian and mean curvatures of the helicoid parametrized by $\mathbf{x}(u, v) = (u \cos v, u \sin v, bv)$.

EXERCISE 2.11. Find the Gaussian and mean curvatures for the saddle surface $M: z = xy$. Take a Monge patch $\mathbf{x}(u, v) = (u, v, uv)$.

EXERCISE 2.12. Compute the Gaussian curvature of **Kuen's Surface** $\mathbf{x}(u, v) =$

$$\left(\frac{2(\cos u + u \sin u) \sin v}{1 + u^2 \sin^2 v}, \; \frac{2(\sin u - u \cos u) \sin v}{1 + u^2 \sin^2 v}, \; \ln(\tan \frac{v}{2}) + \frac{2 \cos v}{1 + u^2 \sin^2 v} \right).$$

EXERCISE 2.13.
(a) Show that a ruled surface $\mathbf{x}(u, v) = \beta(u) + v\delta(u)$ has Gaussian curvature

$$K = \frac{-(\beta' \cdot \delta \times \delta')^2}{W^4}$$

where $W = |\beta' \times \delta + v\delta' \times \delta|$. Using ruling patches,
(b) Compute K for $M: z = xy$.
(c) Compute K for a cone $\mathbf{x}(u, v) = p + v\delta(u)$.
(d) Compute K for a cylinder $\mathbf{x}(u, v) = \beta(u) + vq$.
(e) Compute K for the helicoid.
(f) Compute K for a hyperboloid of one sheet.
(g) Compute K for the hyperbolic paraboloid.
Interpret your computations in terms of the geometry of the surface. Compare your calculations with previous ones.

EXERCISE 2.14. A cone (minus its vertex) is a surface. For the standard cone $z = \sqrt{x^2 + y^2}$, compute K and H using the two patches:
(a) $\mathbf{x}(u, v) = (u, v, \sqrt{u^2 + v^2})$ (i.e. a Monge patch).
(b) $\mathbf{x}(u, v) = v(\cos u, \sin u, 1)$ (i.e. a ruling patch).

EXERCISE 2.15. A ruled surface $M \colon \mathbf{x}(u, v) = \beta(u) + v\delta(u)$ is *developable* if its unit normal U is constant along the rulings. Hence, U does not depend on v. Show that a ruled surface is developable if and only if its Gaussian curvature is zero. See [BM] for an industrial application.

EXERCISE 2.16. Show directly that cones and cylinders have constant normals along rulings and, therefore, are developable.

EXERCISE 2.17. Besides cones and cylinders, typical examples of developable surfaces are ones which arise from curves, the *tangent developables*. Given a curve $\beta(u)$, its tangent developable is defined to be the surface parametrized by $\mathbf{x}(u, v) = \beta(u) + v\beta'(u)$. Show that this surface is, in fact, developable. Graph the tangent developable of the helix $\beta(u) = (\cos u, \sin u, u)$.

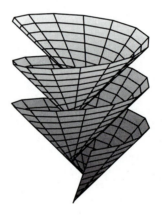

FIGURE 3.1.   Tangent developable of a helix

EXERCISE 2.18. Let $\beta \colon I \to M$ be a curve on a surface M with unit normal U. Show that β is a line of curvature on M if and only if the surface defined by $\mathbf{y}(u, v) = \beta(u) + v\,U(u)$ is developable. Here, $U(u)$ denotes the normal of M along β.

EXERCISE 2.19. For $M \colon z = f(x, y)$
(a) Give a Monge patch for this surface.
(b) Compute E, F, G, l, m, n.
(c) Find formulas for K and H.

(d) Recall that (u_0, v_0) is a *critical point* for f if $f_u(u_0, v_0) = 0$ and $f_v(u_0, v_0) = 0$. Now, (u_0, v_0) may be a maximum, a minimum or a saddle point. The 2nd Derivative Test is designed to decide which of these choices pertains. It goes like this. First, compute $D = f_{uu}(u_0, v_0)f_{vv}(u_0, v_0) - (f_{uv}(u_0, v_0))^2$. If $D = 0$, then the test fails and no information is obtained. If $D < 0$, then (u_0, v_0) is a saddle point. If $D > 0$, then there are two cases:

(I) If $f_{uu}(u_0, v_0) > 0$, then (u_0, v_0) is a minimum.

(II) If $f_{uu}(u_0, v_0) < 0$, then (u_0, v_0) is a maximum.

(In the above, f_{vv} could equally well be used.)

Show that this test works! Hints: (1) How is K related to D? (2) Consider $K > 0$, $K < 0$, $K = 0$. (3) What do these imply about normal and principal curvatures? (4) Recall $k(\mathbf{v}) = S(\mathbf{v}) \cdot \mathbf{v}$ for $\mathbf{v}$ a unit vector. If $\mathbf{w}$ is not a unit vector, then it can be made one by $\mathbf{w}/|\mathbf{w}|$. Thus,

$$S(\mathbf{w}) \cdot \mathbf{w} = (S(\mathbf{w}/|\mathbf{w}|) \cdot w/|\mathbf{w}|)|\mathbf{w}|^2 \qquad \text{Why?}$$
$$= k(\mathbf{w}/|\mathbf{w}|)|\mathbf{w}|^2.$$

(5) $f_{uu}(u_0, v_0) = U(u_0, v_0) \cdot \mathbf{x}_{uu}(u_0, v_0)$.

3.3 SURFACES OF REVOLUTION

We have already seen that many interesting examples of surfaces arise from revolving curves about an axis. These surfaces of revolution have patches of the form $\mathbf{x}(u, v) = (g(u), h(u) \cos v, h(u) \sin v)$.

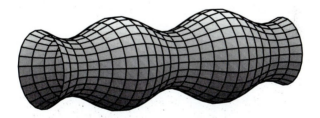

FIGURE 3.2.   A surface of revolution

The curves on the surface which are circles formed by revolving a single point about the axis are called *parallels* (denoted π). The curves which look exactly like the original curve (but rotated out) are called *meridians* (denoted μ). Also, recall that $g(u)$ is the distance along the axis of revolution and $h(u)$ is the radius of a parallel. We have, $\mathbf{x}_u = (g', h' \cos v, h' \sin v)$, $\mathbf{x}_v = (0, -h \sin v, h \cos v)$ and $\mathbf{x}_u \times \mathbf{x}_v = (hh', -g'h \cos v, -g'h \sin v)$. Hence,

$$U = \frac{(h', -g' \cos v, -g' \sin v)}{\sqrt{g'^2 + h'^2}}.$$

Also, the second partial derivatives are $\mathbf{x}_{uu} = (g'', h'' \cos v, h'' \sin v)$, $\mathbf{x}_{uv} = (0, -h' \sin v, h' \cos v)$, and $\mathbf{x}_{vv} = (0, -h \cos v, -h \sin v)$. Hence,

$$E = g'^2 + h'^2 \qquad\qquad l = \frac{(g''h' - h''g')}{\sqrt{g'^2 + h'^2}}$$

$$F = 0 \qquad\qquad\qquad m = 0$$

$$G = h^2 \qquad\qquad\qquad n = \frac{hg'}{\sqrt{g'^2 + h'^2}}.$$

Finally, the Gaussian curvature is computed to be

$$K = \frac{g'(g''h' - h''g')}{h(g'^2 + h'^2)^2}.$$

EXERCISE 3.1. Verify the computations above.

If the original curve $\alpha(t) = (g(t), h(t), 0)$ has $g'(t) \neq 0$ for all t, then g is strictly increasing. Thus, it is one-to-one and has an inverse function g^{-1} which is differentiable as well. Such an inverse allows us to reparametrize the curve α. Define $f = h \circ g^{-1}$ and get

$$\overline{\alpha}(u) = \alpha \circ g^{-1}(u) = (gg^{-1}(u), hg^{-1}(u), 0)$$
$$= (u, f(u), 0).$$

Thus, our calculations become somewhat easier. For instance, the formula for Gaussian curvature becomes $K = -\frac{f''}{f(1+f'^2)^2}$. To avoid confusion, we will still write $\alpha(u) = (u, h(u), 0)$ and $K = -\frac{h''}{h(1+h'^2)^2}$.

Example 3.1: The Torus.

FIGURE 3.3.　A torus

$$\mathbf{x}(u, v) = (\underbrace{(R + r \cos u)}_{h(u)} \cos v, \; (R + r \cos u) \sin v, \; \underbrace{r \sin u}_{g(u)})$$

$$K = \frac{\cos u}{r(R + r \cos u)}.$$

Observe that, for $-\frac{\pi}{2} < u < \frac{\pi}{2}$, we have $\cos u > 0$. Hence, $K > 0$ on the outer half of the torus. For $u = -\frac{\pi}{2}, \frac{\pi}{2}$ we have $\cos u = 0$, so $K = 0$ on the top and bottom of the torus. Finally, for $\frac{\pi}{2} < u < \frac{3\pi}{2}$ we have $\cos u < 0$ and $K < 0$ on the inner half of the torus. Note that the maximum K occurs at $u = 0$,

$$K(0) = \frac{1}{r(R+r)}$$

on the outermost circle, while the minimum K occurs at $u = \pi$,

$$K(\pi) = -\frac{1}{r(R-r)}$$

on the innermost circle.

EXERCISE 3.2. Verify the formula for K above.

EXERCISE 3.3. For a surface of revolution parametrized as above, the v-parameter curves are parallels and the u-parameter curves are meridians. The normal curvatures in the directions of $\mathbf{x}_v$ and $\mathbf{x}_u$ are denoted k_π and k_μ respectively. Show that $k_\mu = l/E$ and $k_\pi = n/G$ by calculating the eigenvectors of S as $\mathbf{x}_u$ and $\mathbf{x}_v$ with eigenvalues l/E, n/G respectively. This shows directly that meridians and parallels are lines of curvature. Hint: $S(\mathbf{x}_u) = a\mathbf{x}_u + b\mathbf{x}_v$. Use $F = 0$, $m = 0$.

EXERCISE 3.4. For a general surface of revolution, show that

$$k_\mu = \frac{g''h' - g'h''}{(g'^2 + h'^2)^{3/2}}, \qquad k_\pi = \frac{g'}{h(g'^2 + h'^2)^{1/2}}$$

and obtain formulas for K and H. What is k_μ on a torus? What is the geometric reason for this?

EXERCISE 3.5. Revolve the catenary $y = c \cosh(\frac{x}{c})$ around the x-axis. Show the following:

$$k_\mu = -\frac{1}{c \cosh^2 \frac{u}{c}}, \qquad k_\pi = -k_\mu, \qquad K = -\frac{1}{c^2 \cosh^4 \frac{u}{c}}, \qquad H = 0.$$

Can you think of another surface of revolution which is minimal?

EXERCISE 3.6. For the following surfaces of revolution, find K and describe where $K > 0$, $K < 0$, $K = 0$.

(a) Revolve $\alpha(u) = (u, e^{-u^2/2}, 0)$ about the x-axis.

(b) **Elliptical Torus.** Revolve the ellipse $\frac{(x-R)^2}{a^2} + \frac{y^2}{b^2} = 1$ about the z-axis.

EXERCISE 3.7. If a surface of revolution is generated by a *unit speed curve* $\alpha(u) = (g(u), h(u), 0)$, then show (a) $E = 1$, $F = 0$, $G = h^2$ and (b) $K = -h''/h$.

Example 3.2. As another example of our computational techniques, we will consider a surface of revolution which is analogous to the sphere in the sense that it has *constant Gaussian curvature*. This surface is called the *Pseudo-sphere* and it is defined by means of a condition on the generating curve. Namely, let α be a curve which begins at $(0, c)$ and traces a path so that its tangent line at any point reaches the x-axis after running a distance exactly equal to c. This means the curve α must decrease and flatten. The curve α is called a *tractrix* and is physically represented by the path a ship would follow when starting at $(0, c)$ and is being pulled by a tugboat which moves along the x-axis.

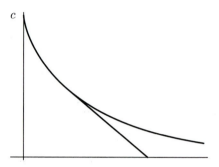

FIGURE 3.4.   Tangent line segment of length c

Write $\alpha(u) = (u, h(u))$ and note $h' < 0$ and $g(u) = u$. Now, $\alpha'(u) = (1, h'(u))$ and a tangent line at $(u, h(u))$ has the equation

$$l(t) = \alpha(u) + t\alpha'(u).$$

The y-coordinate of $l(t)$ is then $h + th'$, so for $t = 0$ we are at $\alpha(u)$ and for $t = -h/h'$ we are on the x-axis. We want the length of this segment to be c, so we have

$$c = |\alpha(u) - (\alpha(u) - \frac{h}{h'}\alpha'(u))|$$

$$= \frac{h}{|h'|}|\alpha'(u)|$$

$$= \frac{h}{|h'|}\sqrt{1 + h'^2}.$$

Then $c^2 = \frac{h^2}{h'^2}(1 + h'^2) = \frac{h^2}{h'^2} + h^2$. Solving for h' and then taking another derivative, we get

$$h' = -\frac{h}{\sqrt{c^2 - h^2}} \qquad h'' = -\frac{h'c^2}{(c^2 - h^2)^{3/2}}.$$

EXERCISE 3.8. Show that, for the surface of revolution obtained by revolving α about the x-axis, we obtain (a) $k_\mu = h'/c$ (b) $k_\pi = -1/ch'$ and (c) $K = -1/c^2$. Note that the Gaussian curvature of the pseudosphere is a constant. However, in contrast to a sphere's constant positive curvature, the pseudosphere's curvature is negative. For $c = 1$, we obtain a surface at each point of which the Gaussian curvature is -1. Later we will meet up with more geometrically natural surfaces of this kind.

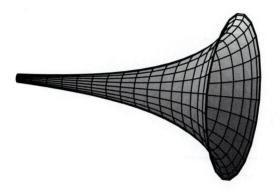

FIGURE 3.5.   The Pseudosphere

EXERCISE 3.9. Solve the following separable differential equation

$$h' = -\frac{h}{\sqrt{1-h^2}}.$$

in two ways to get (1) $u = \ln\left|\frac{\sqrt{1-h^2}}{h} + \frac{1}{h}\right| - \sqrt{1-h^2}$ and (2) $h(u) = \operatorname{sech} w$, $u = w - \tanh w$. For the first, turn the picture over so we revolve the tractrix about the z-axis and the pseudosphere (with $c = 1$) lies over the disk of radius 1. If we write things in cylindrical coordinates (r, θ, z) then the old u becomes z and the old h becomes r. By substitution in the formula above, we get

$$z = -\sqrt{1-r^2} - \ln(r) + \ln(1 + \sqrt{1-r^2}).$$

Of course, rectangular coordinates may now be introduced by $r = \sqrt{x^2 + y^2}$. Use this formula to graph the pseudosphere via MAPLE. For the second, we obtain a patch for the pseudosphere

$$\mathbf{x}(w, v) = (w - \tanh w,\ \operatorname{sech} w \cos v, \operatorname{sech} w \sin v).$$

Plot this patch with MAPLE also. Compare the two plots. Hints: For the first, use trigonometric substitution with $h = \cos w$. For the second, use hyperbolic trigonometric substitution.

The pseudosphere is actually determined by the requirement of constant curvature and certain initial conditions. Suppose we start off by requiring that a surface of revolution M have constant Gauss curvature $K = -1$. We can suppose also that M is parametrized by $\mathbf{x}(u, v) = (g(u), h(u) \cos v, h(u) \sin v)$ with $g'(u)^2 + h'(u)^2 = 1$ (i.e. the profile curve is unit speed). Then, as we have seen, the formula for Gauss curvature reduces to $K = -h''/h$, so the requirement of constant curvature $K = -1$ gives us a linear differential equation to solve,

$$h'' = h.$$

The solution to this differential equation is easy. We obtain

$$h(u) = A\,e^u + B\,e^{-u}.$$

Then, plugging $h(u)$ into the the unit speed relation $g'(u)^2 + h'(u)^2 = 1$ and solving for $g(u)$ gives

$$g(u) = \int_{u_0}^{u} \sqrt{1 - (A\,e^t - B\,e^{-t})^2}\,dt.$$

In choosing A and B, new surfaces of constant Gauss curvature $K = -1$ are created.

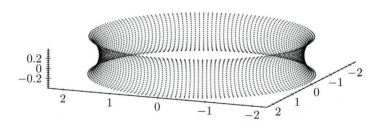

FIGURE 3.6.   A constant negative curvature surface of revolution

EXERCISE 3.10. Let $A = 1$ and $B = 0$. Show that this surface of constant Gauss curvature $K = -1$ is the pseudosphere with $c = 1$. Hint: let $e^t = \operatorname{sech} w$.

EXERCISE 3.11. Carry out the same procedure for the case of positive Gauss curvature $K = +1$. What should A and B be to get the unit sphere? See section 7.

3.4 A FORMULA FOR GAUSSIAN CURVATURE

So far, we have seen that Gaussian curvature tells us a great deal about the geometry of a surface. However, since we compute K using the unit normal U, it seems that the precise way in which we situate a surface in $\mathbb{R}^3$ may well change K while leaving the essential geometry unaltered. This would mean that Gaussian curvature would not be an invariant of the geometry of the surface. We shall now give a formula for K which obviates the use of U and shows that K depends only on E, F, and G, the so-called *metric* of the surface. In fact, although a more general formula exists (see Exercise 4.5 below), we shall restrict ourselves to the case where $F = \mathbf{x}_u \cdot \mathbf{x}_v = 0$. That is, we shall assume that the u and v-parameter curves always meet at right angles. In this case we prove

Theorem 4.1. *The Gauss curvature depends only on the metric E, F and G,*

$$(*) \qquad K = -\frac{1}{2\sqrt{EG}} \left(\frac{\partial}{\partial v} \left(\frac{E_v}{\sqrt{EG}} \right) + \frac{\partial}{\partial u} \left(\frac{G_u}{\sqrt{EG}} \right) \right).$$

Here, we have used the notation

$$E_v = \frac{\partial}{\partial v} E = \frac{\partial}{\partial v} (\mathbf{x}_u \cdot \mathbf{x}_u) \quad \text{and} \quad G_u = \frac{\partial}{\partial u} G = \frac{\partial}{\partial u} (\mathbf{x}_v \cdot \mathbf{x}_v).$$

Of course, to prove the theorem, we will show that our usual formula $K = \frac{ln - m^2}{EG - F^2}$ reduces to the one above. Notice that the formula of the theorem does not depend on the particular patch used, but only on the metric coefficients E, G (and implicitly F). We will consider this a bit more carefully later when we discuss isometries of surfaces, but suffice it to say that, intuitively, a surface may be "bent" without stretching and still have the same Gaussian curvature. The prime example of this phenomenon is the usual right circular cylinder. If the cylinder is cut along a ruling it can be made to lie flat on a plane without stretching. This means that, locally, the cylinder and the plane have the same geometry. This further explains, geometrically, the vanishing of the cylinder's Gaussian curvature. The dependence of K on the metric alone is known as *Gauss's Theorem Egregium*.

We point out two things. First, this result *does not mean* that two surfaces with the same Gaussian curvature are *the same geometrically*. Again, to make this precise we need the notion of isometry, so for the moment we give

EXERCISE 4.1. Show that the following two patches have the same Gaussian curvature:

$$\mathbf{x}(u, v) = (u \cos v, u \sin v, v) \qquad \mathbf{y}(u, v) = (u \cos v, u \sin v, \ln u).$$

These surfaces are not isometric as will be shown later.

Secondly, the Theorem Egregium holds for Gaussian curvature, *but not for mean curvature*. For instance, as we have mentioned, a cylinder and a plane are geometrically the same with the same Gauss curvature $K = 0$, but their mean curvatures are quite different. The mean curvature of the plane is zero while that of the cylinder is one-half the curvature of the base circle. Thus, mean curvature is not truly an invariant of *the surface itself*, but depends on the way in which the surface sits in $\mathbb{R}^3$.

We now turn to the derivation of the formula in Theorem 4.1. While the details of the calculations are quite tedious, keep in mind that we are simply finding coefficients for a vector in terms of a particular basis. Because $l = \mathbf{x}_{uu} \cdot U$, $m = \mathbf{x}_{uv} \cdot U$ and $n = \mathbf{x}_{vv} \cdot U$, we need expressions for $\mathbf{x}_{uu}$, $\mathbf{x}_{uv}$, and $\mathbf{x}_{vv}$ in terms of the basis for 3-space $\{\mathbf{x}_u, \mathbf{x}_v, U\}$. Write,

(**)
$$\mathbf{x}_{uu} = \Gamma^u_{uu}\mathbf{x}_u + \Gamma^v_{uu}\mathbf{x}_v + lU \quad (\text{since } \mathbf{x}_{uu} \cdot U = l \text{ etc.})$$
$$\mathbf{x}_{uv} = \Gamma^u_{uv}\mathbf{x}_u + \Gamma^v_{uv}\mathbf{x}_v + mU$$
$$\mathbf{x}_{vv} = \Gamma^u_{vv}\mathbf{x}_u + \Gamma^v_{vv}\mathbf{x}_v + nU.$$

Our job is to find the Γ's. These are just coefficients in a basis expansion, but they are known by the name *Christoffel Symbols*. We use what we know about dot products to determine the Γ's.

$$\mathbf{x}_{uu} \cdot \mathbf{x}_u = \Gamma^u_{uu}\mathbf{x}_u \cdot \mathbf{x}_u + 0 + 0$$
$$= \Gamma^u_{uu}E \qquad \text{by definition of } E$$

If we can compute $\mathbf{x}_{uu} \cdot \mathbf{x}_u$ then we will know Γ^u_{uu}. This is simply the product rule.

$$E = \mathbf{x}_u \cdot \mathbf{x}_u, \quad \text{so} \quad E_u = \mathbf{x}_{uu} \cdot \mathbf{x}_u + \mathbf{x}_u \cdot \mathbf{x}_{uu} = 2\mathbf{x}_{uu} \cdot \mathbf{x}_u.$$

Thus,

$$\mathbf{x}_{uu} \cdot \mathbf{x}_u = \frac{E_u}{2} \quad \text{and} \quad \Gamma^u_{uu} = \frac{E_u}{2E}.$$

Further, $\mathbf{x}_u \cdot \mathbf{x}_v = 0$ so taking the partial with respect to u gives

$$0 = \mathbf{x}_{uu} \cdot \mathbf{x}_v + \mathbf{x}_u \cdot \mathbf{x}_{uv} \quad \text{or} \quad \mathbf{x}_{uu} \cdot \mathbf{x}_v = -\mathbf{x}_u \cdot \mathbf{x}_{uv}.$$

Also, $E = \mathbf{x}_u \cdot \mathbf{x}_u$, so taking the partial with respect to v gives $E_v = 2\mathbf{x}_u \cdot \mathbf{x}_{uv}$ and, consequently, $E_v/2 = \mathbf{x}_u \cdot \mathbf{x}_{uv} = -\mathbf{x}_{uu} \cdot \mathbf{x}_v$. Moreover,

$$\Gamma^v_{uu} = (\mathbf{x}_{uu} \cdot \mathbf{x}_v)/G = -E_v/2G \quad \text{and} \quad \Gamma^u_{uv} = \mathbf{x}_{uv} \cdot \mathbf{x}_u/E = E_v/2E.$$

Continuing on, $G = \mathbf{x}_v \cdot \mathbf{x}_v$, so $G_u/2 = \mathbf{x}_{uv} \cdot \mathbf{x}_v$. Then, since $0 = \mathbf{x}_v \cdot \mathbf{x}_u$, we have

$$-\mathbf{x}_v \cdot \mathbf{x}_{uv} = \mathbf{x}_{vv} \cdot \mathbf{x}_u \quad \text{with} \quad \Gamma^v_{uv} = \mathbf{x}_{uv} \cdot \mathbf{x}_v/G = G_u/2G$$
$$\text{and} \quad \Gamma^u_{vv} = \mathbf{x}_{vv} \cdot \mathbf{x}_u/E = -G_u/2E.$$

Finally, $\mathbf{x}_v \cdot \mathbf{x}_v = G$, so $\mathbf{x}_{vv} \cdot \mathbf{x}_v = G_v/2$ and $\Gamma_{vv}^v = \mathbf{x}_{vv} \cdot \mathbf{x}_v/G = G_v/2G$. We end up with the following formulas.

$$\mathbf{x}_{uu} = \frac{E_u}{2E}\mathbf{x}_u - \frac{E_v}{2G}\mathbf{x}_v + lU$$

$$\mathbf{x}_{uv} = \frac{E_v}{2E}\mathbf{x}_u + \frac{G_u}{2G}\mathbf{x}_v + mU$$

$$\mathbf{x}_{vv} = -\frac{G_u}{2E}\mathbf{x}_u + \frac{G_v}{2G}\mathbf{x}_v + nU$$

$$U_u = -\frac{l}{E}\mathbf{x}_u - \frac{m}{G}\mathbf{x}_v$$

$$U_v = -\frac{m}{E}\mathbf{x}_u - \frac{n}{G}\mathbf{x}_v$$

EXERCISE 4.2. Compute U_u and U_v as above. Hint: $U_u \overset{\text{def}}{=} \nabla_{\mathbf{x}_u} U = A\mathbf{x}_u + B\mathbf{x}_v$. Find A by $\nabla_{\mathbf{x}_u} U \cdot \mathbf{x}_u = A\mathbf{x}_u \cdot \mathbf{x}_u = AE$ and $0 = \mathbf{x}_u(U \cdot \mathbf{x}_u) = \nabla_{\mathbf{x}_u} U \cdot \mathbf{x}_u + U \cdot \mathbf{x}_{uu}$.)

We know that mixed partial derivatives are equal no matter the order of differentiation, so $\mathbf{x}_{uuv} = \mathbf{x}_{uvu}$, or $\mathbf{x}_{uuv} - \mathbf{x}_{uvu} = 0$. This means that the coefficients of $\mathbf{x}_u$, $\mathbf{x}_v$ and U are *all zero* when $\mathbf{x}_{uuv} - \mathbf{x}_{uvu}$ is written in this basis. Let's concentrate on the $\mathbf{x}_v$-term. (We again use the product rule repeatedly.)

$$\mathbf{x}_{uuv} = \left(\frac{E_u}{2E}\right)_v \mathbf{x}_u + \frac{E_u}{2E}\mathbf{x}_{uv} - \left(\frac{E_v}{2G}\right)_v \mathbf{x}_v - \frac{E_v}{2G}\mathbf{x}_{vv} + l_v U + lU_v.$$

Now replace $\mathbf{x}_{uv}$, $\mathbf{x}_{vv}$ and U_v by their basis expansions to get,

$$\mathbf{x}_{uuv} = [\ ]\mathbf{x}_u + \left[\frac{E_u G_u}{4EG} - \left(\frac{E_v}{2G}\right)_v - \frac{E_v G_v}{4G^2} - \frac{ln}{G}\right]\mathbf{x}_v + [\ ]U$$

$$\mathbf{x}_{uvu} = \left(\frac{E_v}{2E}\right)_u \mathbf{x}_u + \frac{E_v}{2E}\mathbf{x}_{uu} + \left(\frac{G_u}{2G}\right)_u \mathbf{x}_v + \frac{G_u}{2G}\mathbf{x}_{uv} + m_u U + mU_u$$

$$\mathbf{x}_{uvu} = [\ ]\mathbf{x}_u + \left[-\frac{E_v E_v}{4EG} + \left(\frac{G_u}{2G}\right)_u + \frac{G_u G_u}{4G^2} - \frac{m^2}{G}\right]\mathbf{x}_v + [\ ]U.$$

Because the $\mathbf{x}_v$-coefficient of $\mathbf{x}_{uuv} - \mathbf{x}_{uvu}$ is zero, we get,

$$0 = \frac{E_u G_u}{4EG} - \left(\frac{E_v}{2G}\right)_v - \frac{E_v G_v}{4G^2} + \frac{E_v E_v}{4EG} - \left(\frac{G_u}{2G}\right)_u - \frac{G_u G_u}{4G^2} - \frac{ln - m^2}{G}.$$

Notice the last term! Dividing by E, we have

$$\frac{ln - m^2}{EG} = \frac{E_u G_u}{4E^2 G} - \frac{1}{E}\left(\frac{E_v}{2G}\right)_v - \frac{E_v G_v}{4EG^2} + \frac{E_v E_v}{4E^2 G} - \frac{1}{E}\left(\frac{G_u}{2G}\right)_u - \frac{G_u G_u}{4EG^2}.$$

Of course, the lefthand side is K (since $F = 0$) and the righthand side only depends on E and G. Thus, we have a formula for K which does not make explicit use of the normal U.

EXERCISE 4.3. Show that the righthand side above is

$$-\frac{1}{2\sqrt{EG}}\left(\frac{\partial}{\partial v}\left(\frac{E_v}{\sqrt{EG}}\right) + \frac{\partial}{\partial u}\left(\frac{G_u}{\sqrt{EG}}\right)\right).$$

EXERCISE 4.4. Compute the curvature of the sphere of radius R using this formula. Does this agree with what the shape operator tells you?

EXERCISE 4.5. Show, by similar arguments, that, if $F \neq 0$, the following formula holds.

$$K =$$

$$\frac{1}{(EG-F^2)^2}\left(\begin{vmatrix} -\frac{E_{vv}}{2}+F_{uv}-\frac{G_{uu}}{2} & \frac{E_u}{2} & F_u-\frac{E_v}{2} \\ F_v-\frac{G_u}{2} & E & F \\ \frac{G_v}{2} & F & G \end{vmatrix} - \begin{vmatrix} 0 & \frac{E_v}{2} & \frac{G_u}{2} \\ \frac{E_v}{2} & E & F \\ \frac{G_u}{2} & F & G \end{vmatrix}\right)$$

EXERCISE 4.6. Show that the U-terms of $\mathbf{x}_{uuv} - \mathbf{x}_{uvu} = 0$ and $\mathbf{x}_{vvu} - \mathbf{x}_{uvv} = 0$ give

$$l_v - m_u = m\left(\frac{G_u}{2G} - \frac{E_u}{2E}\right) + E_v H$$

$$= l\frac{E_v}{2E} + m\left(\frac{G_u}{2G} - \frac{E_u}{2E}\right) + n\frac{E_v}{2G}$$

and

$$n_u - m_v = m\left(\frac{E_v}{2E} - \frac{G_v}{2G}\right) + G_u H$$

$$= l\frac{G_u}{2E} + m\left(\frac{E_v}{2E} - \frac{G_v}{2G}\right) + n\frac{G_u}{2G}$$

where H is mean curvature (and F is assumed to be zero). These are the Codazzi-Mainardi equations. What do they say in the case where the surface is minimal (i.e. $H = 0$) and the parametrization satisfies $E = G$, $F = 0$?

3.5 SOME EFFECTS OF CURVATURE(S)

In this section we will derive some geometric consequences of various conditions on Gaussian and mean curvatures. Of course, the point of all this is simply to show that we have not wasted our time so far in learning how to compute K and H. These quantities really do tell us about the geometry of a surface. In the next chapter we will focus on mean curvature and see that it is intimately related to certain structures in Nature. For now, however, our goal is simply to see geometry reflected in the linear algebra and calculus represented by K and H.

Let us start by considering a completely impractical situation! Namely, suppose we knew that every point on a surface was an umbilic point. Recall that this means that, at every point, the principal curvatures are equal. Although this doesn't mean that, *a priori*, the principal curvatures are the same at every point, this is precisely what we shall prove below. Of course, a direct verification of such a hypothesis is out of the question, yet we shall see later that this result is useful all the same. We shall use the following

EXERCISE 5.1. Show that two vectors $\mathbf{v} = v_1\mathbf{x}_u + v_2\mathbf{x}_v + v_3U$ and $\mathbf{w} = w_1\mathbf{x}_u + w_2\mathbf{x}_v + w_3U$ are equal if and only if the following conditions hold:

$$\mathbf{v} \cdot \mathbf{x}_u = \mathbf{w} \cdot \mathbf{x}_u, \quad \mathbf{v} \cdot \mathbf{x}_v = \mathbf{w} \cdot \mathbf{x}_v, \quad \mathbf{v} \cdot U = \mathbf{w} \cdot U.$$

Hint: We do *not* assume $\mathbf{x}_u \cdot \mathbf{x}_v = 0$ here. Use $EG - F^2 \neq 0$.

EXERCISE 5.2. Suppose $p \in M$ is an umbilic point (i.e. $k_1 = k_2 = k$ at p). Show that, at p,

$$\frac{l}{E} = \frac{m}{F} = \frac{n}{G}.$$

Hint: Write $S(\mathbf{x}_u) = A\mathbf{x}_u + B\mathbf{x}_v = k\mathbf{x}_u$ since p is umbilic. Compute B another way:

$$l = S(\mathbf{x}_u) \cdot \mathbf{x}_u = AE + BF, \quad m = S(\mathbf{x}_u) \cdot \mathbf{x}_v = AF + BG.$$

Solve for B. Do the same for $S(\mathbf{x}_v)$.

Theorem 5.1. *A surface M consisting entirely of umbilic points is contained in either a plane or a sphere.*

Proof. At each point $\frac{l}{E} = \frac{m}{F} = \frac{n}{G} = C$, but C may vary from point to point in M. Now, by Exercise 5.2 above, we can see that

$$\nabla_{\mathbf{x}_u} U = -C\mathbf{x}_u \quad \text{and} \quad \nabla_{\mathbf{x}_v} U = -C\mathbf{x}_v.$$

For example, $\nabla_{\mathbf{x}_u} U \cdot \mathbf{x}_u = -l = -l/E\mathbf{x}_u \cdot \mathbf{x}_u = -l/EE$ and $\nabla_{\mathbf{x}_u} U \cdot \mathbf{x}_v = -m = -m/FF = -m/F\mathbf{x}_u \cdot \mathbf{x}_u$. Now the mixed covariant derivative may be calculated.

$$\nabla_{\mathbf{x}_v}\nabla_{\mathbf{x}_u} U = \sum_k \sum_i \sum_j \left(\frac{\partial^2 u_k}{\partial x_i \partial x_j}\frac{\partial x_i}{\partial u}\frac{\partial x_j}{\partial v} + \frac{\partial u_k}{\partial x_i}\frac{\partial^2 x_i}{\partial u \partial v}\right)e_k$$
$$= \nabla_{\mathbf{x}_u}\nabla_{\mathbf{x}_v} U$$

so differentiating above by v (left) and u (right), we get, $-C_v\mathbf{x}_u - C\mathbf{x}_{uv} = -C_u\mathbf{x}_v - C\mathbf{x}_{vu}$ or $C_v\mathbf{x}_u = C_u\mathbf{x}_v$. But $\mathbf{x}_u$ and $\mathbf{x}_v$ are linearly independent, so we must have $C_v = 0 = C_u$ and this implies that C *is a constant*.

Case 1. Suppose $C = 0$. Then the shape operator is zero everywhere. By Theorem 2.2.3, M is contained in a plane.

Case 2. Suppose $C \neq 0$. Then consider $\mathbf{x}(u, v) + \frac{1}{c}U$ and compute,

$$\frac{\partial}{\partial u}[\mathbf{x}(u, v) + \frac{1}{C}U] = \mathbf{x}_u + \frac{1}{C}U_u$$

$$= \mathbf{x}_u + \frac{1}{C}[-C\mathbf{x}_u] \qquad \text{by above}$$

$$= \mathbf{x}_u - \mathbf{x}_u$$

$$= 0.$$

In a similar fashion

$$\frac{\partial}{\partial v}[\mathbf{x}(u, v) + \frac{1}{C}U] = 0.$$

Hence, $\mathbf{x}(u, v) + \frac{1}{C}U = p$ constant. Then $|\mathbf{x}(u, v) - p| = |-\frac{1}{C}| = \frac{1}{C}$, so every point $\mathbf{x}(u, v)$ lies a distance $1/C$ from p. Hence, $\mathbf{x}(u, v)$ lies on a sphere of radius $1/C$.

We note that we have proved the result only for a single patch. However, because M is connected, any patch must overlap some other patch. For a patch which overlaps $\mathbf{x}(u, v)$, the points in the overlap lie on the plane or sphere as above. Since the argument above produces a constant C on the new patch, it must, therefore, agree with the C of $\mathbf{x}(u, v)$. Hence, the geometry of $\mathbf{x}(u, v)$ propagates to overlapping patches. This then continues over the whole surface, each patch reproducing the geometry of its neighbors. □

Notice that in the situation above,

$$K = \frac{ln - m^2}{EG - F^2} = \frac{\frac{l}{E}\frac{n}{G}(EG) - \frac{m^2}{F^2}F^2}{EG - F^2} = \frac{\frac{l^2}{E^2}(EG - F^2)}{EG - F^2} = \frac{l^2}{E^2},$$

so $C = \sqrt{K}$ and the radius of the sphere is $1/\sqrt{K}$. The next result restricts the form surfaces in $\mathbb{R}^3$ may take in terms of their possible Gaussian curvatures. Recall that a surface in $\mathbb{R}^3$ is *compact* if it is closed and bounded. Here, "bounded" means that the surface may be enclosed by a sphere of sufficient size. The term "closed" means that any sequence of points on the surface which converges to some point in $\mathbb{R}^3$ actually converges to a point on the surface. In the following result the hypothesis of "compactness" is used to ensure that maximums and minimums of functions are actually attained by the functions.

Theorem 5.2. *On every compact surface $M \subseteq \mathbb{R}^3$ there is some point p with $K(p) > 0$.*

Proof. Let $f \colon M \to \mathbb{R}$ be defined by $f(p) = |p|^2$. This function is continuous, so (since M is *compact*) f attains its max and its min. Let p_0 be a max for f. Now $f(p_0) = |p_0|^2$, so p_0 is the farthest point from the origin on M. If $r = |p_0|$, then M is inside the sphere of radius r as shown:

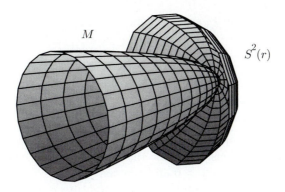

$$M \qquad\qquad\qquad S^2(r)$$

FIGURE 3.7

Therefore, it seems likely that $K(p_0) \geq \frac{1}{r^2}$, the curvature of the r-sphere $S^2(r)$. We verify this and so prove the theorem. Consider a unit tangent vector $\mathbf{u} \in T_{p_0}(M)$ and take a curve α in M with $\alpha(0) = p_0$ and $\alpha'(0) = \mathbf{u}$. Of course the composition $f \circ \alpha$ still has a max at p_0 (since α lies in M), so

$$\frac{d}{dt}(f \circ \alpha)\,|_{t=0} = 0 \qquad \text{since a max is a critical point}$$

$$\frac{d^2}{dt^2}(f \circ \alpha)\,|_{t=0} < 0 \qquad \text{since } p_0 \text{ is a max.}$$

Now $f \circ \alpha(t) = |\alpha(t)|^2 = \alpha(t) \cdot \alpha(t)$, so $\frac{d}{dt}(f \circ \alpha) = 2\alpha \cdot \alpha'$. At $t = 0$,

$$0 = 2\alpha(0) \cdot \alpha'(0) = 2p_0 \cdot \mathbf{u}.$$

But $\mathbf{u}$ is any unit vector in $T_{p_0}M$, so $p_0 \cdot \mathbf{u} = 0$ for all $\mathbf{u} \in T_{p_0}M$. Hence, p_0 is normal to M at p_0. Also, $\frac{d^2}{dt^2}(f \circ \alpha) = 2\alpha' \cdot \alpha' + 2\alpha \cdot \alpha''$ and at $t = 0$,

$$0 \geq \frac{d^2}{dt^2}(f \circ \alpha) = 2\alpha'(0) \cdot \alpha'(0) + 2\alpha(0) \cdot \alpha''(0)$$

$$\geq \mathbf{u} \cdot \mathbf{u} + p_0 \cdot \alpha''(0)$$

$$\geq 1 + p_0 \cdot \alpha''(0)$$

$$-1 \geq p_0 \cdot \alpha''(0).$$

Now, let's calculate normal curvatures at p_0. Since $p_0/r = p_0/|p_0|$ is a unit normal to M at p_0, the calculations above give

$$k(\mathbf{u}) = S(\mathbf{u}) \cdot \mathbf{u}$$

$$= S(\alpha') \cdot \alpha'$$

$$= U(p_0) \cdot \alpha''$$

$$= \frac{p_0}{r} \cdot \alpha''$$

$$\leq -\frac{1}{r}.$$

In particular, both k_1 and k_2 are less than or equal to $-1/r$. Thus, the Gauss curvature has

$$K(p_0) = k_1(p_0)k_2(p_0) \geq \frac{1}{r}\frac{1}{r} = \frac{1}{r^2} > 0.$$

$\square$

Corollary 5.3. *There are no compact surfaces in $\mathbb{R}^3$ with $K \leq 0$. In particular, no minimal surface embedded in $\mathbb{R}^3$ is compact.*

We now come to one of the main results displaying the influence of curvature on geometry. Previously, we saw that an all umbilic surface is either a plane ($K = 0$) or a sphere ($K > 0$ and constant). There is, however, a much stronger result which is based on the Gauss curvature K itself rather than its factors k_1 and k_2. This beautiful result is

Theorem 5.4 (H. Liebmann). *If M is a compact surface of constant Gaussian curvature K, then M is a sphere of radius $1/\sqrt{K}$.*

Proof. By the previous Theorem, there is some $p \in M$ with $K(p) > 0$. Since K is constant, then $K > 0$ everywhere on M. Because K is constant and $K = k_1k_2$, where k_1 and k_2 are the principal curvatures, if p is a point where the function k_1 is a maximum, then p is also a point where k_2 is a minimum. We know that such a p exists since M is compact and k_1, k_2 are continuous functions. We now have two cases to consider.

Case 1. Suppose $k_1 = k_2$ at p. Because k_1 is a maximum and k_2 is a minimum at p, *all* normal curvatures on M must lie between these two values. Therefore, if $k_1 = k_2$ at p, the normal curvature k must be constant on M. Hence, M is all umbilic with $K > 0$, so it is a sphere of radius $1/\sqrt{K}$. (This is what we meant by saying that Theorem 5.1 would prove useful.) We shall now show that Case 2 below is impossible, so that Case 1 proves the theorem.

Case 2. Suppose $k_1 > k_2$ at p. There is actually a small open neighborhood of p in M where $k_1 > k_2$. If not, we could take a decreasing sequence of open "balls" about p and choose points α_i in the balls with $k_1(\alpha_i) = k_2(\alpha_i)$ and $\alpha_i \to p$. Then, by continuity, we would have $k_1(p) = k_2(p)$. This contradicts our assumption, so we must have the desired neighborhood. Although we shall not prove it, on this neighborhood we may choose two principal directions which are perpendicular at each point and use these to provide a patch with $\mathbf{x}_u \cdot \mathbf{x}_u = E$, $\mathbf{x}_u \cdot \mathbf{x}_v = 0$, $\mathbf{x}_v \cdot \mathbf{x}_v = G$ and, since $\mathbf{x}_u$, $\mathbf{x}_v$ are *principal vectors*,

$$S(\mathbf{x}_u) = k_1\mathbf{x}_u, \quad S(\mathbf{x}_u) = k_2\mathbf{x}_v.$$

(A proof of the existence of such a patch $\mathbf{x}$ may be found in [DoC1, p. 185].) Note that $m = S(\mathbf{x}_u) \cdot \mathbf{x}_v = k\mathbf{x}_u \cdot \mathbf{x}_v = 0$ also. Moreover, $l = S(\mathbf{x}_u) \cdot \mathbf{x}_u = k_1\mathbf{x}_u \cdot \mathbf{x}_u = k_1E$, so $k_1 = \frac{l}{E}$ and, similarly, $k_2 = \frac{n}{G}$. Of course, we have the formula,

$$K = -\frac{1}{2\sqrt{EG}}\left[\frac{\partial}{\partial v}\left(\frac{E_v}{\sqrt{EG}}\right) + \frac{\partial}{\partial u}\left(\frac{G_u}{\sqrt{EG}}\right)\right]$$

and from the U-term of $\mathbf{x}_{uuv} - \mathbf{x}_{uvu} = 0$ (see Exercise 4.6), we get

$$0 = \frac{-nE_v}{2G} + l_v - \frac{E_v l}{2E}$$

$$l_v = \frac{E_v}{2}\left[\frac{l}{E} + \frac{n}{G}\right]$$

$$= \frac{E_v}{2}(k_1 + k_2).$$

Similarly, using the U-term of $\mathbf{x}_{vvu} - \mathbf{x}_{vuv} = 0$ we get,

$$n_u = \frac{G_u}{2}\left[\frac{l}{E} + \frac{n}{G}\right] = \frac{G_u}{2}(k_1 + k_2).$$

If we differentiate the formulas $k_1 = \frac{l}{E}$ and $k_2 = \frac{n}{G}$ repeatedly and use the expressions for l_v and n_u we get,

$$k_{1v} = \frac{1}{E^2}(El_v - E_v l) = \frac{E_v}{2E}(k_2 - k_1)$$

$$k_{2u} = \frac{1}{G^2}(Gn_u - G_u n) = \frac{G_u}{2G}(k_1 - k_2)$$

$$k_{1vv} = \frac{E_{vv}}{2E}(k_2 - k_1) + E_v(\ldots)$$

$$k_{2uu} = \frac{G_{uu}}{2G}(k_1 - k_2) + G_u(\ldots).$$

Now, at p, k_1 is a max and k_2 is a min, so

$$k_{1v} = 0 \qquad k_{2u} = 0$$
$$k_{1vv} \leq 0 \qquad k_{2uu} \geq 0.$$

Plugging these estimates into the equations above we get, at p,

$$E_v = 0 \qquad G_u = 0$$
$$E_{vv} \geq 0 \qquad G_{uu} \geq 0.$$

Note that, even though $E_v = 0 = G_u$ at p, the expressions

$$\frac{\partial}{\partial v}\left(\frac{E_v}{\sqrt{EG}}\right) \quad \text{and} \quad \frac{\partial}{\partial u}\left(\frac{G_u}{\sqrt{EG}}\right)$$

may be nonzero at p since the differentiation is done before the evaluation at p. These expressions then give us, at p,

$$K = -\frac{1}{2EG}(E_{vv} + G_{uu}) \leq 0.$$

But this contradicts our hypothesis that $K > 0$, so our assumption $k_1 > k_2$ must be incorrect. Hence Case 1 holds and M is a sphere. □

Notice that in the course of the proof we have discovered a Lemma of David Hilbert which will prove useful in a later exercise.

Lemma 5.5 (Hilbert). *If k_1 has a max at p, k_2 has a min at p and k_1 is strictly greater than k_2 at p, then $K(p) \leq 0$.*

So far, we have concentrated on the effects of Gaussian curvature on geometry. We now balance this somewhat by giving a result which indicates the effect mean curvature has on the geometry of a surface. Recall that a surface is *minimal* if the mean curvature is zero at each point of the surface. Just as for other geometric qualities, we may test the notion of minimality on classes of surfaces which we understand best. In Chapter 4 we shall consider other types of minimal surfaces, but for now let's look at the surfaces of revolution.

Theorem 5.6. *If a surface of revolution M is minimal, then M is contained in either a plane or a catenoid.*

Proof. For simplicity, we will take the special case of a patch $\mathbf{x}(u, v) = (u, h(u) \cos v, h(u) \sin v)$ for M. Then

$$H = \frac{k_\mu + k_\pi}{2} = \frac{1}{2} \left(\frac{-h''}{(1 + h'^2)^{3/2}} + \frac{1}{h(1 + h'^2)^{1/2}} \right) = \frac{1}{2} \left(\frac{-hh'' + 1 + h'^2}{h(1 + h'^2)^{3/2}} \right).$$

The surface M is minimal, so $H = 0$ and, thus, $hh'' = 1 + h'^2$. Let $w = h'$. Then

$$h'' = w' = \frac{dw}{dh} \frac{dh}{du} = \frac{dw}{dh} w.$$

Here we consider w as a function of h on an interval with $h'(u) \neq 0$. We can do this because, on such an interval, h has an inverse function f with $u = f(h)$. Then, taking a derivative and applying the chain rule gives

$$1 = \frac{df}{dh} \frac{dh}{du} = \frac{df}{dh} w \qquad \text{or} \qquad w = \frac{1}{\frac{df}{dh}},$$

so that w is a function of h since f and df/dh are. Thus, $hh'' = 1 + h'^2$ implies $hw \, dw/dh = 1 + w^2$; or $\frac{w}{1+w^2} dw = \frac{1}{h} dh$. We can integrate both sides to get,

$$\ln(\sqrt{1 + w^2}) = \ln h + c$$

$$\sqrt{1 + w^2} = ch$$

$$w = \sqrt{c^2 h^2 - 1}.$$

Now, $w = \frac{\partial h}{\partial u}$, so $\frac{1}{\sqrt{c^2 h^2 - 1}} dh = du$ and integration yields (with $ch = \cosh l$)

$$\frac{1}{c} l = u + D$$

$$\frac{1}{c} \cosh^{-1} ch = u + D$$

$$ch = \cosh(cu + D)$$

$$h = \frac{1}{c} \cosh(cu + D).$$

Therefore, M is part of a catenoid.   □

EXERCISE 5.3. In the proof above we assumed that $h'(u) \neq 0$. What happens if $h'(u) = 0$? Explain.

EXERCISE 5.4. Show that a flat $(K = 0)$ surface of revolution is part of a cone or a cylinder. Hint: Model your proof after the theorem above.

EXERCISE 5.5. Show that if M is a compact, connected (oriented) surface with $K > 0$ and H constant, then M is a sphere of radius $\frac{1}{\sqrt{K}}$.

Hints: (1) Consider the function $H^2 - K = (k_1 - k_2)^2/4$ and let $p \in M$ be a max for it. (2) If $H^2 - K(p) = 0$, show K is constant and use Liebmann's Theorem. (3) Suppose $H^2 - K(p) > 0$ (so $k_1(p) > k_2(p)$). Note $(k_1 - k_2)(p)$ is a max. (4) Use Hilbert's Lemma.

3.6 SURFACES OF DELAUNAY

Just as we saw that surfaces of revolution of constant Gaussian curvature are determined by certain integrals with chosen initial conditions, we can ask whether something similar occurs for constant mean curvature. Let's take a surface of revolution $M\colon \mathbf{x}(u, v)$, where the patch has the form $\mathbf{x}(u, v) = (u, h(u)\cos v, h(u)\sin v)$. The mean curvature is given by

$$H = \frac{1}{2}\frac{-hh'' + 1 + h'^2}{h(1 + h'^2)^{3/2}}.$$

Suppose $H = c/2$ is constant. We then have the differential equation

$$1 + h'^2 - hh'' = ch(1 + h'^2)^{\frac{3}{2}}.$$

First, consider the case when $c = 0$. The differential equation then reduces to $1 + h'^2 - hh'' = 0$. This is precisely the equation we solved to get the catenoid. Of course this must be the case since the condition $c = 0$ is saying that M is a surface of revolution with zero mean curvature. Secondly, suppose $c = \pm 1/a$

with $a > 0$. We get (for $c = -1/a$), $1 + h'^2 - hh'' = \frac{-1}{a} h(1 + h'^2)^{\frac{3}{2}}$, or

$$\frac{a(1 + h'^2) - ahh''}{(1 + h'^2)^{\frac{3}{2}}} + h = 0$$

$$2h' \left[\frac{a(1 + h'^2) - ahh''}{(1 + h'^2)^{\frac{3}{2}}} + h \right] = 0$$

$$\frac{2ah'(1 + h'^2) - 2ahh'h''}{(1 + h'^2)^{\frac{3}{2}}} + 2hh' = 0$$

$$\frac{d}{du} \left[\frac{2ah}{\sqrt{1 + h'^2}} + h^2 \right] = 0$$

$$h^2 + \frac{2ah}{\sqrt{1 + h'^2}} = \pm b^2 \qquad \text{a constant.}$$

If we take all cases into consideration (and note that the steps above may be reversed), we obtain the general form of the differential equation describing surfaces of revolution of constant mean curvature.

Theorem 6.1. *M is a surface of revolution of constant mean curvature parametrized by* $\mathbf{x}(u, v) = (u, h(u) \cos v, h(u) \sin v)$ *if and only if the function* $h(u)$ *satisfies*

$$h^2 \pm \frac{2ah}{\sqrt{1 + h'^2}} = \pm b^2$$

where a and b are constants.

It is an amazing fact discovered by Delaunay that this differential equation arises geometrically. That is to say, there is a geometric construction which produces the differential equation above and, consequently, all surfaces of revolution of constant mean curvature. We shall consider one example below in detail and then simply give the geometric characterization in a theorem.

Example 6.1: The Roulette of an Ellipse.
Suppose that an ellipse rolls without slipping along the x-axis. If we follow the path of one of the foci of the ellipse as it rolls, then this path gives an example of a *roulette* — a curve formed by a point associated to one curve as it rolls upon another. Consider the following "rolling" diagram where F and F' are the foci of the ellipse, K is the point of (tangent) contact with the x-axis and FT is the tangent to the curve traced out by F.

We will use various properties of the ellipse which were, perhaps, better known in years past (see [G], [Ya] or [Zw] for example). Let the ellipse have major axis of length a and minor axis of length b. Then it is known that the sum of the distances from the foci to a given point on the ellipse is constant and equal to $2a$. In particular, $FK + F'K = 2a$. Furthermore, the ellipse has the beautiful (and useful) reflection property that any light ray originating

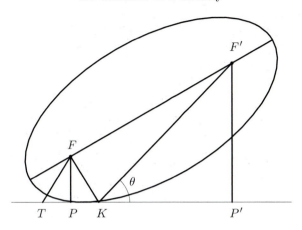

FIGURE 3.8.   Ellipse rolling on the x-axis

from one focus will reflect off the ellipse to the other focus. Since the angles of incidence and reflection are equal, this means that $\theta \overset{\text{def}}{=} \angle FKP = \angle F'KP'$. A more esoteric property of the ellipse (its *pedal equation*) is the fact that the product of the lengths of the line segments from the foci to any tangent line which meet the tangent perpendicularly is a constant equal to the square of the minor axis. In our case this means that $FP \cdot F'P' = b^2$. Finally, a general property of this sort of roulette is the fact that the direction of the normal to the curve traced by F passes through the point of contact with the line. Here, this means that FK is perpendicular to FT.

Now, the x and y coordinates of F are given by OP and FP respectively. Then $y = FK \sin \theta$. Because $\angle KFT = \pi/2$, we have $\angle FTP = \pi/2 - \theta$. Moreover, $\angle FTP$ is the angle between the tangent of the traced curve and the x-axis. That is, for the unit tangent of the traced curve $T = (\frac{dx}{ds}, \frac{dy}{ds})$,

$$\sin \theta = \cos \left(\frac{\pi}{2} - \theta \right) = T \cdot (1,0) = \frac{dx}{ds}.$$

Hence, $y = FK \frac{dx}{ds}$. Also, since $\angle FKP = \angle F'KP'$, we have $y' \overset{\text{def}}{=} F'P' = F'K \frac{dx}{ds}$. Note that

$$y + y' = (FK + F'K) \frac{dx}{ds} = 2a \frac{dx}{ds}$$

by the fundamental property of the ellipse. Also, the pedal equation gives

$$yy' = b^2.$$

Solving for y' in the second equation and plugging into the first equation gives

$$y^2 - 2ay \frac{dx}{ds} + b^2 = 0.$$

Now, the arclength is given by $s = \int \sqrt{1 + y'^2}\, dx$, so $\frac{dx}{ds} = 1/\sqrt{1 + y'^2}$. Replacing $\frac{dx}{ds}$ in the equation above and noting that its sign changes when $\angle FTK$ is large enough, we obtain

$$y^2 \pm \frac{2ay}{\sqrt{1 + y'^2}} + b^2 = 0.$$

This is precisely the equation we obtained from the assumption of constant mean curvature. Therefore, the roulette of an ellipse, an *undulary*, is a profile curve (i.e. meridian) of a surface of revolution of constant mean curvature, an *unduloid*. We note first that, for this case, we must have $a > b$. Also, we note that the companion equation $y^2 \pm \frac{2ay}{\sqrt{1+y'^2}} - b^2 = 0$ is the differential equation describing the roulette of (the focus of) a hyperbola, a *nodary*.

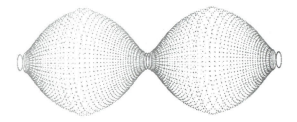

FIGURE 3.9.   The unduloid with a=1.5 and b=1

EXERCISE 6.1. What is the roulette obtained when $b = a$? Hint: think geometrically.

EXERCISE 6.2. What is the roulette obtained when $b = 0$. Hint: solve the differential equation and then interpret geometrically.

EXERCISE 6.3. Find the roulette of the focus of a parabola as it is rolled along the x-axis.

EXERCISE 6.4. Show that the differential equation $y^2 \pm 2ay/\sqrt{1 + y'^2} + b^2 = 0$ may be put in the form

$$\frac{dx}{dy} = \frac{-b^2 - y^2}{\sqrt{4a^2 y^2 - (y^2 + b^2)^2}}.$$

This expression may be used with MAPLE to graph surfaces of Delaunay.

A surface M is *immersed* in $\mathbb{R}^3$ if there is a mapping $f \colon M \to \mathbb{R}^3$ which may not be one-to-one, but which has a one-to-one derivative map f_* at *every point*. Also, although we shall treat the idea of *completeness* in Chapter V, we mention here that one property of a complete surface is that it is not part of some larger surface. Finally, recall that the word "conic" refers to any of circles, lines, parabolas, ellipses and hyperbolas. Now we can state Delaunay's Theorem.

Theorem 6.2 (Delaunay). *A complete immersed surface of revolution of constant mean curvature is a roulette of a conic.*

More comprehensive discussions of the surfaces of Delaunay may be found in [Ee1], [Ee2] (which we have mostly followed above) and, of course, [Del]. When we talk more about surfaces with constant mean curvature in Chapter 4, we shall see that they come about as the result of a variational principle (again, see [Ee1], [Ee2]). That is, informally, the surface takes its shape to "minimize" surface area subject to having fixed volume. This principle has both physical and biological consequences. In particular, if a soap bubble is formed between two rings (with the ends included), then it holds a certain amount of air inside. If the amount of air is just right, then the bubble is a cylinder, the surface of Delaunay associated to the roulette of a circle. If the amount of air is changed or if the rings are pulled apart to lengthen the bubble, then the waist of the bubble narrows and an unduloid results. These types of experiments are discussed in D'Arcy Wentworth Thompson's book *On Growth and Form* [DWT]. In Chapter 5 of this classic text Thompson describes informally how surface tension and pressure combine to determine the shapes of various one-celled creatures — shapes suspiciously Delaunayan!

3.7 Calculating Curvature with MAPLE

As might be expected, MAPLE may be used to calculate Gaussian and mean curvatures as well as such necessary ingredients as the metric coefficients E, F and G, the quantities (sometimes referred to as the coefficients of the second fundamental form) l, m and n and unit normals. The procedures which carry out these calculations are listed below. Of course, for plotting surfaces, we should always start with

```
> with(plots):
```

Also, just as for curves, we can use the procedures for dot and cross products and lengths.

```
> dp := proc(X,Y)
>      X[1]*Y[1]+X[2]*Y[2]+X[3]*Y[3];
>      end:
```

```
> nrm := proc(X)
>      sqrt(dp(X,X));
>      end:

> xp := proc(X,Y)
>      local a,b,c;
>      a := X[2]*Y[3]-X[3]*Y[2];
>      b := X[3]*Y[1]-X[1]*Y[3];
>      c := X[1]*Y[2]-X[2]*Y[1];
>      [a,b,c];
>      end:
```

We also need to compute the Jacobian matrix whose columns comprise the tangent vectors to the parameter curves.

```
> Jacf := proc(X)
>      local Xu,Xv;
>      Xu := [diff(X[1],u),diff(X[2],u),diff(X[3],u)];
>      Xv := [diff(X[1],v),diff(X[2],v),diff(X[3],v)];
>      simplify([Xu,Xv]);
>      end:
```

The following procedures are then self-explanatory once you realize that each calls another procedure (i.e. Jacf or UN) to provide the blocks upon which it will build. For example, in EFG below, the very definition $E = \mathbf{x}_u \cdot \mathbf{x}_u$ requires that Jacf(X) be called to provide $\mathbf{x}_u$. Similarly, EFG calls the dot product procedure dp to carry out the operations on the building blocks. Also, note the form of the output of EFG (for example) is a vector because, later on, we may want to only consider one of the metric coefficients and it is easy to write EFG[1] for E, EFG[2] for F and EFG[3] for G.

```
> EFG := proc(X)
>      local E,F,G,Y;
>      Y := Jacf(X);
>      E := dp(Y[1],Y[1]);
>      F := dp(Y[1],Y[2]);
>      G := dp(Y[2],Y[2]);
>      simplify([E,F,G]);
>      end:
> UN := proc(X)
>      local Y,Z,s;
>      Y := Jacf(X);
>      Z := xp(Y[1],Y[2]);
>      s := nrm(Z);
>      simplify([Z[1]/s,Z[2]/s,Z[3]/s]);
>      end:
```

```
> lmn := proc(X)
>       local Xu,Xv,Xuu,Xuv,Xvv,U,l,m,n;
>       Xu := [diff(X[1],u),diff(X[2],u),diff(X[3],u)];
>       Xv := [diff(X[1],v),diff(X[2],v),diff(X[3],v)];
>       Xuu := [diff(Xu[1],u),diff(Xu[2],u),diff(Xu[3],u)];
>       Xuv := [diff(Xu[1],v),diff(Xu[2],v),diff(Xu[3],v)];
>       Xvv := [diff(Xv[1],v),diff(Xv[2],v),diff(Xv[3],v)];
>       U := UN(X);
>       l := dp(U,Xuu);
>       m := dp(U,Xuv);
>       n := dp(U,Xvv);
>       simplify([l,m,n]);
>       end:
```

Now it's a relatively simple matter to put together procedures to calculate Gauss curvature and mean curvature.

```
> GK := proc(X)
>       local E,F,G,l,m,n,S,T;
>       S := EFG(X);
>       T := lmn(X);
>       E := S[1];
>       F := S[2];
>       G := S[3];
>       l := T[1];
>       m := T[2];
>       n := T[3];
>       simplify((l*n-m^2)/(E*G-F^2));
>       end:

> MK := proc(X)
>       local E,F,G,l,m,n,S,T;
>       S := EFG(X);
>       T := lmn(X);
>       E := S[1];
>       F := S[2];
>       G := S[3];
>       l := T[1];
>       m := T[2];
>       n := T[3];
>       simplify((G*l+E*n-2*F*m)/(2*E*G-2*F^2));
>       end:
```

EXERCISE 7.1. Write a MAPLE procedure to calculate Gauss curvature (when $F = 0$) from the formula

$$K = -\frac{1}{2\sqrt{EG}} \left(\frac{\partial}{\partial v}\left(\frac{E_v}{\sqrt{EG}} \right) + \frac{\partial}{\partial u}\left(\frac{G_u}{\sqrt{EG}} \right) \right).$$

Here's a procedure to plot the Gauss curvature of a surface on a given domain (which could be, say, the parameter domain of the surface). Note that, since dom[1] etc. are used in the procedure, MAPLE is expecting the domain to be given as a vector.

```
> Gaussplot := proc(X,dom)
>      local k;
>      k:= GK(X);
>      plot3d(k,u = dom[1] .. dom[2],v = dom[3] .. dom[4], axes = frame,
                title = 'Gaussian Curvature');
>      end:
```

For example, if I want to plot the Gauss curvature function for the sphere of radius R, I would first enter

```
> Rsphere:=[R*cos(u)*cos(v),R*sin(u)*cos(v),R*sin(v)];
```

and then

```
> Gaussplot(Rsphere,[0,2*Pi,-Pi/2,Pi/2]);
```

and the output will be a three dimensional graph of the curvature function over the rectangular domain $u = 0$ to 2π and $v = -\pi/2$ to $\pi/2$. Of course, what should this plot turn out to be?

EXERCISE 7.2. Write a MAPLE procedure to plot the mean curvature of a surface.

EXERCISE 7.3. Calculate Gauss and mean curvatures for the usual list of surfaces: helicoid, catenoid, torus, Enneper and hyperboloid(s). Also calculate for the pretzel surface

$$\mathbf{x}(u, v) = (\cos(u)\cos(v), \sin(u)\cos(v), \sin(v) + u).$$

With some complicated surfaces such as Henneberg's and Catalan's surfaces (which we shall meet later), you must be careful when asking for curvature calculations. *Always save your worksheet before asking for a complicated calculation.* In particular, Henneberg will either give a mess which MAPLE has difficulty simplifying or simply crash. This probably depends on how much RAM is available. It is also possible to get a picture of Gauss curvature on the surface itself by coloring points with similar Gauss curvature by the same color. For example, the following colors the helicoid according to its Gauss curvature via the option "color=GK(helicoid)."

```
> helicoid:=[u*cos(v),u*sin(v),v];

> plot3d(helicoid,u=0..1.5,v=0..5*Pi,color=GK(helicoid),style=
        patchnogrid,grid=[10,40]);
```

EXERCISE 7.4. Color Enneper's surface according to its Gauss curvature. What happens if you color according to mean curvature?

As a less computational subject amenable to MAPLE, consider the surfaces of revolution of constant Gauss curvature. In Exercise 3.11, you were asked to determine the integral formula for $g(u)$ when $h(u)$ satisfies $h'' = -h$. Of course, the answer to this question is that (with suitable shifting)

$$h(u) = c\cos(u) \qquad \text{and} \qquad g(u) = \int_{u_0}^{u} \sqrt{1 - c^2\sin^2(t)}\, dt.$$

Notice that the second expression will only make sense when the squareroot is defined; that is, when $u < \arcsin(1/c)$. Here is a MAPLE procedure which graphs surfaces of constant positive Gauss curvature given an initial c and n. The n is just a parameter which allows a finer picture. That is, the larger n is, the more filled in the picture will be and the longer it will take to render.

```
> with(plots):

> plotconstcurv:=proc(c,n)
>       local X,listX,k,j,eval,desys,dequ,deqgu ;
>       X:=[h(u)*cos(v),h(u)*sin(v),g(u)];
>       desys:=dsolve({diff(g(u),u)=sqrt(1-c^2*sin(u)^2), g(0)=0},g(u),
>               type=numeric, output=listprocedure):
>       deqgu:=subs(desys,g(u));
>       dequ:=subs(desys,u);
>
>       eval:=proc(t,m,Y,du,dg,s)
>          local i;
>          seq(evalf(subs({h(u)=s*cos(du(t/m)),
>               g(u)=dg(t/m),v=i/m},Y)),i=0..m*6.283):
>          end:
>
>       if c<=1 then k:=n*evalf(Pi/2) else k:=n*evalf(arcsin(1/c)) fi;
>       listX:={seq(eval(j,n,X,dequ,deqgu,c),j=-k+0.00001..k-0.00001)};
>       pointplot(listX,color=red,scaling=constrained);
>       end:
```

We find $g(u)$ above by converting the integral for $g(u)$ into a differential equation by differentiating and using the Fundamental Theorem of Calculus.

MAPLE's "dsolve" command then solves the differential equation numerically and the option "output=listprocedure" allows us to keep track of the u's and $g(u)$'s so that they may be inserted into other procedures. That's exactly what the little embedded procedure "eval" does. Namely, when we do "listX" further on in the procedure, it creates a sequence of points which are the solutions of the differential equation *evaluated* in the parametrization for a surface of revolution. In other words, the command listX:= {seq(eval(j,n,X,dequ,deqgu,c), j=-k+0.00001..k-0.00001)} takes the differential equation solutions "dequ" and "deqgu" and shoves them into the parametrization X. The "k" is the limit of integration $\arcsin(1/c)$ and j is just a parameter going between $-k$ and k. The ± 0.00001 tacked on to k is just a way to avoid singularities in plotting, so there is nothing special about the number itself — feel free to change it. Finally, notice that I had to go through two evaluations; one to create the curve by solving the differential equation numerically and the other to revolve the curve around the axis. Then, once the sequence(s) of points are generated, the MAPLE command "pointplot" may be invoked to plot all the points in the sequences. This is a crazy way to do all this and once MAPLE's elliptic integral package is improved, it will not be necessary. For then, the package will create the curve and you will only have to revolve it. In Chapter 5, geodesics will be drawn *on surfaces* using this simpler approach. Also, see [Gr] for the MATHEMATICA elliptic integral approach to constant curvature surfaces.

EXERCISE 7.5. Try the plots plotconstcurv(1,10), plotconstcurv(0.5,15), plotconstcurv(0.95,12) and plotconstcurv(2,15).

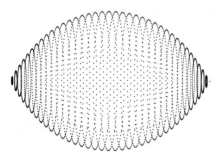

FIGURE 3.10.   Constant positive curvature surface of revolution

EXERCISE 7.6. Write a MAPLE procedure to draw constant negative curvature surfaces of revolution. You may use the parameters A and B as in the text or take cases where $h(u) = c \cosh(u)$ or $h(u) = c \sinh(u)$. Remember though that the squareroot must be defined.

EXERCISE 7.7. Write a MAPLE procedure to graph surfaces of Delaunay. This exercise will surely test your MAPLE skills. Hint: just obtain half of the picture and reflect it to obtain the other half.

Chapter 4

CONSTANT MEAN
CURVATURE SURFACES

4.1 INTRODUCTION

Much of science has to do with determining mathematical principles which underlie the phenomena of our universe. Although Kant seems to have overstated the case in declaring that humans are a priori *Euclidean* geometers, there can be no denying the proliferation of physical theories based on geometry. From Newton, who gave only geometric arguments in the *Principia*, to Einstein, who formulated "gravitation" as the geometry of space-time, to the present-day creators of Grand Unified Field Theories, the underlying mechanisms of the universe are geometric in nature.

Of course, in saying that the great forces of nature are consequences of some geometric structure, we lose sight of the fact that phenomena which we observe every day often display the same geometric dependence. Furthermore, we have already seen that nuclear cooling towers take the shape of hyperboloids of one sheet in order to prevent unnecessary stresses, so it should not be surprising that other types of construction might be similarly geometrically affected. Indeed, the Frei Otto school of architecture makes extensive use of minimal surfaces in its designs.

In this chapter we shall give a brief introduction to *minimal surfaces* and *surfaces of constant mean curvature*. While these surfaces are intrinsically mathematically interesting, they also are objects of study and application in the sciences. In a later chapter we will consider this subject from a more mathematically advanced point of view (involving complex analysis).

Surface tension is that quality responsible for the cohesiveness of liquids. Think of a liquid as a collection of (polar) molecules each of which exerts attractive forces of equal strength on the others. Deep inside the liquid a molecule feels equal forces from all directions, but near the surface of the liquid a molecule feels more of a force from inside the liquid than it does from the small number of molecules between it and the surface. Hence, those molecules near the surface are drawn into the liquid and the surface of the liquid displays a "curvature." Thus, we see liquids form into drops or bubbles because surface tension acts as a skin holding the liquid together. But a soap bubble doesn't shrink away to nothing, so at some point surface tension must be balanced by an internal pressure. In the case of the soap bubble, of course, the atmospheric pressure inside the bubble is larger than that outside, so an eventual equilibrium is attained. Indeed, this war between surface tension and pressure is brought to a cease-fire by the mediating effect of geometry (in the form of curvature).

Around 1800, Thomas Young (who is best known for his diffraction experiments establishing the wave nature of light) and Pierre Simon Laplace analyzed the phenomenon of surface tension in the following way. Consider the diagram below, which we take to be an "infinitesimal" piece of surface area expanded outward by an increased pressure. Note that, since the piece is very tiny, we may assume that the front and side curves are pieces of circles of radii R_1 and R_2 respectively.

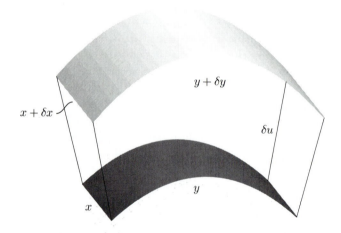

FIGURE 4.1.   Work done on a surface

Let's compute the work done to expand the area shown (which we denote by S). Remember that work is force × distance and pressure is force per unit area. We then get,

$$\begin{aligned} W &= F \cdot D \\ &= p \cdot S \cdot \delta u \\ &= p \cdot x\, y \cdot \delta u. \end{aligned}$$

But we can compute work in another way. Ask yourself, how could a soap film do work? Try this. Take a wire in the shape of a horseshoe and attach (loosely) a string to the two ends by loops. Hold the loops and dip the whole object in a soap solution. Now let the loops go — what happens? The string is pulled to the top of the wire by the film! This means that the film has the potential to do work. Moreover, the quality responsible for contracting the string and the film itself is *surface tension*. Also, from this description we see that the amount of work done must be equal to the surface tension T (the units of which are force per unit length) multiplied by the change in surface area ΔS. Of course,

$$\Delta S = (x + \delta x)(y + \delta y) - x\,y$$

from the picture (approximating the areas appropriately). Now, the length x is part of a circle of radius R_1 and $x + \delta x$ is part of a circle of radius $R_1 + \delta u$ with *the same included angle*. Therefore, we have the equality

$$\frac{x + \delta x}{R_1 + \delta u} = \frac{x}{R_1}.$$

Treating y similarly gives

$$x + \delta x = x\left(1 + \frac{\delta u}{R_1}\right)$$

$$y + \delta y = y\left(1 + \frac{\delta u}{R_2}\right).$$

Plug these into ΔS to obtain,

$$\Delta S = x\left(1 + \frac{\delta u}{R_1}\right)y\left(1 + \frac{\delta u}{R_2}\right) - xy$$

$$= x\,y\,\delta u\left(\frac{1}{R_1} + \frac{1}{R_2}\right) + x\,y\,\frac{(\delta u)^2}{R_1 R_2}.$$

If δu is small, we may neglect the last term. Now put this expression for ΔS into the work formula $W = T\Delta S$ and equate it to our previous work calculation. We end up with,

$$p\,x\,y\,\delta u = T\,x\,y\,\delta u\left(\frac{1}{R_1} + \frac{1}{R_2}\right)$$

$$p = T\left(\frac{1}{R_1} + \frac{1}{R_2}\right).$$

This is the Laplace-Young equation. It says that a pressure difference on either side of a bubble or film is given by the product of the surface tension of the bubble or film and a quantity which is clearly related to the shape of the bubble or film. In fact, notice that the only requirement (implicitly) placed on our infinitesimal piece of area was that the curves meet at right angles (so $S \approx xy$). Also, notice that the quantities $\frac{1}{R_1}$ and $\frac{1}{R_2}$ are the normal curvatures of the surface in those perpendicular directions. But, then (by Exercise 3.1.4) we recognize the quantity $2H = 1/R_1 + 1/R_2$ as twice the *mean curvature* of the surface.

Now consider the case of a soap film bounded by a wire say. Since there is no enclosed volume, the pressure is the same on both sides of the film. Hence, $p = 0$ and (since T is constant) $H = 0$ as well. Therefore, we see that

Theorem 1.1. *Every soap film is a physical model of a minimal surface.*

It is clear that surface tension tries to make the surface as "taut" as possible. That is, the surface should have the least surface area among surfaces satisfying certain constraints like having fixed boundaries or fixed volumes. In fact, this area-minimizing property implies the curvature property above and gives rise, in the case of films, to the term *minimal*. Let's see how this works.

4.2 First Notions in Minimal Surfaces

To begin, let's understand (in a somewhat informal way) how to integrate over a surface. Of course the simplest surface is the plane and here we do ordinary double integration. Our whole philosophy in dealing with surfaces has been to do all calculus computations in the plane and then transport them to a given surface via a patch $\mathbf{x}(u, v)$. Therefore, it's no surprise that we do the same thing for this situation. In Chapter 1 we noted that the length of the cross product of two vectors equals the area of the parallelogram which they span. Now, what is the area of a tiny parallelogram-like piece of a surface? Since it is very small, we may think of the parallelogram as spanned by tangent vectors $\mathbf{x}_u$ and $\mathbf{x}_v$. Hence, the area is given by $|\mathbf{x}_u \times \mathbf{x}_v|$. This quantity then approximates the area of that tiny piece of the surface. Of course this is the usual *modus operandi* of calculus; approximate a small piece and then add up continuously (i.e. integrate) over the entire region. So, let $\mathbf{x}(u, v)$ be a patch for a surface M. Then we make the

Definition 2.1. The area of the patch $\mathbf{x}$, denoted $A_{\mathbf{x}}$, is

$$A_{\mathbf{x}} = \int \int |\mathbf{x}_u \times \mathbf{x}_v| \, du \, dv,$$

where the limits of integration are the defining limits of the patch.

Furthermore, although we will not prove it, any compact oriented surface may be cut-up into a finite number of patches $\mathbf{x}$ (which, in fact may be taken to be oriented triangles) meeting only along their boundary curves with the opposite orientation.

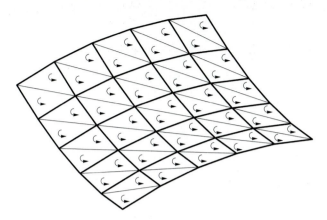

FIGURE 4.2.   Oriented triangulation of a patch

Therefore, we may calculate the whole surface area of a compact oriented surface M as the sum of the areas of the individual patches

$$A = \sum_{\mathbf{x}} \int \int |\mathbf{x}_u \times \mathbf{x}_v| \, du \, dv.$$

Note also that if we miss a few points, this does not affect the integral. So, all of the patches we have considered up to this point allow surface area to calculated in a single integration. We shall return to this later and give some exercises, but right now our only aim is to use this idea to understand something about minimal surfaces.

Finally, we will need Green's theorem from calculus. Suppose P and Q are two real valued (smooth) functions of two variables x and y defined on a simply connected region of the plane. Then Green's Theorem says that

$$\int_y \int_x \frac{\partial P}{\partial x} + \frac{\partial Q}{\partial y} dx \, dy = \int_C P dy - Q dx$$

where the righthand-side is the line integral around the boundary of the region C. Because we pull all integrals back to the plane for computation, we will find Green's Theorem particularly useful.

Now we may begin our analysis of minimal surfaces. Let $z = f(x, y)$ be a function of two variables and take a Monge patch for its graph: $\mathbf{x}(u, v) = (u, v, f(u, v))$. From Exercise 3.2.19 we have

$$\mathbf{x}_u = (1, 0, f_u) \qquad\qquad \mathbf{x}_{uu} = (0, 0, f_{uu})$$
$$\mathbf{x}_v = (0, 1, f_v) \qquad\qquad \mathbf{x}_{uv} = (0, 0, f_{uv})$$
$$\mathbf{x}_{vv} = (0, 0, f_{vv})$$

$$\mathbf{x}_u \times \mathbf{x}_v = (-f_u, -f_v, 1) \qquad\qquad U = \frac{(-f_u, -f_v, 1)}{\sqrt{1 + f_u^2 + f_v^2}}$$

$$E = 1 + f_u^2 \qquad\qquad F = f_u f_v \qquad\qquad G = 1 + f_v^2$$

$$\ell = \frac{f_{uu}}{\sqrt{1+f_u^2+f_v^2}} \qquad m = \frac{f_{uv}}{\sqrt{1+f_u^2+f_v^2}} \qquad n = \frac{f_{vv}}{\sqrt{1+f_u^2+f_v^2}}$$

$$K = \frac{f_{uu}f_{vv} - f_{uv}^2}{(1 + f_u^2 + f_v^2)^2}$$

$$H = \frac{(1 + f_v^2)f_{uu} + (1 + f_u^2)f_{vv} - 2f_u f_v f_{uv}}{2(1 + f_u^2 + f_v^2)^{\frac{3}{2}}}.$$

The expression for H immediately gives the following

Proposition 2.1. *M is minimal if and only if*

$$f_{uu}(1 + f_v^2) - 2f_u f_v f_{uv} + f_{vv}(1 + f_u^2) = 0.$$

This partial differential equation is called the *minimal surface equation*. By placing algebraic or geometric constraints on the function f, we determine various types of minimal surfaces.

EXERCISE 2.1. Suppose we require the algebraic condition that $f(x,y) = g(x) + h(y)$. Show that $H = 0$ implies $(1 + h'^2(y))g''(x) + (1 + g'^2(x))h''(y) = 0$. Separate variables and solve. Infer that

$$f(x, y) = \frac{1}{a} \ln \left(\frac{\cos ax}{\cos ay} \right).$$

The surface $z = f(x,y)$ is called *Scherk's minimal surface*. Note that Scherk's surface is only defined for $\frac{\cos ax}{\cos ay} > 0$. For example, a piece of Scherk's surface is given over the square $-\frac{\pi}{2} < ax < \frac{\pi}{2}$, $-\frac{\pi}{2} < ay < \frac{\pi}{2}$. In Chapter 7 we will discuss how different pieces of Scherk's surface fit together. Surprisingly, only the catenoid and helicoid were known to be minimal in the 1700's. Scherk's surface was the next example of a minimal surface and it was discovered by Scherk in 1835. It is a wonder that the simple algebraic condition above was not considered earlier.

FIGURE 4.3.   Scherk's surface

EXERCISE 2.2. Suppose we require the geometric condition that every level curve $f(x, y) = c$ be a line. What types of minimal surfaces do we now obtain? Hints:

1. In general, if a curve is given implicitly as $f(x, y) = c$, then the curvature is given by

$$\kappa = \frac{-f_{xx} f_y^2 + 2 f_x f_y f_{xy} - f_{yy} f_x^2}{|\nabla f|^3}.$$

Show this by invoking the implicit function theorem and writing $f(x, g(x)) = c$ with parametrization $\alpha(x) = (x, g(x), c)$. Then,

$$\kappa = \frac{|g''|}{(1 + g'^2)^{\frac{3}{2}}}.$$

Show $g' = -\frac{f_x}{f_y}$ by using $\nabla f \cdot \alpha' = 0$. Now find g'' by implicit differentiation.

2. Show $H = 0 \Leftrightarrow f_{xx} + f_{yy} = \kappa |f_x^2 + f_y^2|^{\frac{3}{2}}$.

3. Now $\kappa = 0$ (why?), so $f_{xx} + f_{yy} = 0$.

4. Show that a solution to Laplace's equation above with the given geometric constraint is

$$f(x, y) = A \arctan(\frac{y - y_0}{x - x_0}) + B.$$

If we let $x - x_0 = u \cos v, y - y_0 = u \sin v, z - B = Av$, then we see that the surface must be a plane or a helicoid. In fact, the proof that $f(x, y)$ is the only such solution requires much harder analysis and is known as Hamel's Theorem (see [Grau]).

EXERCISE 2.3. A unit speed curve on a surface $\alpha: I \to M$ is said to be an *asymptotic curve* if $\alpha'' \cdot U = 0$ for each point on the curve. By our formula $\alpha'' \cdot U = S(\alpha') \cdot \alpha' = k(\alpha')$, we see that asymptotic curves are those curves which travel in directions of zero normal curvature. Show that a surface M is minimal if and only if through each point there are two *orthogonal* asymptotic curves. Verify that the parameter curves for the helicoid $\mathbf{x}(u, v) = (u \cos v, u \sin v, v)$ are asymptotic and orthogonal at each point. Hint: use $K \leq 0$, Euler's formula for normal curvature and $H = 0$.

We have seen in the last chapter that *a nonplanar minimal surface of revolution must be a catenoid*. Naturally, we can ask similar questions by imposing other geometric hypotheses. In Chapter 7 we shall consider the question of whether a minimal surface can also possess constant Gaussian curvature. Here, we confine ourselves to the following

Theorem 2.2 (Catalan's Theorem). *Any ruled minimal surface in $\mathbb{R}^3$ is part of a plane or a helicoid.*

Proof. (after [BC]) Let $M: \mathbf{x}(u, v) = \beta(u) + v\delta(u)$ where, without loss of generality, we make the following assumptions: (1) by Exercise 2.1.17, we may take β perpendicular to the rulings of M (i.e. $\beta' \cdot \delta = 0$) with unit speed (i.e. $\beta' \cdot \beta' = 1$).   (2) δ is a unit vector field of directions for the rulings along β (i.e. $\delta \cdot \delta = 1$). Hence, we also have $\delta \cdot \delta' = 0$. From these assumptions we derive the data,

$$\mathbf{x}_u = \beta' + v\delta' \qquad \mathbf{x}_v = \delta \qquad \mathbf{x}_u \times \mathbf{x}_v = \beta' \times \delta + v\delta' \times \delta$$
$$E = 1 + 2v\beta' \cdot \delta' + v^2|\delta'|^2 \qquad F = 0 \qquad G = 1$$
$$\mathbf{x}_{uu} = \beta'' + v\delta'' \qquad \mathbf{x}_{uv} = \delta' \qquad \mathbf{x}_{vv} = 0$$

$$\ell = \frac{\beta'' \cdot \beta' \times \delta + v\beta'' \cdot \delta' \times \delta + v\delta'' \cdot \beta' \times \delta + v^2\delta'' \cdot \delta' \times \delta}{\sqrt{E}} \qquad m =? \quad n = 0.$$

Then $H = \frac{\ell}{2E} = 0$ since M is minimal, so $\ell = 0$. Now, the numerator of ℓ is a polynomial in v,

$$\beta'' \cdot \beta' \times \delta + v\left[\beta'' \cdot \delta' \times \delta + \delta'' \cdot \beta' \times \delta\right] + v^2\delta'' \cdot \delta' \times \delta,$$

so for it to be zero, we must have each coefficient zero. Set the first term to zero, $\beta'' \cdot \beta' \times \delta = 0$. This equation says that β'' is contained in the plane spanned by β' and δ, denoted $\langle\beta', \delta\rangle$. Since β is unit speed, β' and β'' are perpendicular. But by assumption, β' is perpendicular to δ as well. Since all three vectors are in the plane $\langle\beta', \delta\rangle$, it must be true that β'' is parallel to δ. In fact, since the length of δ is 1 and the length of β'' is the curvature κ_β of β, we must have $\beta'' = \pm\kappa_\beta\delta$. At any rate, we certainly have $\beta'' \cdot \delta' \times \delta = 0$.

Now look at the coefficient of v set to zero and use $\beta'' \cdot \delta' \times \delta = 0$ to infer $\delta'' \cdot \beta' \times \delta = 0$. This says that $\delta'' \in \langle\beta', \delta\rangle$ also. Now, the coefficient of v^2

set to zero is $\delta'' \cdot \delta' \times \delta = 0$ and this implies that $\delta'' \in \langle \delta', \delta \rangle$ as well. Hence, $\delta'' \in \langle \delta', \delta \rangle \cap \langle \beta', \delta \rangle$. Now there are two possibilities.

First, this intersection may be an equality $\langle \delta', \delta \rangle = \langle \beta', \delta \rangle$. This follows if δ'' is not parallel to δ at a point (so in some neighborhood as well). In this case, $\beta' = a\delta'$ since β', δ' and δ are in the same plane and both β' and δ' are perpendicular to δ. But then the unit normal of M would take the form

$$U = \frac{(1 \pm av)(\beta' \times \delta)}{|1 \pm av| \, |\beta' \times \delta|} = \pm \beta' \times \delta,$$

since $|\beta' \times \delta| = 1$. But then, taking the derivative of U with respect to u, we have $U' = \beta'' \times \delta + \beta' \times \delta' = 0$ since $\beta'' = \kappa_\beta \delta$ and $\beta' = a\delta'$. Hence, U is constant, so M is part of a plane.

Secondly, if $\langle \delta', \delta \rangle \neq \langle \beta', \delta \rangle$, then the intersection is the line $\langle \delta \rangle$. Hence, $\delta'' = a\delta$ and, consequently, $\beta' \cdot \delta'' = 0$. But $\beta' \cdot \delta = 0$ implies $\kappa_\beta = \kappa_\beta \delta \cdot \delta = \beta'' \cdot \delta = -\beta' \cdot \delta'$ and this, in turn, gives

$$\frac{d\kappa_\beta}{du} = -\beta'' \cdot \delta' - \beta' \cdot \delta'' = 0$$

since β'' and δ'' are parallel to δ and β' and δ' are perpendicular to δ. Hence, the curvature κ_β is a constant. Now consider the torsion τ_β of β. The usual formula becomes $\tau_\beta = \beta' \times \delta \cdot \delta'$ since $\beta'' = \kappa_\beta \delta$ and, because κ is constant, $\beta''' = \kappa_\beta \delta'$.

EXERCISE 2.4. Show that $\frac{d\tau_\beta}{du} = 0$.

Therefore, τ_β is a constant. Recall that a curve with constant curvature and torsion is a *circular helix*. Up to a rigid motion of $\mathbb{R}^3$, we may then parametrize β by

$$\beta(u) = (A \cos u, A \sin u, Bu)$$

with $A^2 + B^2 = 1$. Also, δ is parallel to β'', so $\delta(u) = (\cos u, \sin u, 0)$ since δ has unit speed. Let $A + v = \bar{v}$. We then obtain a parametrization for the helicoid, $\mathbf{x}(u, v) = (\bar{v} \cos u, \bar{v} \sin u, Bu)$. $\qquad\qquad\square$

So, we see that the two most natural geometric conditions to impose along with minimality, being a surface of revolution or being ruled, lead to the minimal surfaces of the 18th century, the catenoid and the helicoid.

EXERCISE 2.5. Plot the following surfaces. Show that Helcat, Enneper, Scherk and Planar Lines of Curvature are minimal and compute their Gaussian curvatures. You may find MAPLE to be of use here, but be careful. If you try Henneberg or Catalan, save your worksheet before actually computing!

1. **Helcat.** $\mathbf{x}(u,v) = (x^1(u,v), x^2(u,v), x^3(u,v))$, for any fixed t, where
$$x^1(u,v) = \cos(t)\sinh(v)\sin(u) + \sin(t)\cosh(v)\cos(u)$$
$$x^2(u,v) = -\cos(t)\sinh(v)\cos(u) + \sin(t)\cosh(v)\sin(u)$$
$$x^3(u,v) = u\cos(t) + v\sin(t).$$

2. **Henneberg's Surface.** $\mathbf{x}(u,v) = (x^1(u,v), x^2(u,v), x^3(u,v))$, where
$$x^1(u,v) = 2\sinh(u)\cos(v) - \frac{2}{3}\sinh(3u)\cos(3v)$$
$$x^2(u,v) = 2\sinh(u)\sin(v) - \frac{2}{3}\sinh(3u)\sin(3v)$$
$$x^3(u,v) = 2\cosh(2u)\cos(2v).$$

3. **Catalan's Surface.**
$$\mathbf{x}(u,v) = (u - \sin(u)\cosh(v), 1 - \cos(u)\cosh(v), 4\sin(u/2)\sinh(v/2)).$$

4. **Enneper's Surface.**
$$\mathbf{x}(u,v) = (u - u^3/3 + uv^2, v - v^3/3 + vu^2, u^2 - v^2).$$

5. **Scherk's Fifth Surface.** This surface is often written in nonparametric form $\sin z = \sinh x \sinh y$. Parametrically,
$$\mathbf{x}(u,v) = (\operatorname{arcsinh}(u), \operatorname{arcsinh}(v), \arcsin(uv)).$$

6. **Planar lines of curvature surface.**
$$\mathbf{x}(u,v) = \left(\frac{cu \pm \sin u \cosh v}{\sqrt{1-c^2}}, \frac{v \pm c\cos u \sinh v}{\sqrt{1-c^2}}, \pm\cos u \cosh v \right).$$

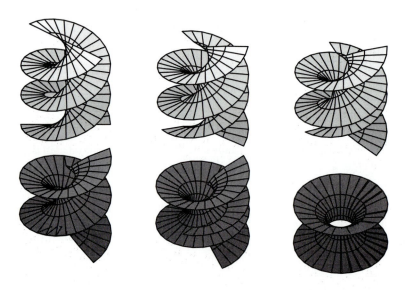

FIGURE 4.4.　A one-parameter family of minimal surfaces joining the helicoid and the catenoid

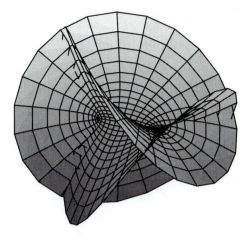

FIGURE 4.5. Henneberg's surface

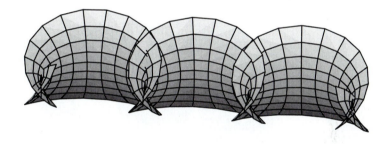

FIGURE 4.6. Catalan's surface

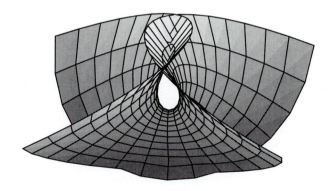

FIGURE 4.7. Enneper's surface

FIGURE 4.8.   Scherk's fifth surface

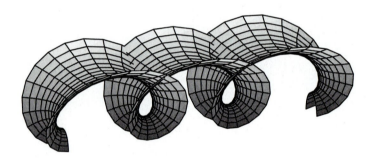

FIGURE 4.9.   Planar lines of curvature surface

4.3 AREA MINIMIZATION

Now let's try to understand this word "minimal" a little better. In the mid-1800's the Belgian physicist Plateau set the following problem: given a curve C, find a minimal surface M having C as boundary. Plateau was interested in thin films (e.g. soap films) and his problem was a natural outgrowth of his physical experiments. As we shall see below, *least-area* surfaces are minimal. In fact, it was eventually realized that hopes of finding a general method of solution for Plateau's problem rested on producing least-area surfaces. Thus, another version of Plateau's problem is to find a least-area surface having C as boundary. But even the *existence* of area minimizing surfaces is not automatic. In fact, it was only in the 1920's and 1930's that Plateau's problem was first solved by J. Douglas and T. Rado (see [Dou] and [R]). They proved

Theorem 3.1. *There exists a least-area disk-like minimal surface spanning any given Jordan curve.*

(A minimal surface is *disk-like* if its parameter domain is the unit disk $D = \{(u, v)|u^2 + v^2 \leq 1\}$ and the boundary circle maps to the given Jordan curve.) To show the delicateness of this problem, consider

EXERCISE 3.1. Let C be the unit circle in the xy plane. Show that the following variant of Plateau's problem has no solution: Find the surface of least area which has C as boundary and passes through $(0, 0, 1)$. Note that this problem begs the question: exactly what is a surface? Does a disk with a spike qualify (see Figure 4.10)? This (along with an appropriate definition of area) was one of the main difficulties in considering Plateau's problem. Of course from our very "differentiable" point of view, we obviate such considerations.

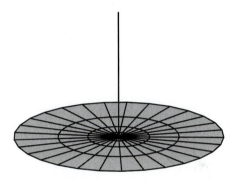

FIGURE 4.10.   Disk with spike

Let's look at a simpler question. What is a necessary condition that M have least area among surfaces with boundary C? The answer may be found through a simplified version of the calculus of variations as follows.

Suppose $M: z = f(x, y)$ is a surface of least area with boundary C. Consider the nearby surfaces which look like slightly deformed versions of M,

$$M^t: z^t(x, y) = f(x, y) + tg(x, y).$$

Here, g is a function on the domain of f which has the effect, when multiplied by a small t and added to f, of moving points of M a small bit and leaving C fixed. That is, $g|_{\tilde{C}} = 0$, where $\tilde{C}$ is the boundary of the domain of f and $f(\tilde{C}) = C$. A Monge patch for M^t is given by

$$\mathbf{x}^t(u, v) = (u, v, f(u, v) + tg(u, v)).$$

Immediately we compute

$$|\mathbf{x}_u^t \times \mathbf{x}_v^t| = \sqrt{1 + f_u^2 + f_v^2 + 2t(f_u g_u + f_v g_v) + t^2(g_u{}^2 + g_v{}^2)}.$$

By the definition of area, we see that the area of M^t is

$$A(t) = \int_v \int_u \sqrt{1 + f_u^2 + f_v^2 + 2t(f_u g_u + f_v g_v) + t^2(g_u{}^2 + g_v{}^2)} \, du \, dv.$$

Now take the derivative with respect to t (which then passes inside the integral),

$$A'(t) = \int_v \int_u \frac{f_u g_u + f_v g_v + t(g_u{}^2 + g_v{}^2)}{\sqrt{1 + f_u^2 + f_v^2 + 2t(f_u g_u + f_v g_v) + t^2(g_u{}^2 + g_v{}^2)}} \, du \, dv.$$

We assumed $z = z_0$ was a minimum, so $A'(0) = 0$. Therefore, setting $t = 0$ in the equation above, we get

$$\int_v \int_u \frac{f_u g_u + f_v g_v}{\sqrt{1 + f_u^2 + f_v^2}} \, du \, dv = 0.$$

Now, let

$$P = \frac{f_u g}{\sqrt{1 + f_u^2 + f_v^2}} \qquad Q = \frac{f_v g}{\sqrt{1 + f_u^2 + f_v^2}}.$$

EXERCISE 3.2. Compute $\frac{\partial P}{\partial u}$ and $\frac{\partial Q}{\partial v}$ and apply Green's Theorem.

We then get

$$\int_v \int_u \frac{f_u g_u + f_v g_v}{\sqrt{1 + f_u^2 + f_v^2}} \, du \, dv$$
$$+ \int_v \int_u \frac{g[f_{uu}(1 + f_v^2) + f_{vv}(1 + f_u^2) - 2f_u f_v f_{uv}]}{(1 + f_u^2 + f_v^2)^{\frac{3}{2}}} \, du \, dv$$
$$= \int_C \frac{f_u g \, dv}{\sqrt{1 + f_u^2 + f_v^2}} - \frac{f_v g \, du}{\sqrt{1 + f_u^2 + f_v^2}} = 0$$

since $g|_{\tilde{C}} = 0$. Of course the first integral is zero as well, so we end up with

$$\int_v \int_u \frac{g[f_{uu}(1 + f_v^2) + f_{vv}(1 + f_u^2) - 2f_u f_v f_{uv}]}{(1 + f_u^2 + f_v^2)^{\frac{3}{2}}} \, du \, dv = 0.$$

Since this is true for all such g, we must have

$$f_{uu}(1 + f_v^2) + f_{vv}(1 + f_u^2) - 2f_u f_v f_{uv} = 0.$$

But this is the minimal surface equation. Therefore, we have shown the following *necessary* condition for a surface to be area minimizing.

Theorem 3.2. *If M is area minimizing, then M is minimal.*

EXERCISE 3.3. Let M be a graph $z = f(x, y)$ above a domain $D \subseteq \mathbb{R}^2$ with boundary a closed curve C. Show that, if f satisfies the minimal surface equation, then M has least area among all graphs of functions $z = g(x, y)$ with $g|_C = f|_C$. Hints: (1) a surface Z is formed by adjoining the two graphs along $g(C) = f(C)$, (2) take the normal vector field for M defined in all of $D \times \mathbb{R}$ by $U_f = \left(\frac{-f_x}{\sqrt{1+f_x^2+f_y^2}}, \frac{-f_y}{\sqrt{1+f_x^2+f_y^2}}, \frac{1}{\sqrt{1+f_x^2+f_y^2}} \right)$, (3) show $\operatorname{div}(U_f) = 0$, apply the divergence theorem and estimate each surface integral obtained by using the fact that U_f is a normal vector field for one of the pieces of Z, but not the other!

EXERCISE 3.4. Determine the parameter a for the catenary $y = a \cosh(\frac{x}{a})$ which passes through the points $(-.6, 1)$ and $(.6, 1)$. Compute the surface area of the catenoid generated by revolving the catenary about the x-axis. Show that this surface area is greater than the surface area of two disks of radius 1. Hence, if the original boundary curve consisted of two circles perpendicular to the x-axis centered at $(-.6, 1)$ and $(.6, 1)$, then the catenoid would be a minimal, but non-area-minimizing, surface spanning the boundary. Hint: MAPLE (e.g. "fsolve" and numeric integration).
In fact, more can be shown. Let $x_0, -x_0$ be the points on the x-axis which are centers of the given circles of radius y_0.
1. If $\frac{x_0}{y_0} > .528$ (approximately), then the two disks give an absolute minimum for surface area. This is the so-called Goldschmidt discontinuous solution.
2. If $\frac{x_0}{y_0} < .528$ (approximately), then a catenoid is the absolute minimum and the Goldschmidt solution is a local minimum.
3. If $.528 < \frac{x_0}{y_0} < .663$ (approximately), then the catenoid is only a local minimum.
4. If $\frac{x_0}{y_0} > .663$ (approximately), then there is no catenoid joining the points. This can be seen using soap films by forming a catenoid between two rings and slowly pulling the rings apart. Measure $\frac{x_0}{y_0}$ and you will see that the catenoid spontaneously jumps to the two disk solution at approximately $\frac{x_0}{y_0} = .663$. For an informal discussion of this problem, see [Isen]. For a formal discussion, see [Bli].

4.4 Constant Mean Curvature

Now let's look at situations where mean curvature is non-zero, but constant. Before we begin, we present a very beautiful formula which will be the foundation for our study. Let M be a compact oriented immersed surface in $\mathbb{R}^3$ with unit normal $U = \frac{\mathbf{x}_u \times \mathbf{x}_v}{|\mathbf{x}_u \times \mathbf{x}_v|}$, mean curvature H and area $A = \int \int |\mathbf{x}_u \times \mathbf{x}_v| \, du \, dv$. Recall that when we say that M is *immersed*, we mean that we allow self intersections, but not locally. That is, around each point in M there is a small open neighborhood for which there are no self intersections. In terms of a patch $\mathbf{x}(u, v)$ for M, an immersion requires $\mathbf{x}_u$ and $\mathbf{x}_v$ to be linearly independent at every point. Thus, a nonzero normal $\mathbf{x}_u \times \mathbf{x}_v$ is well defined at every point of the surface. Enneper's surface is an example of an immersed surface.

To continue, just as in our derivation of the minimal surface equation, we perturb M a bit by a vector field V on M. Note that we use a vector field here instead of a function g because M is not defined as the graph over some domain.

$$M^t : \mathbf{y}^t(u,v) = \mathbf{x}(u,v) + tV(u,v)$$

and write the area of M^t as

$$A(t) = \int\int |\mathbf{y}_u \times \mathbf{y}_v| \, du \, dv$$

$$= \int\int \sqrt{|\mathbf{x}_u \times \mathbf{x}_v|^2 + 2t(\mathbf{x}_u \times \mathbf{x}_v)(\mathbf{x}_u \times V_v + V_u \times \mathbf{x}_v) + O(t^2)} \, du \, dv.$$

Then, taking the derivative with respect to t and evaluating at $t = 0$, we obtain

$$A'(0) = \int\int \frac{\mathbf{x}_u \times \mathbf{x}_v}{|\mathbf{x}_u \times \mathbf{x}_v|}(\mathbf{x}_u \times V_v + V_u \times \mathbf{x}_v) \, du \, dv$$

$$= \int\int \frac{\mathbf{x}_u \times \mathbf{x}_v}{|\mathbf{x}_u \times \mathbf{x}_v|}(\mathbf{x}_u \times V_v - \mathbf{x}_v \times V_u) \, du \, dv$$

$$= \int\int V_v \cdot U \times \mathbf{x}_u - V_u \cdot U \times \mathbf{x}_v \, du \, dv.$$

EXERCISE 4.1. Let $P = -V \cdot U \times \mathbf{x}_v$ and $Q = V \cdot U \times \mathbf{x}_u$, apply Green's Theorem and the shape operator formulas of Chapter 3.1 to get

$$\int\int V_v \cdot U \times \mathbf{x}_u - V_u \cdot U \times \mathbf{x}_v + V(2H\mathbf{x}_u \times \mathbf{x}_v) \, du \, dv$$

$$= -\int_C V \cdot U \times \mathbf{x}_v \, dv + V \cdot U \times \mathbf{x}_u \, du.$$

When we integrate over the entire collection of patches on M, the line integrals around the boundaries all cancel due to the orientation of M. Hence, over M, the lefthandside above is zero and we have

$$A'(0) = \int\int V_v \cdot U \times \mathbf{x}_u - V_u \cdot U \times \mathbf{x}_v \, du \, dv$$

$$= -\int\int V(2H\mathbf{x}_u \times \mathbf{x}_v) \, du \, dv$$

$$= -\int\int 2HU \cdot V \, dA \quad \text{since, by definition, } dA = |\mathbf{x}_u \times \mathbf{x}_v| \, du \, dv.$$

Let's do two quick applications of this formula. First, let $V = \mathbf{x}(u,v)$ so that $\mathbf{y}^t = (1+t)\mathbf{x}$ and $A(t) = (1+t)^2 A$ since M is simply stretched uniformly. Clearly, $A'(0) = 2A$ so we have

$$A = -\int\int HU \cdot \mathbf{x} \, dA.$$

Thus, we get a formula connecting surface area to mean curvature. We will exploit this below. Secondly, let $V = fU$, a normal vector field. Of course, this is the type of force field we would expect to arise from pressure exerted uniformly on a surface. Now, $U \cdot U = 1$, so we have $A'(0) = -2 \int\int fH \, dA$. Also, the geometric interpretation of the divergence of a vector field as a rate of change of volume V (see [MT, p. 227]) together with the divergence theorem itself gives

$$V'(0) = \int\int\int \operatorname{div} f U \, dx \, dy \, dz$$
$$= \int\int fU \cdot U \, dA$$
$$= \int\int f \, dA .$$

(Recall that the divergence (or Gauss's) theorem says that, for a surface M enclosing a volume Ω, the volume integral of the divergence $\operatorname{div} V = \partial V^1/\partial x^1 + \partial V^2/\partial x^2 + \partial V^3/\partial x^3$ of a vector field $V = (V^1, V^2, V^3)$ is equal to the surface integral of the component of V normal to M. Formally,

$$\int\int\int_{\Omega} \operatorname{div} V \, d\Omega = \int\int_{M} V \cdot U \, dA$$

where U is the unit outward normal of M.) So, the problem of minimizing surface area (i.e. $A'(0) = 0$) subject to having a fixed volume (i.e. $V'(0) = 0$) is equivalent to finding H satisfying

$$\int\int fH \, dA = 0 \quad \text{for all } f \text{ with} \quad \int\int f \, dA = 0.$$

EXERCISE 4.2. Show that this condition implies $H = c$, a constant. Hints: (1) Write $H = c + (H - c) = c + J$ where $c = \frac{\int\int H \, dA}{A}$. (2) Show $\int\int J^2 \, dA = \int\int J(H - c) \, dA = 0$. (Use the hypothesis.) (3) Then $J = 0$. (4) Hence, $H = c$.

Of course, the surface tension of a bubble works to achieve exactly the situation above. Therefore, we have shown that

Theorem 4.1. *A soap bubble must always take the form of a surface of constant mean curvature.*

But what are these surfaces then? We have seen in Chapter 3 that, in the 1800's, Delaunay classified all surfaces of revolution with constant mean curvature. Of these, however, only the sphere is compact (i.e. closed and bounded) and embedded in $\mathbb{R}^3$. The question arose whether there are any other compact embedded surfaces of constant mean curvature? In other words, must non-self-intersecting soap bubbles always be spheres? The answer, originally due to Alexandrov, is *yes*. To prove Alexandrov's result we make use of our formula above together with a beautiful estimate due to Antonio Ros. In these results notice the essential connections between geometry (in the forms of area, volume and curvature) and vector calculus (divergence theorem, curvature). Of course, *curvature* is the link which binds these subjects together. We begin by giving Ros's estimate, then prove Alexandrov's theorem and, finally, prove Ros's theorem. Ros' theorem is pretty heavy going, but an understanding of the ideas involved definitely has its geometric rewards. (The following is adapted from [Oss2].)

Theorem 4.2 [Ros]. *Let M be a compact embedded surface in $\mathbb{R}^3$ bounding a domain D of volume V. If $H > 0$ on M, then*

$$\int_M \frac{1}{H}\, dA \geq 3V.$$

Furthermore, equality holds if and only if M is the standard sphere.

Theorem 4.3 (Alexandrov). *If M is a compact embedded surface of constant mean curvature, then M is a standard sphere.*

Proof. Suppose H is constant and use the area formula above and the divergence theorem to get (where U is the interior normal)

$$A = -\int\int HU \cdot \mathbf{x}\, dA$$

$$= H \int\int\int \operatorname{div} \mathbf{x}\, dx\, dy\, dz$$

$$= 3HV\ .$$

But then we have

$$\int\int \frac{1}{H}\, dA = \frac{1}{H}\int\int dA = \frac{A}{H} = \frac{A}{\frac{A}{3V}} = 3V$$

and by Ros' Theorem, M must be a sphere. $\qquad\qquad\square$

Proof of Ros's Theorem. First, let's fatten M up a bit by taking a shell S on M's inside given by

$$S = \{p + \epsilon\, h(p)\, U \mid p \in M\}$$

where U is the unit normal of M, $0 \leq \epsilon \leq 1$ and $h(p)$ is defined by

$$h(p) = \sup\{r \mid \text{ the point } p \text{ is the unique nearest point on } M \text{ to the point } q$$

$$\text{at a distance } r \text{ from } p \text{ along } U\}.$$

This definition means the following. For any $p \in M$, go in the direction of the inner normal U and consider each of the points q along the way. For *some* of these points q on the line $p + tU$, p is the nearest point of M to q. Now just focus on *these* points and take the distance r of the farthest such point. (In fact, there is a small technical point here. It is possible that a point q which is farthest does not have a unique nearest point on M. Think of the center of the sphere! This is why we take the *supremum* of the distances of the q's for which p is the unique nearest point.)

We consider this shell S because we use the volume V of D in our estimate *and* we will see that

$$\text{Vol}(D) = \text{Vol}(S).$$

Moreover, the volume of S may be calculated in terms of curvatures in a convenient way. In order to see the volume equality, first note that there can be *no overlap* on the interior of the shell. This follows from the definition of $h(p)$ since every point of the shell lies on a unique normal line to M. Of course, there may be overlap on the inside boundary of the shell (as the sphere example shows), but this will have no effect on volume since the boundary is only 2-dimensional.

Secondly, let $q \in D$ be arbitrary and let d be the distance from q to M (i.e. to M's nearest point). Of course such a distance is obtained along a line perpendicular to M. At any rate, if $B(q, d)$ denotes the closed ball of radius d about q, then the interior of $B(q, d)$ contains no points of M and the boundary of $B(q, d)$, denoted $\partial B(q, d)$, contains at least one $p \in M$. Consider the radius of $B(q, d)$ from q to such a p and take *any* q' on this radius. Let r be the distance along the radius from q' to p and note that $B(q', r)$ lies in the interior of $B(q, d)$ *except at* p. Therefore, p is the unique point of M realizing distance r from q' to M and *all of these* q''s are in the defining set for $h(p)$. Now, the supremum of the distances of the q' is d since the q' get arbitrarily close to q. Of course, there may be other points in the set further out, but nevertheless, we see that $d \leq h(p)$. This means, by definition of the shell S, that $q \in S$. Thus, every point of D is in the shell S and the only overlap of S occurs on the 2-dimensional boundary. Therefore, the volumes of D and S must be the same.

Now, if M has a local parametrization $\mathbf{x}(u, v)$, then we may parametrize the shell S by

$$\mathbf{y}(u, v, t) = \mathbf{x}(u, v) + t\, U(u, v).$$

Note that we need three parameters u, v and t since the shell is 3-dimensional. Here, t varies as $0 \leq t \leq h(p) = h(u, v)$ where $p = \mathbf{x}(u, v)$. Now, a small chunk of S may be thought of as a parallelepiped determined by $\mathbf{y}_u$, $\mathbf{y}_v$ and $\mathbf{y}_t$. Consequently, the small chunk has volume $|\mathbf{y}_u \times \mathbf{y}_v \cdot \mathbf{y}_t|$. The volume of S is then obtained by integration

$$\text{Vol}(S) = \int \int \int_0^{h(u,v)} |\mathbf{y}_u \times \mathbf{y}_v \cdot \mathbf{y}_t|\, dt\, du\, dv.$$

To determine the integrand, for the moment assume that the parameter curves $\mathbf{x}(u, v_0)$ and $\mathbf{x}(u_0, v)$ are *lines of curvature*. That is, $k_1\, \mathbf{x}_u = S(\mathbf{x}_u) =$

$-\nabla_{\mathbf{x}_u}U = -U_u$ and $k_2\,\mathbf{x}_v = S(\mathbf{x}_v) = -\nabla_{\mathbf{x}_v}U = -U_v$. Then,

$$\mathbf{y}_u = \mathbf{x}_u + tU_u = (1 - k_1t)\mathbf{x}_u \quad \text{and} \quad \mathbf{y}_v = \mathbf{x}_v + tU_v = (1 - k_2t)\mathbf{x}_v.$$

This means $\mathbf{y}_u \times \mathbf{y}_v = (1 - k_1t)(1 - k_2t)\,\mathbf{x}_u \times \mathbf{x}_v = (1 - 2Ht + Kt^2)\,\mathbf{x}_u \times \mathbf{x}_v$ since $k_1 + k_2 = 2H$ and $k_1 k_2 = K$. *But H and K don't depend on the parametrization chosen.* So, even though we assumed $\mathbf{x}_u$ and $\mathbf{x}_v$ were principal vectors to obtain the formula, in fact the formula must hold in general! Of course, $\mathbf{y}_t = U$, so this means $|\mathbf{y}_u \times \mathbf{y}_v \cdot \mathbf{y}_t| = |1 - 2Ht + Kt^2||\mathbf{x}_u \times \mathbf{x}_v|$ and

$$\mathrm{Vol}(S) = \int\int\int_0^{h(u,v)} |\mathbf{y}_u \times \mathbf{y}_v \cdot \mathbf{y}_t| dt\, du\, dv$$

$$= \int\int \left\{ \int_0^{h(u,v)} |1 - 2Ht + Kt^2|\, dt \right\} |\mathbf{x}_u \times \mathbf{x}_v|\, du\, dv$$

$$= \int_M \left\{ \int_0^{h(u,v)} |1 - 2Ht + Kt^2|\, dt \right\} dA$$

$$= \int_M \int_0^{h(u,v)} |(1 - k_1t)(1 - k_2t)|\, dt\, dA.$$

From this expression for the volume of the shell we see that the principal curvatures come into the scheme of things. In fact, they are related to the function $h(p)$ by the following estimate:

$$\frac{1}{h(p)} \geq \max\{k_1(p), k_2(p)\}$$

for all $p \in M$. Let's see why this is so. Let q be the point a distance $h(p)$ along the normal from p. Then the *open* ball $B(q, h(p))$ cannot contain any points of M, for if there were a $p' \in M \cap B(q, h(p))$, its distance to q would be less than $h(p)$ and for all q' close to q on the radius from q to p the distance from q' to p' would be less than the distance from q' to p. This contradicts the definition of $h(p)$.

EXERCISE 4.3. Prove this last statement. Hints:(1) $h(p)$ is a supremum, so there is a sequence $q' \to q$ with p the nearest point of M to q' along a normal, (2) if a p' exists, then the distance of q' to M is less than the distance of q' to p, (3) since this is true for all q' close to q, the sequence $q' \to q$ can't exist.

Hence, $B(q, h(p))$ contains no points of M and, therefore, the sphere $S(q, h(p))$ which bounds $B(q, h(p))$ lies completely inside M only touching at p. But this means that this sphere of radius $h(p)$ has *larger* normal curvature at p *in every direction* than M. (Compare the proof of Theorem 3.5.2 for the opposite estimate.) Of course, the normal curvatures on a sphere are constant and equal to the reciprocal of the radius. In our case then, the normal curvature is always $\frac{1}{h(p)}$ and we have the required inequality.

Now, the inequality above says that $k_1 h(p) \leq 1$ and $k_2 h(p) \leq 1$. Hence, $(1 - k_1 t)$ and $(1 - k_2 t)$ are nonnegative for $0 \leq t \leq h(p)$ and

$$\int_0^{h(p)} |(1 - k_1 t)(1 - k_2 t)| \, dt = \int_0^{h(p)} (1 - k_1 t)(1 - k_2 t) \, dt.$$

The normal curvature inequality and the geometric-arithmetic mean inequality $ab \leq (\frac{a+b}{2})^2$ imply

$$\begin{aligned}
(1 - \frac{1}{h(p)} t)^2 &= (1 - \frac{1}{h(p)} t)(1 - \frac{1}{h(p)} t) \\
&\leq (1 - k_1 t)(1 - k_2 t) \\
&\leq (1 - Ht)^2.
\end{aligned}$$

Hence, $\frac{1}{h(p)} \geq H(p)$ or $\frac{1}{H(p))} \geq h(p)$. This gives

$$\int_0^{h(p)} (1 - k_1 t)(1 - k_2 t) \, dt \leq \int_0^{\frac{1}{H}} (1 - Ht)^2 \, dt = \frac{1}{3H}$$

and consequently, for $V = \text{Vol}(S)$,

$$V = \int_M \int \int_0^{h(p)} (1 - k_1 t)(1 - k_2 t) \, dt \, dA \leq \int_M \frac{1}{3H} \, dA$$

as was desired. Finally, note that for equality to hold, it must hold in $(1 - k_1 t)(1 - k_2 t) \leq (1 - Ht)^2$.

EXERCISE 4.4. Show that $(1 - k_1 t)(1 - k_2 t) = (1 - Ht)^2$ if and only if $k_1 = k_2$ everywhere.

Hence, equality holds in the estimate if and only if M is all umbilic. By Theorem 3.5.1, this is true if and only if M is a sphere. $\qquad\square$

Results such as Alexandrov's Theorem led H. Hopf to conjecture that all *immersed* surfaces of constant mean curvature are spheres. It has only been in recent years that this conjecture has been shown to be false. H. Wente [We] was the first to construct tori of constant mean curvature which are immersed in $\mathbb{R}^3$. Since Wente's work, many such have been constructed with important consequences for mechanics (for example). We have by no means given an exhaustive study of mean curvature. There are many interesting and important questions which remain and we hope that the brief outline above will lead the reader to further study.

4.5 HARMONIC FUNCTIONS

Before we leave this chapter, we want to indicate some of the connections between minimal surfaces and other areas of Mathematics. We will examine these connections more closely in a later chapter, but it is worthwhile to understand, from the outset, that this subject touches, and is touched by, many areas of Mathematics and Science.

A very important partial differential equation studied in mathematical physics is the *Laplace equation or potential equation*. In two-dimensional Cartesian coordinates, the *Laplace equation* is

$$\Delta \Phi \stackrel{\text{def}}{=} \frac{\partial^2 \Phi}{\partial x^2} + \frac{\partial^2 \Phi}{\partial y^2} = 0.$$

(An analogous version of this equation is found in higher dimensions as well as in other coordinate systems.) This equation describes, among other things, steady-state temperature distribution, equilibrium displacements of a membrane and gravitational and electrostatic potentials.

Definition 5.1. A real-valued function $\Phi(x, y)$ is said to be *harmonic* if all of its second-order partial derivatives are continuous and each point in its domain satisfies the Laplace equation.

Although in a later chapter we shall be precise in specifying the particulars of the relationship between harmonic and complex analytic functions, we mention here

Theorem 5.1. *If $f(z) = u(x, y) + iv(x, y)$ is complex analytic (i.e. the derivative $f'(z)$ exists at all points of a region of the complex plane), then each of $u(x, y)$ and $v(x, y)$ is harmonic.*

Example 5.2. Let $f(z) = z^2$, where $z = x + iy$. Then

$$z^2 = (x + iy)^2$$
$$= x^2 + 2ixy - y^2$$
$$= (x^2 - y^2) + i(2xy).$$

So, in this case, $\text{Re}(z^2) = u(x,y) = x^2 - y^2$ and $\text{Im}(z^2) = v(x,y) = 2xy$ are the particular harmonic functions (and consequently, harmonic conjugates) produced from $f(z) = z^2$. Thus, since the partial derivatives for u are $u_x = 2x$, $u_y = -2y$, $u_{xx} = 2$ and $u_{yy} = -2$, then u satisfies the Laplace equation, $u_{xx} + u_{yy} = 2 + (-2) = 0$. Similarly, the partial derivatives for v are $v_x = 2y$, $v_y = 2x$, $v_{xx} = 0$ and $v_{yy} = 0$, so v satisfies the Laplace equation, $v_{xx} + v_{yy} = 0 + 0 = 0$ as well.

EXERCISE 5.1. Let $f(z) = z^3$, where $z = x + iy$. Calculate the real part of f. The harmonic function obtained produces the surface known as the monkey saddle. Calculate the monkey saddle's curvature functions. Graph the monkey saddle.

It is often useful to visualize what harmonic functions look like in $\mathbb{R}^3$ in order to discuss some of their properties. The graphs of the two harmonic functions in the previous example are the standard "saddle" surfaces. It is not always a simple task, though, to recognize the graphs of more complicated harmonic functions. This is where our soap film can be used, since an informal physical interpretation can be simply made using soap film. However, it cannot be said in general that graphs of harmonic functions are minimal surfaces or vice versa. Consider the harmonic function $u(x,y) = x^2 - y^2$ from the previous example. Calculating H for the corresponding Monge patch yields

$$H = \frac{2(1 + 4y^2) - 2(1 + 4x^2)}{2(1 + 4x^2 + 4y^2)} = \frac{4y^2 - 4x^2}{1 + 4x^2 + 4y^2}.$$

Clearly, the mean curvature for such a surface is not identically zero. Similarly, in general, minimal surfaces are not graphs of harmonic functions. Unfortunately, calculating comparative information from a minimal surface can be very difficult. Very little is known about the parametrizations for an arbitrary soap film. Finding explicit parametrizations for minimum area surfaces, even when given relatively simple boundaries, has proved to be a nearly impossible task since Plateau's day.

A natural question to then ask is, how closely do minimal surfaces approximate harmonic functions? First, consider the plane. The function $z = \Phi(x,y)$ would have the form $ax + by + c$, which is clearly harmonic. (The variable z is not a complex variable here, but is the vertical coordinate in Cartesian 3-space.) Now, what if the surface were nearly planar? Using an argument similar to the one we used above to show area-minimizing implies minimal, it can be shown that in this case the function Φ which minimizes area across a loop is harmonic as well.

EXERCISE 5.2. Suppose $z = f(x,y)$ is a function whose graph spans a given curve and has least area among surfaces spanning the curve. Further, suppose all higher powers ≥ 4 of partials of f are negligible. That is, f_x^4, f_y^4 and $f_x^2 f_y^2$ are so small that the integrand $\sqrt{1 + f_x^2 + f_y^2}$ of the area integral is well approximated by

$\sqrt{(1 + \frac{1}{2}[f_x^2 + f_y^2])^2} = 1 + \frac{1}{2}[f_x^2 + f_y^2]$. Then, up to a constant provided by the term 1 and the factor $\frac{1}{2}$, the relevant area integral is

$$A = \int f_x^2 + f_y^2 \, dx \, dy.$$

Now vary $z = f(x, y)$ to get a nearby surface $z^t(x, y) = f(x, y) + tg(x, y)$ and carry through the derivation of Theorem 3.2 as before using the area integral above to show that f is harmonic. See [CH].

Unfortunately in most cases, as the minimal surface becomes less planar, it begins to vary from the graph of a true harmonic function. The argument used above breaks down when fourth-order terms are no longer negligible. The soap film is less and less a good approximation of the graph of a harmonic function.

Another interesting relationship between minimal surfaces and harmonic functions comes about when the surface is parametrized by *isothermal coordinates*. A parametrization $\mathbf{x}(u, v)$ is called *isothermal* if $E = G$ and $F = 0$. We will show later that every minimal surface has a parametrization with isothermal coordinates. When isothermal parameters are used, there is a close tie between the Laplace operator $\Delta \mathbf{x} = \mathbf{x}_{uu} + \mathbf{x}_{vv}$ and mean curvature. Recall that we have the following formulas for an orthogonal coordinate system:

$$\mathbf{x}_{uu} = \frac{E_u}{2E}\mathbf{x}_u - \frac{E_v}{2G}\mathbf{x}_v + l\, U,$$

$$\mathbf{x}_{uv} = \frac{E_v}{2E}\mathbf{x}_u + \frac{G_u}{2G}\mathbf{x}_v + m\, U,$$

$$\mathbf{x}_{vv} = -\frac{G_u}{2E}\mathbf{x}_u + \frac{G_v}{2G}\mathbf{x}_v + n\, U.$$

Theorem 5.2. *If the patch* $\mathbf{x}$ *is isothermal, then* $\Delta \mathbf{x} \overset{\text{def}}{=} \mathbf{x}_{uu} + \mathbf{x}_{vv} = (2\, E\, H)\, U.$

Proof.

$$\mathbf{x}_{uu} + \mathbf{x}_{vv} = \left(\frac{E_u}{2E}\mathbf{x}_u - \frac{E_v}{2G}\mathbf{x}_v + l\, U\right) + \left(-\frac{G_u}{2E}\mathbf{x}_u + \frac{G_v}{2G}\mathbf{x}_v + n\, U\right)$$

$$= \frac{E_u}{2E}\mathbf{x}_u - \frac{E_v}{2E}\mathbf{x}_v + l\, U - \frac{E_u}{2E}\mathbf{x}_u + \frac{E_v}{2E}\mathbf{x}_v + n\, U \qquad \text{since } E = G$$

$$= (l + n)\, U$$

$$= 2\, E \left(\frac{l + n}{2E}\right) U.$$

By examining the formula for mean curvature when $E = G$ and $F = 0$, we see that

$$H = \frac{El + En}{2E^2} = \frac{l + n}{2E}.$$

Therefore,

$$\mathbf{x}_{uu} + \mathbf{x}_{vv} = (2\,E\,H)\,U.$$

$\square$

Corollary 5.3. *A surface* $M \colon \mathbf{x}(u,v) = (x^1(u,v), x^2(u,v), x^3(u,v))$, *with isothermal coordinates is minimal if and only if* x^1, x^2 *and* x^3 *are harmonic functions.*

Proof. If M is minimal, then $H = 0$ and, by the previous theorem, $\mathbf{x}_{uu} + \mathbf{x}_{vv} = 0$. Therefore, $\mathbf{x}$ is harmonic. On the other hand, suppose x^1, x^2, and x^3 are harmonic functions. Then $\mathbf{x}$ is harmonic, so $\mathbf{x}_{uu} + \mathbf{x}_{vv} = 0$ and, by the previous theorem, $(2\,E\,H)\,U = 0$. Therefore, since U is the unit normal and $E = \mathbf{x}_u \cdot \mathbf{x}_u \neq 0$, then $H = 0$ and M is minimal. $\square$

This result will have a major impact on our later discussion of minimal surfaces. It is, in fact, the link between the geometry of minimal surface theory and the powerful methods of complex analysis. For more on minimal surfaces and their relation to the sciences, see Chapter 7 as well as the wonderful books [Ni], [DHKW], [Oss1], [Bo], [DWT], [HT] and [Isen]. In particular, [Ni, Introduction] and [HT] provide histories and overviews of the calculus of variations, minimal surfaces and the uses of minimal surfaces in science, engineering and architecture.

Chapter 5

GEODESICS, METRICS, AND ISOMETRIES

5.1 INTRODUCTION

In axiomatic geometry the axioms focus on the characteristics of the fundamental objects of geometry, points and lines. Moreover, in order to test these axioms, certain models of geometry are constructed. For example, Riemann's non-Euclidean geometry is modelled by the sphere with lines being the great circles (i.e. circles on the sphere having the center of the sphere as their own center). With these definitions, the sphere models a geometry where there do *not* exist *any* parallels to a given line through a given point. Of course the ramifications of such a result are tremendous. In particular, the angle sum of a triangle is greater than 180°.

If differential geometry is to somehow connect with traditional axiomatic geometry, then it must produce its own abstract definition of line (as opposed to the Euclidean axiomatic nondefinition of it) and show that this works for the typical models of Euclidean and non-Euclidean geometry. In particular, if we have the correct notion of line, then we should be able to prove (rather than assume or define) that the lines of the sphere are great circles!

So what should the differential geometric notion of line be? Previously we showed that a straight line in the plane gives the shortest distance between two points. We could attempt to use this "distance-minimization" criterion as our definition, but it is a hard condition to check. Instead, let's pick out another property of a straight line — a property which is readily calculable and which also characterizes the line — *the vanishing of its second derivative*.

Now, we cannot take a curve α on a surface M in $\mathbb{R}^3$ and require $\alpha'' = 0$, since this would just give us a straight line in $\mathbb{R}^3$ and, for an arbitrary M (such as the sphere), there is no reason why such a line should remain on M. But, let's think of all this from the point of view of a resident of M. That is, let's take the viewpoint of a creature who lives on M and has no perception of the 3rd dimension given by the unit normal U. Because the creature cannot see anything outside of the tangent plane T_pM, it cannot see the normal component of acceleration. Suppose α has unit speed. Then we

150

have two perpendicular unit vectors $T = \alpha'$ and U (the unit normal of M). A third such vector may be obtained by taking $T \times U$. Now, three vectors which are mutually perpendicular form a basis of $\mathbb{R}^3$, so any vector is a linear combination of them. Hence, we may write $\alpha'' = AT + B(T \times U) + CU$ and calculate the coefficients as

$$\alpha'' \cdot T = A, \qquad \alpha'' \cdot T \times U = B, \qquad \alpha'' \cdot U = C,$$

where we use the fact that T, $T \times U$ and U are unit vectors. Hence, we may write

$$\alpha'' = (\alpha'' \cdot T)T + (\alpha'' \cdot T \times U)(T \times U) + (\alpha'' \cdot U)U.$$

Further, $\alpha' \cdot \alpha' = 1$ since α has unit speed, so the product rule for differentiation gives $\alpha'' \cdot \alpha' + \alpha' \cdot \alpha'' = 2\alpha' \cdot \alpha'' = 0$. Hence, $\alpha' \cdot \alpha'' = T \cdot \alpha'' = 0$ and we have no T-component for α'',

$$\alpha'' = (\alpha'' \cdot T \times U)(T \times U) + (\alpha'' \cdot U)U.$$

The usual identities involving dot product and cross product give $U \cdot T \times U = 0$ and $\alpha'' \cdot T \times U = U \cdot \alpha'' \times \alpha'$, so $T \times U$ is in $T_p M$ for all $p \in M$ and

$$\begin{aligned}
\alpha'' \cdot T \times U &= U \cdot \alpha'' \times \alpha' \\
&= |U|\,|\alpha'' \times \alpha'|\cos\theta \\
&= |\alpha'' \times \alpha'|\cos\theta \\
&= \kappa_\alpha \cos\theta
\end{aligned}$$

where κ_α is the curvature of α and θ is the angle between $\alpha'' \times \alpha'$ and U. This quantity is often called *the geodesic curvature* of α and is denoted

$$\kappa_g = \kappa_\alpha \cos\theta.$$

EXERCISE 1.1. Show that the following relation holds between the curvature κ_α of α, the normal curvature $k(\alpha')$ of α' and the geodesic curvature κ_g of α:

$$\kappa_\alpha^2 = k(\alpha')^2 + \kappa_g^2.$$

Hint: how is the normal curvature of α' related to the curvature of α?

EXERCISE 1.2. For the topmost parallel (i.e. $u = \frac{\pi}{2}$) on the torus, compute κ, κ_g and k and show that the relation above holds in this case.

So now we see that we are able to decompose acceleration into tangential and normal components

$$\alpha''_{\text{tan}} = \kappa_g\, T \times U \qquad \text{and} \qquad \alpha''_{\text{normal}} = (\alpha'' \cdot U)U.$$

In keeping with our viewpoint as a resident of M, we see no acceleration exactly when $\alpha''_{\text{tan}} = 0$. Therefore we make the following definition. A curve α in M is a *geodesic* if $\alpha''_{\text{tan}} = 0$.

EXERCISE 1.3. For a non-unit speed curve $\alpha(t)$ with speed ν, show that

$$\alpha'' = \frac{d\nu}{dt} T + \kappa_g \nu^2 T \times U + (\alpha'' \cdot U) U.$$

Hints: (1) Recall $T = \frac{\alpha'}{\nu}$ and differentiate $\alpha' \cdot \alpha' = \nu^2$ to get $\alpha'' \cdot T = \frac{d\nu}{dt}$. (2) Use the formula for the curvature of a non-unit speed curve to show that $\alpha'' \cdot \frac{\alpha'}{\nu} \times U = \kappa_g \nu^2$.

The geodesics of a surface will be the lines of our geometry. We first note a very simple property of a geodesic (which also follows from the exercise above).

Lemma 1.1. *A geodesic has constant speed.*

Proof. The speed of α is $\nu = |\alpha'|$, so $\nu^2 = \alpha' \cdot \alpha'$. Differentiation yields $2\nu \frac{d\nu}{dt} = \alpha'' \cdot \alpha' + \alpha' \cdot \alpha'' = 2\alpha' \cdot \alpha'' = 0$ since $\alpha'' = \alpha''_{\text{normal}}$ and $\alpha''_{\text{normal}} \cdot \alpha' = 0$. Hence, $\frac{d\nu}{dt} = 0$, so ν is constant. $\qquad\square$

Note also that a straight line $\alpha(t) = p + tq$ lying in M must be a geodesic because $\alpha'' = 0$. Of course, as we expect, this condition characterizes geodesics in our familiar Euclidean geometry. To see this from our definition, suppose P is a plane with normal U and α is a geodesic in P. By definition, $\alpha''_{\text{tan}} = 0$, so $\alpha'' = (\alpha'' \cdot U) U$. But $\alpha' \cdot U = 0$ since α lies in P and the product rule gives,

$$0 = (\alpha' \cdot U)' = \alpha'' \cdot U + \alpha' \cdot U' = \alpha'' \cdot U$$

since U is constant. Hence, since both the tangential and normal components of α'' vanish, we have $\alpha'' = 0$. Therefore, α must be a straight line.

Another class of examples is provided by surfaces of revolution. Let $\alpha(u) = (g(u), h(u), 0)$ be a curve with unit speed parametrization. Recall that the associated surface of revolution (about the x-axis) has parametrization $\mathbf{x}(u, v) = (g(u), h(u) \cos v, h(u) \sin v)$ with

$$E = \mathbf{x}_u \cdot \mathbf{x}_u = 1, \qquad F = 0, \qquad G = \mathbf{x}_v \cdot \mathbf{x}_v = h^2 > 0.$$

Let's differentiate E and F with respect to u. Recall that this means we take the directional derivative of these functions in the $\mathbf{x}_u$ direction. Since $E = 1$ and $F = 0$, we have

$$
\begin{aligned}
0 = \mathbf{x}_u[E] \\
= \mathbf{x}_u[\mathbf{x}_u \cdot \mathbf{x}_u] \\
= \mathbf{x}_{uu} \cdot \mathbf{x}_u + \mathbf{x}_u \cdot \mathbf{x}_{uu} \\
= 2\mathbf{x}_u \cdot \mathbf{x}_{uu}
\end{aligned}
\qquad\qquad
\begin{aligned}
0 = \mathbf{x}_u[F] \\
= \mathbf{x}_u[\mathbf{x}_u \cdot \mathbf{x}_v] \\
= \mathbf{x}_{uu} \cdot \mathbf{x}_v + \mathbf{x}_u \cdot \mathbf{x}_{uv} \\
= \mathbf{x}_{uu} \cdot \mathbf{x}_v
\end{aligned}
$$

since $\mathbf{x}_u \cdot \mathbf{x}_{uv} = (g', h'\cos v, h'\sin v) \cdot (0, -h'\sin v, h'\cos v) = 0$. Hence, $\mathbf{x}_{uu} \cdot \mathbf{x}_u = 0$ and $\mathbf{x}_{uu} \cdot \mathbf{x}_v = 0$, so $\mathbf{x}_{uu}$ is perpendicular to $T_p(M)$. That is, $(\mathbf{x}_{uu})_{\tan} = 0$. Hence, *meridians of surfaces of revolution are geodesics*. We shall generalize this statement shortly when we consider Clairaut parametrizations.

Example 1.1: Geodesics on S_R^2. Let S_R^2 be an R-sphere with parametrization

$$\mathbf{x}(u, v) = (R\cos u \cos v,\ R\sin u \cos v,\ R\sin v).$$

Here we see that the parametrization gives a surface of revolution with $g(v) = R\sin v$ and $h(v) = R\cos v$, so the v-parameter curves are geodesics. That is, longitudes from the North pole to the South pole are geodesics. Since the sphere is symmetric about its center, by an appropriate rotation we may consider it as a surface of revolution about any line which passes through the center. The v-parameter curves are then circles on the sphere which connect the points of intersection of the line with the sphere. Since these points correspond to the North and South poles under the rotation, the circles connecting them are great circles on S_R^2. Recall that a *great circle* is a circle of radius R on an R-sphere and, therefore, has center the center of the sphere. This discussion shows that all great circles on S_R^2 are geodesics. To see that great circles are the only geodesics on S_R^2, suppose α is a unit speed geodesic. Then $\alpha''_{\tan} = \alpha'' - (\alpha'' \cdot U)\,U = 0$ by definition, so $\alpha'' = (\alpha'' \cdot U)\,U$. But, on S_R^2, we know that $U(\alpha(t)) = \alpha(t)/R$, so $\alpha'' = \frac{1}{R^2}(\alpha'' \cdot \alpha)\,\alpha$. Also then,

$$(\alpha' \times U)' = \frac{1}{R}(\alpha' \times \alpha)'$$

$$= \frac{1}{R}(\alpha'' \times \alpha + \alpha' \times \alpha')$$

$$= 0$$

since α'' is parallel to α and α' is parallel to itself. Then $\alpha' \times U \stackrel{\text{def}}{=} \mathcal{N}$ is a constant vector with $\alpha \cdot \mathcal{N} = 0$ since $\mathcal{N}$ is perpendicular to $U = \alpha/R$. But this means that α lies in the plane having $\mathcal{N}$ as normal vector. Further, since $U = \alpha/R$ lies in the plane, from any $\alpha(t)$ we can get to the origin $(0,0,0)$ and stay in the plane. Hence, the plane passes through $(0,0,0)$ as well. Since α also lies on S_R^2, it is in the intersection of the sphere with a plane passing through the origin. Therefore, it is (a part of) a great circle. The following exercise provides another approach, in terms of the Frenet Formulas, to characterizing geodesics on the sphere.

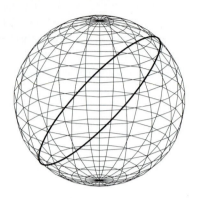

FIGURE 5.1.   Geodesic on a sphere

EXERCISE 1.4. Show that a curve α on S_R^2 is a geodesic if and only if it is a great circle. Hints: (1) suppose α is a unit speed geodesic with $\alpha' = T$ and $\alpha'' = T' = \kappa N$. Show $S(T) = \kappa_\alpha T - \tau_\alpha B = T/R$, where S is the shape operator of S_R^2. (Note that the first part of the formula for S holds in general.) (2) What does this imply about κ_α and τ_α?

EXERCISE 1.5. Show that a geodesic on M, α, which is contained in a plane is also a line of curvature. Hint: $\tau_\alpha = 0$ and the binormal B is the plane's normal.

EXERCISE 1.6. Show that a curve α in a surface M is both a geodesic and a line of curvature in M if and only if α lies in a plane P which is perpendicular to M everywhere along their intersection. Hint: Frenet Formulas.

EXERCISE 1.7. Find the geodesics on the cylinder $M \colon x^2 + y^2 = R^2$. Hints: (1) Write $\alpha(t) = \mathbf{x}(u(t), v(t)) = (R\cos u(t), R\sin u(t), bv(t))$ and take α''. (2) A unit normal is $U = (\cos u, \sin u, 0)$. Decompose $\alpha'' = -R\left(\frac{du}{dt}\right)^2 U + \alpha''_{\text{tan}}$, find α''_{tan} and set $\alpha''_{\text{tan}} = 0$. (3) What conditions are imposed on $\frac{d^2u}{dt^2}$ and $\frac{d^2v}{dt^2}$.

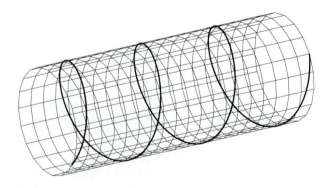

FIGURE 5.2.   Geodesic on a cylinder

Another way to see that our notion of geodesic is correct is to consider the following special situation. Suppose M is a surface with a patch (near a given point) $\mathbf{x}$ having $E = 1$, $F = 0$ and $G > 0$. (In fact, such a patch near a point may always be found. It is called a *geodesic polar coordinate patch*. See Chapter 6.) For instance, a surface of revolution generated by a unit speed curve has $E = 1$, $F = 0$ and $G = h(u)^2 > 0$. Let $\alpha\colon [s_0, s_1] \to M$ be a unit speed geodesic. Then the arclength of α is

$$L(\alpha) = \int_{s_0}^{s_1} |\alpha'|\, ds = \int_{s_0}^{s_1} 1\, ds = s_1 - s_0.$$

Now take another curve $\beta\colon [s_0, s_1] \to M$ with $\beta(s_0) = \alpha(s_0)$ and $\beta(s_1) = \alpha(s_1)$. Note that we can always take the same interval $[s_0, s_1]$ by reparametrizing. Also, *in an appropriate region*, the implicit function theorem allows us to write, $\beta(s) = \mathbf{x}(s, g(s))$ for some function g of s. We then obtain

$$\begin{aligned} \beta'(s) &= \mathbf{x}_u \frac{ds}{ds} + \mathbf{x}_v \frac{dg}{ds} \\ &= \mathbf{x}_u + \mathbf{x}_v g' \end{aligned}$$

and,

$$\begin{aligned} |\beta'(s)| &= \sqrt{\mathbf{x}_u \cdot \mathbf{x}_u + 2\mathbf{x}_u \cdot \mathbf{x}_v g' + \mathbf{x}_v \cdot \mathbf{x}_v g'^2} \\ &= \sqrt{E + Gg'^2} \\ &= \sqrt{1 + Gg'^2} \\ &> 1 \end{aligned}$$

since $G > 0$. Hence, we have the following estimate for the length of β,

$$\begin{aligned} L(\beta) &= \int_{s_0}^{s_1} |\beta'|\, ds \\ &= \int_{s_0}^{s_1} \sqrt{1 + Gg'^2}\, ds \\ &> \int_{s_0}^{s_1} 1\, ds \\ &= s_1 - s_0 = L(\alpha). \end{aligned}$$

Thus, geodesics minimize the distance between two points in (an appropriate neighborhood contained in) M.

EXERCISE 1.8. Take the points $(0, 1, 0)$ and $(0, 1, u_0)$ on the cylinder $x^2 + y^2 = 1$ and consider both the straight line and helical geodesics joining them. What goes wrong in the argument above since it is clear that the helical geodesic is *not* shortest length. Hint: here is a situation where the fact that a surface may not be completely covered by a single patch has a true geometric implication.

5.2 THE GEODESIC EQUATIONS AND THE CLAIRAUT RELATION

Now let's try to get a handle on the calculation of geodesics in a more general way. In the following, we only consider orthogonal patches $\mathbf{x}(u, v)$ (i.e. $F = \mathbf{x}_u \cdot \mathbf{x}_v = 0$). Let α be a geodesic in the patch $\mathbf{x}$. Then $\alpha = \mathbf{x}(u(t), v(t))$ and $\alpha' = \mathbf{x}_u u' + \mathbf{x}_v v'$ with

$$\alpha'' = \mathbf{x}_{uu} u'^2 + \mathbf{x}_{uv} v' u' + \mathbf{x}_u u'' + \mathbf{x}_{vu} u' v' + \mathbf{x}_{vv} v'^2 + \mathbf{x}_v v''.$$

Using the formulas for $\mathbf{x}_{uu}$, $\mathbf{x}_{uv}$ and $\mathbf{x}_{vv}$ obtained in Chapter 3, we obtain

$$\alpha'' = \mathbf{x}_u \left[u'' + \frac{E_u}{2E} u'^2 + \frac{E_v}{E} u'v' - \frac{G_u}{2E} v'^2 \right]$$
$$+ \mathbf{x}_v \left[v'' - \frac{E_v}{2G} u'^2 + \frac{G_u}{G} u'v' + \frac{G_v}{2G} v'^2 \right]$$
$$+ U \left[l u'^2 + 2m u'v' + n v'^2 \right]$$

where the first two terms give the tangential part of α''. For α to be a geodesic then, it is both necessary and sufficient that the following *geodesic equations* are satisfied.

(Geodesic Equations)
$$u'' + \frac{E_u}{2E} u'^2 + \frac{E_v}{E} u'v' - \frac{G_u}{2E} v'^2 = 0$$
$$v'' - \frac{E_v}{2G} u'^2 + \frac{G_u}{G} u'v' + \frac{G_v}{2G} v'^2 = 0.$$

EXERCISE 2.1. Find the geodesics on the cylinder $x^2 + y^2 = 1$ by using the geodesic equations.

EXERCISE 2.2. Suppose that a curve α satisfies the first geodesic equation and is also constant speed (i.e. $|\alpha'|^2 = u'^2 E + v'^2 G = c$). Differentiate the constant speed relation, replace u'' by the first geodesic equation and show that you obtain the second geodesic equation. Therefore, the constant speed relation takes the place of the second geodesic equation in the case of a constant speed curve. *Now*, reverse your calculations to show that the two geodesic equations imply that the curve must be constant speed. This is a fact that is hidden by the geodesic equations, but which is essential. Hint: in the first part, be very careful with the chain rule.

The geodesic equations are a system of 2nd order differential equations. Given initial data consisting of a point on M and a tangent vector at the point, the theory of ordinary differential equations guarantees the existence and uniqueness of a geodesic on M through the point and having velocity vector equal to the given tangent vector. We state this formally as

Theorem 2.1. *Let $p = \mathbf{x}(u_0, v_0)$ be a point on a surface $M: \mathbf{x}(u, v)$ and let $\mathbf{v} \in T_pM$. Then there is a unique geodesic $\alpha: (-r, r) \to M$ with $\alpha(0) = p$ and $\alpha'(0) = \mathbf{v}$.*

Proof. We want to obtain a geodesic $\alpha(t) = \mathbf{x}(u(t), v(t))$ with the property that $\alpha'(0) = u'(0)\mathbf{x}_u(u_0, v_0) + v'(0)\mathbf{x}_v(u_0, v_0) = \mathbf{v}$. Since $\mathbf{v}$ is fixed, this gives prescribed values to $u'(0)$ and $v'(0)$. Together with the initial values $u(0) = u_0$ and $v(0) = v_0$ and by the basic existence and uniqueness theorems of differential equations, this is precisely enough information to determine a unique solution (in some interval about 0) to the geodesic equations. $\qquad\square$

EXERCISE 2.3. The plane and the sphere have the property that all geodesics are plane curves. Show that this property characterizes the plane and the sphere. That is, show that if M is a surface such that every geodesic is a plane curve, then M is a part of a plane or a sphere. Hint: Use Exercise 1.5, Theorem 2.1 and Theorem 3.5.1.

Example 2.1: The Unit Sphere S^2. Take the standard patch

$$\mathbf{x}(u, v) = (\cos u \cos v,\ \sin u \cos v,\ \sin v)$$

and calculate $E = \cos^2 v$, $F = 0$ and $G = 1$. The geodesic equations become,

$$u'' - 2 \tan v\, u'v' = 0 \qquad\qquad v'' + \sin v \cos v\, u'^2.$$

As it stands, this is a formidable system of nonlinear differential equations. In fact, it is hardly ever the case that the geodesic equations are solved directly. The sphere provides a situation where we can use a standard trick to determine geodesics. First, assume without loss of generality that $\alpha(t) = \mathbf{x}(u(t), v(t))$ is a *unit speed* geodesic. Then, besides having the geodesic equations, we also have $\alpha' = u'\mathbf{x}_u + v'\mathbf{x}_v$, which leads to the unit speed relation $1 = Eu'^2 + Gv'^2$. On the unit sphere this is simply $1 = \cos^2 v\, u'^2 + v'^2$. Solve the first geodesic equation as follows:

$$\int \frac{u''}{u'} = \int 2 \tan v\, v'$$
$$\ln u' = -2 \ln \cos v + c$$
$$u' = \frac{c}{\cos^2 v}.$$

Now replace u' in the unit speed relation by $\frac{c}{\cos^2 v}$ to get

$$1 = \frac{c^2}{\cos^4 v} \cos^2 v + v'^2$$

$$v'^2 = 1 - \frac{c^2}{\cos^2 v}$$

$$v' = \sqrt{\frac{\cos^2 v - c^2}{\cos^2 v}}.$$

Dividing u' by v' produces the separable differential equation

$$\frac{du}{dv} = \frac{c}{\cos v \sqrt{\cos^2 v - c^2}}$$

which may be integrated (by making the substitutions $w = \frac{c}{\sqrt{1-c^2}} \tan v$ and $w = \sin \theta$ in steps 4 and 5 below respectively) to produce

$$u = \int \frac{c}{\cos v \sqrt{\cos^2 v - c^2}}\, dv$$

$$= \int \frac{c \sec^2 v}{\sqrt{1 - c^2 \sec^2 v}}$$

$$= \int \frac{c \sec^2 v}{\sqrt{1 - c^2 - c^2 \tan^2 v}}$$

$$= \int \frac{dw}{\sqrt{1 - w^2}}$$

$$= \int d\theta$$

$$= \arcsin\left(\frac{c \tan v}{\sqrt{1 - c^2}} \right) + d.$$

Therefore, we have

$$\sin(u - d) = \lambda \tan v$$

where $\lambda = \frac{c}{\sqrt{1-c^2}}$. We may expand $\sin(u - d) = \sin u \cos d - \sin d \cos u$ and find a common denominator $\cos v$ to get

$$\frac{\sin u \cos v}{\cos v} \cos d - \sin d \frac{\cos u \cos v}{\cos v} - \frac{\lambda \sin v}{\cos v} = 0.$$

If we make the substitutions $x = \cos u \cos v$, $y = \sin u \cos v$, $z = \sin v$ and only consider the numerator, we get

$$y \cos d - x \sin d - \lambda z = 0.$$

Hence, the geodesic equations imply that α lies on a plane $ax + by + cz = 0$ through the origin. Just as in our previous discussion of the sphere, this means that α is contained in the intersection of a plane through the origin with the sphere — and this is a great circle.

Before we leave the sphere, let's look at one other piece of information provided by the geodesic equations. Let ϕ denote the (smaller) angle between α' and $\mathbf{x}_v$ at any point along the unit speed curve α. Then (using the angle $\frac{\pi}{2} - \phi$ between α' and $\mathbf{x}_u$) we get

$$\sin \phi = \cos(\frac{\pi}{2} - \phi) = \frac{\alpha' \cdot \mathbf{x}_u}{|\alpha'||\mathbf{x}_u|} = \frac{1}{\sqrt{E}}[(u'\mathbf{x}_u + v'\mathbf{x}_v) \cdot \mathbf{x}_u] = u'\sqrt{E} = u' \cos v.$$

But $u' = \frac{c}{\cos^2 v}$ from the geodesic equations, so we obtain

$$\sin \phi = \frac{c}{\cos v}.$$

This is a special case of what is known as *Clairaut's relation*. We shall discuss this in general shortly. For the moment, let's look at what the relation implies about the behavior of geodesics (i.e. great circles) on the sphere. Suppose a geodesic α starts out parallel to a latitude circle (i.e. a u-parameter curve) $\mathbf{x}(u, v_0)$. This means that $\alpha'(0)$ is parallel to $\mathbf{x}_u(u_0, v_0)$, so the angle ϕ is $\frac{\pi}{2}$. The Clairaut relation says that $1 = \sin \frac{\pi}{2} = \frac{c}{\cos v_0}$, so $c = \cos v_0$. Along α we then have $\cos v \sin \phi = \cos v_0$ and, since $|\sin \phi| \leq 1$, $\cos^2 v \geq \cos^2 v_0$. Because $-\frac{\pi}{2} \leq v \leq \frac{\pi}{2}$, this means $\cos v \geq \cos v_0$. We then have the cases

$$\begin{cases} v \leq v_0 & \text{for} \quad 0 \leq v, v_0 \leq \frac{\pi}{2} \\ v \geq v_0 & \text{for} \quad -\frac{\pi}{2} \leq v, v_0 \leq 0. \end{cases}$$

Since v represents latitude, these inequalities mean that the great circle α is pinched between the latitudes v_0 above and v_0 below the equator.

Example 2.2: The Torus. Take the torus M with parametrization

$$\mathbf{x}(u, v) = ((R + r \cos u) \cos v, \ (R + r \cos u) \sin v, \ r \sin u).$$

We know $E = r^2$, $F = 0$ and $G = (R + r \cos u)^2$, so the geodesic equations are

$$u'' + \frac{(R + r \cos u)}{r} \sin u \, v'^2 = 0 \qquad v'' - 2\frac{r \sin u}{(R + r \cos u)} u'v' = 0.$$

The second equation is separable and we have $v' = \frac{c}{(R+r \cos u)^2}$. Again assuming α is unit speed, we replace v' in the unit speed relation by $\frac{c}{(R+r \cos u)^2}$ to get

$$1 = r^2 u'^2 + \frac{c^2}{(R + r \cos u)^2}$$

$$u' = \frac{1}{r}\sqrt{1 - \frac{c^2}{(R + r \cos u)^2}}.$$

Now we divide v' by u' to obtain

$$\frac{dv}{du} = \frac{cr\sqrt{R + r\cos u}}{\sqrt{(R + r\cos u)^2 - c^2}}$$

$$v = \int \frac{cr\sqrt{R + r\cos u}}{\sqrt{(R + r\cos u)^2 - c^2}}\, du.$$

Unfortunately, unlike the sphere, we cannot integrate the righthand side explicitly. Therefore, the Clairaut relation takes on extra importance since it helps us to visualize the path of a geodesic. For instance, the Clairaut relation for the torus is $(R + r\cos u)\sin\phi = c$, where ϕ is the angle between α' and $\mathbf{x}_u$. Suppose α starts out parallel to the topmost parallel circle $\mathbf{x}(\frac{\pi}{2}, v) = (R\cos v,\ R\sin v,\ r)$. Then $\phi = \frac{\pi}{2}$ and the Clairaut relation gives $R = c$. Again since $|\sin\phi| \leq 1$, we have $R + r\cos u \geq R$ all along α. This implies that $\cos u \geq 0$ and, consequently, $-\frac{\pi}{2} \leq u \leq \frac{\pi}{2}$. Hence, the geodesic is confined to the outside of the torus and, in fact, bounces between the topmost and the bottommost parallels.

FIGURE 5.3.   Geodesic on a torus

What do these two examples have in common? For both the sphere and the torus, the patches are orthogonal and E and G depend only on u or only on v. We say that an orthogonal patch $\mathbf{x}(u, v)$ is a *Clairaut parametrization in* u if $E_v = 0$ and $G_v = 0$. The patch is *Clairaut in* v if $E_u = 0$ and $G_u = 0$. Of course, as we have seen, the sphere is Clairaut in v and the torus is Clairaut in u. The geodesic equations simplify in these cases to

(Clairaut in u)
$$u'' + \frac{E_u}{2E}u'^2 - \frac{G_u}{2E}v'^2 = 0$$
$$v'' + \frac{G_u}{G}u'v' = 0$$

(Clairaut in v)
$$u'' + \frac{E_v}{E}u'v' = 0$$
$$v'' - \frac{E_v}{2G}u'^2 + \frac{G_v}{2G}v'^2 = 0.$$

In the following we shall focus on u-Clairaut parametrizations, but everything we say also applies to the v-Clairaut case as well. Take a u-parameter curve $\mathbf{x}(u, v_0)$ for a u-Clairaut patch $\mathbf{x}$ and assume it is unit speed. In particular $E = \mathbf{x}_u \cdot \mathbf{x}_u = 1$. Then, clearly, $\mathbf{x}(u, v_0)$ satisfies the Clairaut geodesic equations since $v = v_0$ and $E_u = 0$. Hence, for a u-Clairaut patch, u-parameter curves are geodesics. Of course this is a generalization of our previous discussion on surfaces of revolution. Also, note that we assumed the u-parameter curve is unit speed. While every curve can be reparametrized to have unit speed, it is not often the case that u-parameter curves naturally come with a unit speed parametrization. Some authors give the name *pregeodesic* to curves which are not constant speed, but when reparametrized to have constant speed, become geodesics in the sense above. We shall simply continue to use the term *geodesic* for these curves.

Since u-parameter curves of u-Clairaut patches are geodesics, it is natural to ask whether the same is true for v-parameter curves $\mathbf{x}(u_0, v)$. The geodesic equations become

$$0 = -\frac{G_u}{2E}v'^2 = -\frac{G_u}{2E},$$

where $v' = 1$ and G_u is evaluated at u_0. So, it is clear that the v-parameter curve $\mathbf{x}(u_0, v)$ is a geodesic if and only if $G_u(u_0) = 0$. In particular, this applies to surfaces of revolution where $G(u) = h(u)^2$. Putting this together with the discussion above gives

Theorem 2.2. *Let $M \colon \mathbf{x}(u, v)$ be a surface with u-Clairaut patch $\mathbf{x}$. Then every u-parameter curve is a geodesic and a v-parameter curve with $u = u_0$ is a geodesic precisely when $G_u(u_0) = 0$.*

Corollary 2.3. *For a surface of revolution having parametrization $\mathbf{x}(u, v) = (g(u), h(u) \cos v, h(u) \sin v)$, any meridian is a geodesic and a parallel is a geodesic precisely when $h'(u_0) = 0$.*

EXERCISE 2.4. Give an example of a function $h(u)$ so that the surface of revolution parametrized by $\mathbf{x}(u, v) = (u, h(u) \cos v, h(u) \sin v)$ has a geodesic parallel at $u = u_0$, but u_0 is not a maximum or minimum for $h(u)$.

In general, for a u-Clairaut parametrization $\mathbf{x}(u, v)$ and a unit speed geodesic α, we can reduce the second geodesic equation quite easily to first order. This follows just as for the examples of the sphere and torus above. The equation $v'' + \frac{G_u}{G} u'v' = 0$ becomes

$$\frac{v''}{v'} = -\frac{G_u}{G} u'$$

$$\int \frac{v''}{v'} \, dt = -\int \frac{G_u}{G} u' \, dt$$

$$\ln v' = -\ln G + c$$

$$v' = \frac{c}{G}.$$

Again, as in the examples, v' in the unit speed relation may be replaced by c/G to give

$$1 = Eu'^2 + Gv'^2$$

$$1 = Eu'^2 + G\frac{c^2}{G^2}$$

$$1 = Eu'^2 + \frac{c^2}{G}$$

$$u'^2 = \frac{G - c^2}{EG}$$

$$u' = \pm\sqrt{\frac{G - c^2}{EG}}.$$

EXERCISE 2.5. Differentiate either of the last two equations and show that you obtain the first geodesic equation. Therefore, again, the unit speed relation takes the place of this geodesic equation in the case of a Clairaut parametrization.

Now if we divide v' by u', we obtain a single integral which serves to characterize geodesics for a u-Clairaut parametrization.

$$\frac{dv}{du} = \frac{v'}{u'}$$

$$= \frac{\frac{c}{G}}{\pm\sqrt{\frac{G-c^2}{EG}}}$$

$$= \frac{\pm c\sqrt{E}}{\sqrt{G}\sqrt{G - c^2}}$$

$$v = \pm\int \frac{c\sqrt{E}}{\sqrt{G}\sqrt{G - c^2}} \, du.$$

EXERCISE 2.6. Suppose $\alpha(t) = \mathbf{x}(u(t), v(t))$ is a unit speed geodesic and $\mathbf{x}$ is u-Clairaut. Show that the Clairaut relation $\sqrt{G} \sin \phi = c$ holds, where c is a constant and ϕ is the angle from $\mathbf{x}_u$ to α'. Hence, show that α cannot leave the region of the surface for which $G \geq c^2$.

EXERCISE 2.7. Show that geodesics in the plane are straight lines. Hint: Use polar coordinates $\mathbf{x}(u, v) = (u \cos v, u \sin v)$.

EXERCISE 2.8. Let M denote the cone with parametrization $\mathbf{x}(u, v) = (u \cos v, u \sin v, au)$. Show that a unit speed geodesic $\alpha(t) = \mathbf{x}(u(t), v(t))$ on the cone is characterized by the equation $u = c \sec\left(\frac{v}{\sqrt{1+a^2}} + D\right)$. For $a = 1$, determine c and D for the geodesic connecting $(1, 0, 1)$ and $(0, 1, 1)$ and compare the arclength of this geodesic between these points to the arclength of the parallel circle joining the points. Hint: MAPLE.

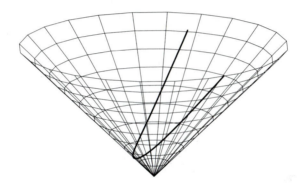

FIGURE 5.4.   Geodesic on a cone

EXERCISE 2.9. Figure 5.5 shows a geodesic on the surface of revolution obtained from the Witch of Agnesi. The geodesic is parallel to the parallel circle with $u = \pi/3$. Explain the behavior of the geodesic in terms of the Clairaut relation.

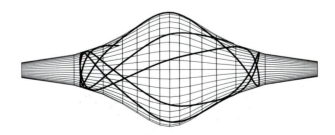

FIGURE 5.5.   Geodesic on the whirling Witch of Agnesi

Before we go on, let's look at the Clairaut relation from a physical viewpoint. Consider a surface M and suppose a particle is constrained to move on M, but is otherwise free of external forces. D'Alembert's Principle in Mechanics (see [Arn]) states that the constraint force F is normal to the surface, so Newton's Law becomes $|F|U = m\alpha''$, where $|F|$ denotes the magnitude of F, U is M's unit normal and $\alpha(t)$ is the motion curve of the particle. Taking $m = 1$, Newton's Law tells us that the $\mathbf{x}_u$ and $\mathbf{x}_v$ components of acceleration vanish and, as we have seen, this leads to the geodesic equations. Hence, a freely moving particle constrained to move on a surface moves along geodesics. Now suppose M is a surface of revolution parametrized by $\mathbf{x}(u,v) = (h(u)\cos v, h(u)\sin v, g(u))$. Along the surface, the radial vector $\mathbf{r}$ from the origin is simply $\mathbf{r} = (h(u)\cos v, h(u)\sin v, g(u))$ and the momentum vector of the particle's trajectory is given by $\mathbf{p} = \alpha' = u'\mathbf{x}_u + v'\mathbf{x}_v$ (since $m = 1$). Note that, since α is a geodesic, we may assume its speed is constant. The angular momentum of the particle about the origin may be calculated to be

$$\mathbf{L} = \mathbf{r} \times \mathbf{p}$$
$$= ((g'h - gh')u'\sin v - ghv'\cos v, -(g'h - gh')u'\cos v - ghv'\sin v, h^2v').$$

Because no external forces (or, more appropriately, torques) act on the particle, angular momentum, and in particular its z-component, is conserved. Therefore, we see that $h^2v' = C$, where C is a constant. Further, if ϕ is the angle from $\mathbf{x}_u$ to α', then

$$|\alpha'|\,h\sin\phi = \alpha' \cdot \mathbf{x}_v$$
$$= (u'\mathbf{x}_u + v'\mathbf{x}_v) \cdot \mathbf{x}_v$$
$$= h^2v'$$
$$= C$$

since $\cos(\frac{\pi}{2} - \phi) = \sin\phi$ and $\mathbf{x}_v \cdot \mathbf{x}_v = h^2$. Because $|\alpha'|$ is constant as well, we have

$$h\sin\phi = \text{constant}.$$

But this is precisely the Clairaut relation since $h = \sqrt{G}$. In this way the Clairaut relation is a physical phenomenon as well as a mathematical one.

EXERCISE 2.10. Let M denote the elliptic paraboloid $z = x^2 + y^2$ parametrized as a surface of revolution by $\mathbf{x}(u,v) = (u\cos v, u\sin v, u^2)$ and let $\alpha = \mathbf{x}(u(t), v(t))$ be a unit speed geodesic. Find v in terms of u and show that the Clairaut relation is $u\sin\phi = c$. Show that a nonmeridan geodesic spirals up the paraboloid and crosses every meridian an infinite number of times. Hints: (1) use the Clairaut relation (2) the integral for v diverges, so α can't approach a meridian as a limit.

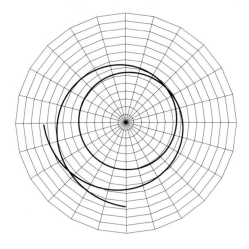

FIGURE 5.6. Geodesics on a paraboloid

EXERCISE 2.11. For the paraboloid M and geodesic α above, determine how low on M the geodesic α can go in terms of the constant c in the Clairaut relation. What happens to the geodesic when it reaches its lowest height? From this, show that a nonmeridian geodesic intersects itself an infinite number of times. Hints: You may use the following result which we will not prove. **Theorem** [DoC1, section 4-7, p. 302]. If a geodesic approaches a parallel as a limit, then the parallel is a geodesic. What parallels are geodesics on the paraboloid?

EXERCISE 2.12. Determine v as an integral for a geodesic on the catenoid. Show that the Clairaut relation $\cosh u \sin \phi = c$ determines the behavior of geodesics approaching the parallel circle with $u_0 = 0$. Hint: take a geodesic starting at $\mathbf{x}(u_0, v_0)$ with $\phi = \phi_0$ and consider the three cases $\cosh u_0 \sin \phi_0 = c < 1$, $\cosh u_0 \sin \phi_0 = c = 1$ and $\cosh u_0 \sin \phi_0 = c > 1$.

EXERCISE 2.13. Determine v as an integral for a geodesic on the hyperboloid of one sheet $x^2 + y^2 - z^2 = 1$ and give the Clairaut relation. Show directly that the ruling of the hyperboloid which passes through $(1, 0, 0)$ satisfies the Clairaut relation. Find a closed geodesic on the hyperboloid. Hints: (1) the ruling may be parametrized by $\alpha(v) = (1, \tan v, \tan v)$ (why?) and a parallel circle may be parametrized by $\beta(v) = (\cosh u_0 \cos v, \cosh u_0 \sin v, \sinh u_0)$. What happens on the central parallel $(\cos v, \sin v, 0)$? (2) A geodesic $\alpha \colon [a, b] \to M$ is closed if $\alpha(a) = \alpha(b)$. When are parallels geodesics?

5.3 A BRIEF DIGRESSION ON COMPLETENESS

In the discussion above, at various points, we have implicitly assumed that geodesics "run forever." That is, we have taken for granted that a geodesic is a unit speed curve $\alpha \colon \mathbb{R} \to M$ whose domain is the real numbers. That this is *not* always the case is apparent from looking at an example such as the plane minus the origin, $M = \mathbb{R}^2 - \{(0,0)\}$. We know that geodesics in the plane are straight lines, so the same must be true for M since the same geodesic equations hold. Suppose a geodesic starts at (r, s) in the direction $(-r, -s)$. We know by Theorem 2.1 that there is a unique geodesic of this type which heads toward $(0,0)$. Since the geodesic is unit speed, its arclength (i.e. distance travelled) corresponds to the length of the interval $[a, b]$ on which it is defined. Because the geodesic cannot pass through the missing origin, it can only go a distance of less than $\sqrt{r^2 + s^2}$. Therefore, the interval $[a, b]$ cannot be all of $\mathbb{R}$. What goes wrong here?

In order to answer this question, let's make a definition. Say that a surface M is *geodesically complete* if every (unit speed) geodesic has domain $\mathbb{R}$. The importance of this definition is reflected in the following beautiful result which we will not prove.

Theorem 3.1 (Hopf-Rinow). *If M is geodesically complete, then any two points of M may be joined by a geodesic which has the shortest length of any curve between the two points.*

When we speak of being a resident of some surface M, we naturally assume that we can get from any point to any other by some shortest-distance path. If we live in a nongeodesically complete surface however, this is just not so. In $M = \mathbb{R}^2 - \{(0,0)\}$, we cannot go from $(-1, 0)$ to $(1, 0)$, say, by a geodesic. Indeed, there is *no* shortest path between these points. So, the question now becomes, how do we recognize geodesically complete surfaces? The example of the plane minus the origin gives a clue. Recall that a subset $M \subseteq \mathbb{R}^3$ is *closed* if any convergent sequence $z_j \to z$ with $z_j \in M$ also has $z \in M$. A sufficient condition for a surface to be geodesically complete is given by

Theorem 3.2. *A closed surface $M \subseteq \mathbb{R}^3$ is geodesically complete.*

Proof. Let α be a unit speed geodesic on M. If $\alpha(s)$ is defined, then the fact that α arises as a solution to the geodesic equations guarantees that α is defined on an open interval $(s - r, s + r)$, for some r. This means that every $s \in \mathbb{R}$ for which α is defined has a small interval about it of other points for which α is defined. Hence, the subset of $\mathbb{R}$ consisting of points for which α is defined is an open set — that is, a union of open intervals. Let $I = (a, b)$ be one of these intervals where we assume $b < \infty$ and b is a least upper bound for I. We shall show that α is, in fact, defined for b as well. Hence, I has no least upper bound (i.e. $b = \infty$). A similar argument for a then shows that α is defined on the whole real line.

Let $z_j \to b$ be a sequence of points in I converging to the least upper bound b. Since the sequence converges, it is also Cauchy. That is, given $\epsilon > 0$, there exists an integer N such that for all $n, m > N$, $|z_n - z_m| < \epsilon$. Now choose $\epsilon > 0$ and look at $\alpha(z_n)$ and $\alpha(z_m)$. Since α is unit speed, its arclength from $\alpha(z_n)$ to $\alpha(z_m)$ is simply $|z_n - z_m| < \epsilon$. Also, as we saw in Chapter 1, the (shortest) distance between two points p, $q \in \mathbb{R}^3$ is given by the line joining them and is $|q - p|$. Hence we have,

$$|\alpha(z_n) - \alpha(z_m)| \leq |z_n - z_m| < \epsilon.$$

This means that the sequence $\{\alpha(z_j)\}$ is Cauchy as well. Now, it is a fact of analysis that Euclidean space $\mathbb{R}^n$ for any n is *complete* in the sense that every Cauchy sequence converges, so we have $\alpha(z_j) \to w$ for some $w \in \mathbb{R}^3$. But $\alpha(z_j) \in M$ for all j and M is closed by hypothesis, so $w \in M$. Hence, we may define $\alpha(b) = w$, showing that α may be defined at b as well and contradicting b's definition. We note that we have only shown that α may be extended continuously to b. Some rather technical arguments indeed show that α extends as a smooth geodesic. $\square$

EXERCISE 3.1. Which of the surfaces we have studied up to now are geodesically complete?

5.4 Surfaces Not in $\mathbb{R}^3$

Up to this point we have taken surfaces in 3-space together with the inner product structure naturally associated to any Euclidean space, the dot product. But there are other inner products which are available. Indeed, any nonsingular symmetric $n \times n$-matrix A gives an inner product on $\mathbb{R}^n$ by taking the matrix multiplication

$$\langle \mathbf{x}, \mathbf{y} \rangle = \mathbf{x}^t A \mathbf{y}$$

for $\mathbf{x}$, $\mathbf{y} \in \mathbb{R}^n$. Here $\mathbf{x}^t$ denotes the transpose of the column vector $\mathbf{x}$ to a row vector. The dot product is simply the inner product which takes A to be the identity matrix. In some sense, the geometry which we see on surfaces in 3-space reflects the geometry of 3-space itself. If we wish to extend our notion of geometry beyond 3-space we must somehow rid ourselves of our dependence on the dot product — or, at least, modify this dependence. Recall the definition of the dot product in $\mathbb{R}^2$. (We use vectors in $\mathbb{R}^2$ because tangent planes of surfaces are 2-dimensional.) Let $\mathbf{x} = (x_1, x_2)$, $\mathbf{y} = (y_1, y_2) \in \mathbb{R}^2$ and define

$$\mathbf{x} \cdot \mathbf{y} = (x_1, x_2) \begin{bmatrix} 1 & 0 \\ 0 & 1 \end{bmatrix} \begin{pmatrix} y_1 \\ y_2 \end{pmatrix} = x_1 y_1 + x_2 y_2.$$

Now, change the matrix to $\begin{bmatrix} 1/a & 0 \\ 0 & 1/a \end{bmatrix}$ for $a > 0$ and find

$$\begin{aligned} \mathbf{x} \circ \mathbf{y} &= (x_1, x_2) \begin{bmatrix} 1/a & 0 \\ 0 & 1/a \end{bmatrix} \begin{pmatrix} y_1 \\ y_2 \end{pmatrix} = \frac{x_1 y_1 + x_2 y_2}{a} \\ &= \frac{\mathbf{x} \cdot \mathbf{y}}{a} \end{aligned}$$

where "$\cdot$" is the dot product and "$\circ$" is the modified dot product. All the usual properties of the dot product (such as symmetry, bilinearity etc.) work for the modified version as well. For an inner product on the tangent planes of a surface M the number a may change at each point $p \in M$. That is, we may take a to be a function on M,

$$a = f(p)^2 \quad \text{for } f \colon M \to \mathbb{R} \text{ and } p \in M.$$

We take the square of the function $f(p)$ to ensure that a is positive at each point. Thus, for two tangent vectors $v, w \in T_p(M)$ we have,

$$v \circ w = \frac{v \cdot w}{f(p)^2}.$$

We say that $\cdot$ and $\circ$ are *metrics* of M since they allow us to calculate E, F, and G, the basic components of distance along M. The dot product is usually called the *Euclidean Metric* and a metric such as $\circ$ is said to be *conformal* (to the Euclidean metric) with scaling factor f.

Now, if these surfaces do not use the Euclidean metric, then where are they? While they may sit in $\mathbb{R}^3$ as sets, their different metrics show that they don't inherit geometry from $\mathbb{R}^3$. So, as surfaces with a metric, they do not sit inside $\mathbb{R}^3$! But if we cannot use the structure of $\mathbb{R}^3$ — for example, the unit normal — how can we understand the geometry of such a surface? As is typical in mathematics, when we extend beyond our usual situation, we use previous theorems as definitions. Because we no longer can use the unit normal U to define Gaussian curvature K, instead we invoke Theorem 3.4.1 to define K. Note that this only makes sense because Theorem 3.4.1 ensures that K depends only on the metric. Therefore,

Important Definition 4.1. For a surface with orthogonal metric (i.e. $F = 0$), the Gaussian curvature K is defined to be

$$K = -\frac{1}{2\sqrt{EG}} \left(\frac{\partial}{\partial v} \left(\frac{E_v}{\sqrt{EG}} \right) + \frac{\partial}{\partial u} \left(\frac{G_u}{\sqrt{EG}} \right) \right).$$

Of course, because this formula was a theorem for surfaces in $\mathbb{R}^3$, this definition agrees with our original definition in case M has a metric induced by the Euclidean metric of $\mathbb{R}^3$.

EXERCISE 4.1. For a conformal metric with scaling factor f, show that

$$K = f(f_{uu} + f_{vv}) - (f_u^2 + f_v^2).$$

A function f is *harmonic* if $f_{uu} + f_{vv} = 0$. In this case, clearly, $K \le 0$ and $K = 0$ only at critical points of f. Define a conformal metric on $\mathbb{R}^2 - \{(0,0)\}$ by taking $f = \frac{1}{2} \ln(u^2 + v^2)$. Compute K and draw the curves in the plane where K is constant.

As the following examples show, we can still define surfaces by taking patches $\mathbf{x}$ in $\mathbb{R}^3$, but with a conformal metric rather than the induced metric. We shall see that we can do geometry in this situation just as before. Understanding this abstract differential geometry is the first step to understanding the geometry of higher dimensions.

Example 4.2: The Poincaré Plane P.
Define P to be the upper half-plane $P = \{(x,y) \in \mathbb{R}^2 \mid y > 0\}$ with the patch $\mathbf{x}(u,v) = (u,v)$ and with the conformal metric,

$$\mathbf{w}_1 \circ \mathbf{w}_2 = \frac{\mathbf{w}_1 \cdot \mathbf{w}_2}{v^2} \quad \text{where } \mathbf{w}_1, \mathbf{w}_2 \in T_p(P) \text{ and } p = (u,v).$$

This definition of the metric means that the usual dot product is scaled down by the height of p (i.e. v) above the x-axis. Let's compute E, F and G (at p). We obtain

$$\mathbf{x}_u = (1,0) \qquad\qquad \mathbf{x}_v = (0,1)$$

$$E = \mathbf{x}_u \cdot \mathbf{x}_u = \frac{1}{v^2} \qquad F = \mathbf{x}_u \cdot \mathbf{x}_v = 0 \qquad G = \mathbf{x}_v \cdot \mathbf{x}_v = \frac{1}{v^2}.$$

Note that $G_u = 0$ and $E_v = \dfrac{-2}{v^3}$. Hence

$$K = -\frac{1}{2\sqrt{1/v^4}} \left(\frac{\partial}{\partial v} \left(\frac{-2/v^3}{\sqrt{1/v^4}} \right) \right)$$

$$= -\frac{v^2}{2} \left(\frac{\partial}{\partial v} \left(-\frac{2}{v} \right) \right)$$

$$= -\frac{v^2}{2} \cdot \frac{2}{v^2}$$

$$= -1.$$

Thus, P has constant curvature equal to -1 at each point. P is then a "negative curvature" analogue of the unit sphere in $\mathbb{R}^3$.

EXERCISE 4.2. Define M to be the upper half-plane $M = \{(x,y) \in \mathbb{R}^2 \mid y > 0\}$ with the patch $\mathbf{x}(u,v) = (u,v)$ and with the conformal metric,

$$\mathbf{w}_1 \circ \mathbf{w}_2 = \frac{\mathbf{w}_1 \cdot \mathbf{w}_2}{v} \quad \text{where } \mathbf{w}_1, \mathbf{w}_2 \in T_p(P) \text{ and } p = (u,v).$$

Calculate the Gaussian curvature K of M.

Example 4.3: The Hyperbolic Plane H.

Define H to be the disk of radius 2 in the plane minus the bounding circle with patch $\mathbf{x}(u, v) = (u \cos v, u \sin v)$ $0 \le u < 2$, $0 \le v < 2\pi$ and conformal metric,

$$\mathbf{w}_1 \circ \mathbf{w}_2 = \frac{\mathbf{w}_1 \cdot \mathbf{w}_2}{(1 - u^2/4)^2} \quad \text{where} \quad \mathbf{w}_1, \mathbf{w}_2 \in T_p(H) \quad \text{and} \quad p = (u \cos v, u \sin v).$$

EXERCISE 4.3. Show that $K = -1$ at each point.

Example 4.4: The Stereographic Sphere S^2_N.

Let S^2_N denote the unit sphere minus the North pole $S^2 - \{(0,0,1)\}$ and define a map $\text{St} \colon S^2_N \to \mathbb{R}^2$ (where $\mathbb{R}^2$ is the xy-plane) by taking the point in $\mathbb{R}^2$ which is the intersection of $\mathbb{R}^2$ and the line in $\mathbb{R}^3$ determined by a point on the sphere and the North pole N. Formally, given a point $p = (\cos u \cos v, \sin u \cos v, \sin v)$ on S^2_N, the line joining p and N is given by

$$\gamma(t) = (0, 0, 1) + t \, (\cos u \cos v, \sin u \cos v, \sin v - 1).$$

The line γ intersects $\mathbb{R}^2$ when the third coordinate is zero. This occurs when $1 + t(\sin v - 1) = 0$ or $t = 1/(1 - \sin v)$. Hence,

$$\text{St}(\cos u \cos v, \; \sin u \cos v, \; \sin v) = \left(\frac{\cos u \cos v}{1 - \sin v}, \; \frac{\sin u \cos v}{1 - \sin v}, \; 0 \right).$$

Clearly, St is a one-to-one onto map from S^2_N to $\mathbb{R}^2$. Recall that the induced linear transformation of tangent vectors St_* may be calculated by simply taking appropriate derivatives of the image u and v curves. For example, fixing a particular v and differentiating gives

$$\text{St}_*(\mathbf{x}_u) = \frac{d}{du} \left(\frac{\cos u \cos v}{1 - \sin v}, \; \frac{\sin u \cos v}{1 - \sin v}, \; 0 \right)$$

$$= \left(\frac{-\sin u \cos v}{1 - \sin v}, \; \frac{\cos u \cos v}{1 - \sin v}, \; 0 \right).$$

Similarly, differentiating with respect to v gives

$$\text{St}_*(\mathbf{x}_v) = \left(\frac{\cos u}{1 - \sin v}, \; \frac{\sin u}{1 - \sin v}, \; 0 \right).$$

We may now define a new metric on S^2_N by saying

$$\mathbf{w}_1 \circ \mathbf{w}_2 = \text{St}_*(\mathbf{w}_1) \cdot \text{St}_*(\mathbf{w}_2)$$

where $\cdot$ denotes the sphere's induced $\mathbb{R}^3$ metric.

EXERCISE 4.4. In the metric $\circ$ on S_N^2, show

$$E = \frac{\cos^2 v}{(1 - \sin v)^2} \qquad F = 0 \qquad G = \frac{1}{(1 - \sin v)^2}$$

and calculate the Gaussian curvature K. Explain your answer.

Example 4.5: The Flat Torus T_{flat}.

Let T_{flat} be defined by the patch with range $\mathbb{R}^4$,

$$\mathbf{x}(u, v) = (\cos u, \sin u, \cos v, \sin v) \;\; 0 \le u \le 2\pi, \; 0 \le v \le 2\pi.$$

and with the induced metric (i.e. dot product) of $\mathbb{R}^4$. Just as before, we may calculate $\mathbf{x}_u$ and $\mathbf{x}_v$ componentwise to get $E = 1$, $F = 0$ and $G = 1$. Hence, $K = 0$ and T_{flat} is flat. Note that this example does not come from a conformal metric, but rather from a Euclidean metric of a higher dimension.

EXERCISE 4.5. Verify the calculations for the flat torus T_{flat} and then show that the name is apt by defining a one-to-one onto map from T_{flat} to a usual torus. Since the flat torus is closed and bounded in R^4, it is compact. Can the flat torus be embedded as a surface in $\mathbb{R}^3$ (i.e. $T_{\text{flat}} \subseteq \mathbb{R}^3$ with the induced metric of $\mathbb{R}^3$)? Explain.

Beyond having a definition of Gaussian curvature, however, we would like to understand the nature of these surfaces in terms of their geodesics. Again we are faced with the problem of removing all traces of the unit normal U from our previous discussions of geodesics. One way is easy. When we calculated the Christoffel symbols for $\mathbf{x}_{uu}$, $\mathbf{x}_{uv}$ and $\mathbf{x}_{vv}$, we never used U. Hence, the same formulas hold for surfaces not in $\mathbb{R}^3$. In particular, the calculation of α'' remains the same, but without the final term involving U. That is,

$$\alpha'' = \mathbf{x}_u \left[u'' + \frac{E_u}{2E} u'^2 + \frac{E_v}{E} u'v' - \frac{G_u}{2E} v'^2 \right]$$
$$+ \mathbf{x}_v \left[v'' - \frac{E_v}{2G} u'^2 + \frac{G_u}{G} u'v' + \frac{G_v}{2G} v'^2 \right].$$

For α to be a geodesic then, just as before, it is both necessary and sufficient that the geodesic equations are satisfied. Now let's look at the original way we defined geodesics in terms of geodesic curvature. Recall that, for a unit speed curve α, we had

$$\alpha'' = \kappa_g \, T \times U + (\alpha'' \cdot U) \, U.$$

Of course, we do not have a U now, so we must drop off the last term and somehow make sense of the first. The tangent vector $T \times U$ was used before to obtain an orthonormal basis for $\mathbb{R}^3$, $\{T, T \times U, U\}$, where T and $T \times U$

provided an orthonormal basis for the tangent plane at every point of the surface. Although we do not have a unit normal to work with for surfaces not in $\mathbb{R}^3$, we may still find an appropriate tangent vector in each tangent plane which is perpendicular to T. The way to do this is as follows. Given a patch $\mathbf{x}(u, v)$, define a linear transformation J of each tangent plane by

$$J\left(\frac{\mathbf{x}_u}{\sqrt{E}}\right) = \frac{\mathbf{x}_v}{\sqrt{G}} \qquad J\left(\frac{\mathbf{x}_v}{\sqrt{G}}\right) = -\frac{\mathbf{x}_u}{\sqrt{E}}.$$

Note that defining J on a basis for each tangent plane suffices to define J completely. Also note that J is defined on unit vectors and has unit vectors as outputs. This is to ensure that J has no effect on lengths, but, rather, only rotates vectors on each tangent plane.

EXERCISE 4.6. Suppose $\alpha \colon I \to M$ is a curve on a surface M in $\mathbb{R}^3$ (i.e. with $\mathbb{R}^3$'s induced metric). For convenience, take α to have unit speed. Recall that patches are now assumed to be orthogonal; that is, $F = 0$. Show that $J(T) = T \times U$. Hint: $\alpha' = u' \mathbf{x}_u + v' \mathbf{x}_v$ and J is linear.

Since $J(T) = T \times U$ for surfaces in $\mathbb{R}^3$, we can take the equation

$$\alpha'' = \kappa_g \, J(T)$$

as the definition of *geodesic curvature* for a unit speed curve $\alpha(t)$. Of course the nonunit speed case is just as before, but without the U term. Now,

$$\begin{aligned} J(T) &= J(u' \mathbf{x}_u + v' \mathbf{x}_v) \\ &= u' J(\mathbf{x}_u) + v' J(\mathbf{x}_v) \\ &= u' \sqrt{\frac{E}{G}} \mathbf{x}_v - v' \sqrt{\frac{G}{E}} \mathbf{x}_u \end{aligned}$$

so

$$\alpha'' = -\kappa_g \, v' \sqrt{\frac{G}{E}} \mathbf{x}_u + \kappa_g \, u' \sqrt{\frac{E}{G}} \mathbf{x}_v.$$

If we equate the $\mathbf{x}_u$ coefficient of this expression for α'' and the one we used to define geodesic equations, we have

$$-\kappa_g \, v' \sqrt{\frac{G}{E}} = u'' + \frac{E_u}{2E} u'^2 + \frac{E_v}{E} u' v' - \frac{G_u}{2E} v'^2$$

$$\kappa_g = \frac{-u''}{v'} \sqrt{\frac{E}{G}} - \frac{E_u}{2\sqrt{EG}} \frac{u'^2}{v'} - \frac{E_v}{\sqrt{EG}} u' + \frac{G_u}{2\sqrt{EG}} v'.$$

We may write all this in a much more meaningful way. Let θ denote the angle between α' and $\mathbf{x}_u$. Since α is unit speed, we have

$$\alpha' = u'\mathbf{x}_u + v'\mathbf{x}_v$$

$$= \cos\theta \, \frac{\mathbf{x}_u}{\sqrt{E}} + \sin\theta \, \frac{\mathbf{x}_v}{\sqrt{G}}.$$

Therefore, $u'\sqrt{E} = \cos\theta$ and $v'\sqrt{G} = \sin\theta$. If we differentiate the first expression and substitute in the second, we obtain

$$u''\sqrt{E} + u'\frac{E_u u' + E_v v'}{2\sqrt{E}} = \frac{d\cos\theta}{dt}$$

$$u''\sqrt{E} + \frac{E_u u'^2}{2\sqrt{E}} + \frac{E_v u'v'}{2\sqrt{E}} = -\sin\theta \, \frac{d\theta}{dt}$$

$$u''\sqrt{E} + \frac{E_u u'^2}{2\sqrt{E}} + \frac{E_v u'v'}{2\sqrt{E}} = -\sqrt{G}\, v' \, \frac{d\theta}{dt}$$

$$\frac{-u''}{v'}\sqrt{\frac{E}{G}} - \frac{E_u u'^2}{2v'\sqrt{EG}} - \frac{E_v u'}{2\sqrt{EG}} = \frac{d\theta}{dt}.$$

Comparing this with the expression for κ_g above gives the following

Theorem 4.1. *Let $\alpha(t)$ be a unit speed curve in a surface $M: \mathbf{x}(u,v)$ and let θ denote the angle between α' and $\mathbf{x}_u$. Then the geodesic curvature of α is given by*

$$\kappa_g = \frac{d\theta}{dt} + \frac{1}{2\sqrt{EG}} \left[G_u v' - E_v u' \right].$$

Note that we have not specified a range for the patch $\mathbf{x}(u,v)$. This is simply to indicate that the expression for κ_g holds for surfaces not in $\mathbb{R}^3$ as well as those in $\mathbb{R}^3$. We shall meet up with geodesic curvature again in Chapter 6 where we shall derive the above formula for κ_g in a different way.

Now let's find some geodesics. Our previous approach in terms of the geodesic equations still works here of course. In particular, the notions of u and v-Clairaut still have meaning and still simplify the geodesic equations greatly. Note, however, that the property of being a Clairaut patch is much more dependent on the definition of the metric now than was the case when we could safely rely on $\mathbb{R}^3$'s dot product.

Example 4.6: The Poincaré plane P.

The patch $\mathbf{x}(u, v) = (u, v)$ with the conformal metric,

$$\mathbf{w}_1 \circ \mathbf{w}_2 = \frac{\mathbf{w}_1 \cdot \mathbf{w}_2}{v^2} \quad \text{where} \quad \mathbf{w}_1, \mathbf{w}_2 \in T_p(P) \quad \text{and} \quad p = (u, v)$$

is v-Clairaut, so, for $\alpha = \mathbf{x}(u(t), v(t))$, the geodesic equations give

$$u'' - \frac{2}{v} u'v' = 0 \qquad\qquad v'' + \frac{1}{v} u'^2 - \frac{1}{v} v'^2 = 0.$$

Since the patch is v-Clairaut, the usual procedure for calculating geodesics gives, from the first geodesic equation,

$$\frac{u''}{u'} = \frac{2}{v} v'$$

$$\ln(u') = 2\ln(v) + c$$

$$u' = cv^2.$$

Plugging this into the unit speed relation $1 = u'^2(1/v^2) + v'^2(1/v^2)$ then gives $v' = v\sqrt{1 - c^2 v^2}$. Dividing v' by u' and integrating produces

$$\int du = \int \frac{cv}{\sqrt{1 - c^2 v^2}} \, dv$$

$$u - d = \frac{1}{c} \int \sin w \, dw \qquad \text{where } v = \frac{1}{c} \sin w$$

$$= \frac{-1}{c} \cos w$$

$$= \frac{-1}{c} \sqrt{1 - c^2 v^2}$$

$$c^2 (u - d)^2 = 1 - c^2 v^2$$

$$(u - d)^2 + v^2 = \frac{1}{c^2}.$$

This is the equation of a circle centered on the u-axis. Also, since the patch is v-Clairaut, vertical lines (i.e. v-parameter curves) are geodesics as well. So, the Poincaré plane P has as its geodesics arcs of circles centered on the u-axis and vertical lines. These are the "straight lines" of P.

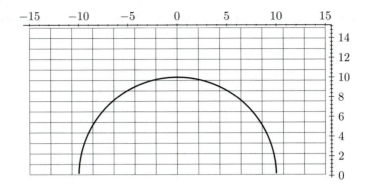

FIGURE 5.7.   Geodesic on the Poincaré plane

EXERCISE 4.7. Show that the parallel postulate does not hold for the Poincaré plane. The parallel postulate says that, given a line and a point not on the line, there is precisely one line through the point which is parallel to the given line. The Poincaré plane is a model for the non-Euclidean geometry of Gauss, Lobachevsky and Bolyai.

EXERCISE 4.8. Find the geodesics for the surface M of Exercise 4.2.

Example 4.7: The Hyperbolic Plane H.
We compute geodesics as follows: (1) Since $\mathbf{x}$ is u-Clairaut, u-parameter curves are geodesics. Hence, radial lines through $(0,0)$ are geodesics. For non-u-parameter geodesics, apply the Clairaut integral formula (after simplification) to get

$$v = \int \frac{c(1-u^2/4)}{u^2 \sqrt{1 - \left(\frac{c(1-u^2/4)}{u}\right)^2}} \, du.$$

This appears formidable, but it can be handled as follows.

EXERCISE 4.9. Let $w = \dfrac{c}{u\sqrt{1+c^2}} \left(1 + u^2/4\right)$ and show $dv = \dfrac{dw}{\sqrt{1-w^2}}$.

Then, $\displaystyle\int dv = \int \frac{dw}{\sqrt{1-w^2}}$ and $v - v_0 = \cos^{-1} w$. We then write $\cos(v - v_0) = w$ or

$$\cos(v - v_0) = \frac{c}{u\sqrt{1+c^2}} \left(1 + u^2/4\right)$$

$$\frac{4u\sqrt{1+c^2}}{c} \cos(v - v_0) = 4 + u^2$$

$$u^2 + 4 - \frac{4u\sqrt{1+c^2}}{c} \cos(v - v_0) = 0.$$

To interpret this equation, replace polar coordinates (u, v) by rectangular coordinates (x, y), $x = u \cos v$ and $y = u \sin v$, expand $\cos(v - v_0)$ to $\cos v \cos v_0 + \sin v \sin v_0$ and complete squares. We have,

$$u^2 + 4 - \frac{4\sqrt{1+c^2}}{c} \cos v_0 \, u \cos v - \frac{4\sqrt{1+c^2}}{c} \sin v_0 \, u \sin v = 0$$

$$x^2 + y^2 + 4 - \frac{4\sqrt{1+c^2}}{c} \cos v_0 \, x - \frac{4\sqrt{1+c^2}}{c} \sin v_0 \, y = 0$$

$$(x - \frac{2\sqrt{1+c^2}}{c} \cos v_0)^2 + (y - \frac{2\sqrt{1+c^2}}{c} \sin v_0)^2 - \frac{4(1+c^2)}{c^2} + 4 = 0$$

$$(x - \frac{2\sqrt{1+c^2}}{c} \cos v_0)^2 + (y - \frac{2\sqrt{1+c^2}}{c} \sin v_0)^2 = \frac{4}{c^2}.$$

This equation represents a circle centered outside H (since $\frac{2\sqrt{1+c^2}}{c} > 2$). Moreover, this circle intersects the boundary circle of H, $x^2 + y^2 = 4$, at *right angles*. Therefore, the geodesics of the hyperbolic plane are radial lines through the origin and arcs of circles centered outside radius 2 which meet H orthogonally.

EXERCISE 4.10. Show that the circle above meets $x^2 + y^2 = 4$ orthogonally. Hint: $(\frac{2\sqrt{1+c^2}}{c})^2 = 2^2 + (\frac{2}{c})^2$.

Note, again, that, given a geodesic and a point off it, there are an infinite number of parallel geodesics through the point. The hyperbolic plane is also a model for the Gauss-Lobachevsky-Bolyai non-Euclidean geometry. Indeed, the hyperbolic plane is *isometric* to the Poincaré plane. That is, from the viewpoint of differential geometry they are indistinguishable. We shall consider this more extensively shortly.

EXERCISE 4.11. Find the geodesics on the stereographic sphere S_N^2. Hints: (1) the patch is v-Clairaut, (2) a convenient substitution may be $\cos v/(1 - \sin v) = c \sec \theta$, (3) just as for the ordinary sphere, geodesics lie on certain kinds of planes — planes which always pass through a particular point of the sphere.

EXERCISE 4.12. Let $\mathbf{x}(r, \theta) = (r \cos \theta, r \sin \theta)$ be the polar coordinate patch for the xy-plane. Show that the inverse to stereographic projection is given by

$$\text{St}^{-1}(r \cos \theta, r \sin \theta) = \left(\frac{2r \cos \theta}{1 + r^2}, \frac{2r \sin \theta}{1 + r^2}, \frac{r^2 - 1}{1 + r^2} \right).$$

Define a metric on the plane by taking $\mathbf{w}_1 \circ \mathbf{w}_2 = \text{St}_*^{-1}(\mathbf{w}_1) \cdot \text{St}_*^{-1}(\mathbf{w}_2)$ where $\cdot$ stands for the usual metric on the sphere. This surface is called the *Stereographic Plane*. Show that

$$E = \frac{4}{(1 + r^2)^2} \qquad F = 0 \qquad G = \frac{4r^2}{(1 + r^2)^2}$$

and $K = +1$. Find the geodesics of the stereographic plane by imitating the approach for the hyperbolic plane. In particular, the substitution $w = c(1 - r^2)/2r\sqrt{1-c^2}$ may be useful. The unit circle and the lines through the origin are geodesics. Why? Finally, show that all geodesics intersect the unit circle, the angle of intersection β obeys $\cos\beta = c$, where c is the Clairaut constant and the formula for a geodesic becomes $(x + \tan\beta \cos\theta_0)^2 + (y + \tan\beta \sin\theta_0)^2 = \sec^2\beta$.

5.5 ISOMETRIES AND CONFORMAL MAPS

Suppose we have two surfaces $M: \mathbf{x}(u,v)$ and $N: \mathbf{y}(r,s)$ given by the indicated patches. If there are smooth functions r and s from the domain of $\mathbf{x}$ to the domain of $\mathbf{y}$, then we may define a map of surfaces $I: M \to N$ by $I(\mathbf{x}(u,v)) = \mathbf{y}(r(u,v), s(u,v))$. We say that I is a *(local) isometry* if

$$E_{\mathbf{x}} = \mathbf{x}_u \circ_{\mathbf{x}} \mathbf{x}_u = I_*(\mathbf{x}_u) \circ_{\mathbf{y}} I_*(\mathbf{x}_u), \quad G_{\mathbf{x}} = \mathbf{x}_v \circ_{\mathbf{x}} \mathbf{x}_v = I_*(\mathbf{x}_v) \circ_{\mathbf{y}} I_*(\mathbf{x}_v).$$

Here, as usual, we assume that orthogonal patches are given, so we implicitly assume that $0 = F_{\mathbf{x}} = I_*(\mathbf{x}_u) \circ_{\mathbf{y}} I_*(\mathbf{x}_v)$. By the chain rule, we know that $I_*(\mathbf{x}_u) = \mathbf{y}_r\, r_u + \mathbf{y}_s\, s_u$ and $I_*(\mathbf{x}_v) = \mathbf{y}_r\, r_v + \mathbf{y}_s\, s_v$, so we get a very explicit check on whether or not a mapping is an isometry. Namely, we must have

$$E_{\mathbf{x}} = \mathbf{x}_u \circ_{\mathbf{x}} \mathbf{x}_u = E_{\mathbf{y}} r_u^2 + G_{\mathbf{y}} s_u^2 \qquad G_{\mathbf{x}} = \mathbf{x}_v \circ_{\mathbf{x}} \mathbf{x}_v = E_{\mathbf{y}} r_v^2 + G_{\mathbf{y}} s_v^2.$$

Of course this is what we obtained in Exercise 3.2.2 for the special case of two parametrizations of the same surface (with I the identity mapping). Moreover, the very same calculations (restricted to our situation where $F = 0$) give $E_{\mathbf{x}} G_{\mathbf{x}} = [r_u s_v - r_v s_u] E_{\mathbf{y}} G_{\mathbf{y}}$, so that plugging into the formula which defines K (and much tedious computation) produces the fundamental

Theorem 5.1. *If $I: M \to N$ is an isometry, then the Gaussian curvatures at corresponding points are equal. That is, $K_M(p) = K_N(I(p))$ for all $p \in M$.*

EXERCISE 5.1. In Exercise 3.4.1 it was shown that the surfaces

$$\mathbf{x}(u,v) = (u\cos v, u\sin v, v) \qquad \mathbf{y}(r,s) = (r\cos s, r\sin s, \ln r)$$

have the same formula for Gaussian curvature, $K = 1/(1+w^2)^2$ where w stands for u and r respectively. Show that these surfaces cannot be isometric however. Hint: an isometry would require $r(u,v) = u$ (up to sign). What goes wrong with $E_{\mathbf{x}} = E_{\mathbf{y}} r_u^2 + G_{\mathbf{y}} s_u^2$?

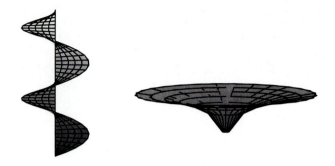

FIGURE 5.8.   Nonisometric surfaces with the same K

If the parameter domains are the same and the functions r and s are the projections $r(u,v) = u$ and $s(u,v) = v$, then an isometry simply means that the metric is preserved

$$E_\mathbf{x} = E_\mathbf{y} \qquad F_\mathbf{x} = F_\mathbf{y} \qquad G_\mathbf{x} = G_\mathbf{y}$$

In fact, many authors say that the functions $r(u,v)$ and $s(u,v)$ give a new parametrization $\bar{\mathbf{y}}(u,v) = \mathbf{y}(r(u,v), s(u,v))$ of the surface N, so that a mapping may be defined by $I(\mathbf{x}(u,v)) = \bar{\mathbf{y}}(u,v)$ and the question of whether I is an isometry is reduced to the simpler criterion above. Of course, the particular mapping must be checked to be a parametrization. In order to avoid this and to make the examples and exercises below more transparent, we have chosen to make the functions r and s explicit.

Clearly, an isometry preserves the lengths of all tangent vectors (since each is a linear combination of $\mathbf{x}_u$ and $\mathbf{x}_v$). Indeed, the word "isometry" comes from the Greek for "same measurement." The Inverse Function Theorem then says that each point of M has a small open neighborhood around it so that I has a smooth inverse function on that neighborhood. In the case where I is one-to-one, onto and has a smooth global inverse $I^{-1}\colon N \to M$, we say that I is a *global* isometry. Although our definition of "isometry" is not the most general one which can be given, it will suffice for all the examples in this book. Let's examine some of these now.

Example 5.1: The Helicoid.
Let two surfaces M and N be defined by patches $\mathbf{x}(u,v) = (u\cos v, u\sin v, v)$ and $\mathbf{y}(r,s) = (\sinh r \cos s, \sinh r \sin s, s)$ respectively. The first patch we recognize as the standard helicoid. Define a map $I\colon M \to N$ by $I(\mathbf{x}(u,v)) = \mathbf{y}(\sinh^{-1} u, v)$. Now, $r = \sinh^{-1} u$ and $s = v$, so the relevant nonzero partials are $r_u = 1/\cosh r = 1/\sqrt{1+u^2}$ and $s_v = 1$. Then we calculate $E_\mathbf{x} = 1 = (1/\cosh^2 r)\cosh^2 r = E_\mathbf{y} r_u^2$ and $G_\mathbf{x} = 1 + u^2 = \cosh^2 r = G_\mathbf{y} s_v^2$. Hence, I is an isometry. Clearly, I is one-to-one and onto with smooth inverse $I^{-1}(\mathbf{y}(r,s)) = \mathbf{x}(\sinh r, s)$, so I is in fact a global isometry.

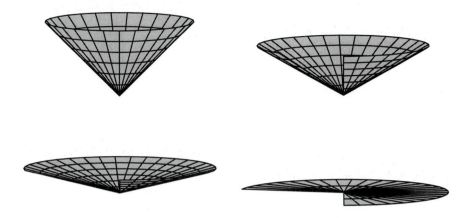

FIGURE 5.9. Unrolling a cone

EXERCISE 5.2. Let $z = a\sqrt{x^2 + y^2}$ be the cone with the parametrization $\mathbf{x}(u, v) = (u \cos v, u \sin v, au)$. Let ϕ denote the vertex angle between the z-axis and the cone. Show that $\sin \phi = 1/\sqrt{1 + a^2}$. If we cut the cone along a ruling line, then we can unroll it onto the plane to get a pie wedge. Let the vertex angle of the pie wedge be θ and show that $\theta = 2\pi \sin \phi$. The unrolling process is given by $\mathbf{y}^t(u, v) =$

$$\left(u\sqrt{1 + ta^2} \, \cos\left(\frac{v}{\sqrt{1 + ta^2}} \right), \; u\sqrt{1 + ta^2} \, \sin\left(\frac{v}{\sqrt{1 + ta^2}} \right), \; au\sqrt{1 - t} \right)$$

for $0 \leq t \leq 1$. Check that this map gives the unrolling above. Show that the map defined by fixing any t, $I(u \cos v, u \sin v, au) =$

$$\left(u\sqrt{1 + ta^2} \, \cos\left(\frac{v}{\sqrt{1 + ta^2}} \right), \; u\sqrt{1 + ta^2} \, \sin\left(\frac{v}{\sqrt{1 + ta^2}} \right), \; au\sqrt{1 - t} \right)$$

is an isometry. Finally, show explicitly that I takes geodesics on the cone to geodesics (i.e. lines) in the plane. Hint: for the last part, a substitution $w = AB/(A - B \tan(v/\sqrt{1 + a^2}))$ may be useful, where $A = c/\sin D$, $B = c/\cos D$ and c and D come from the formula for geodesics on the cone.

EXERCISE 5.3. Invert a cone with vertex angle ϕ so that it sits on the xy-plane. Throw a lasso around the cone with a loop of fixed length L and pull down on the free end of the lasso to tighten the loop around the cone. What happens? More specifically, does the lasso slip or does it stay fixed? Does your answer depend on the vertex angle? Explain. Hint: (1) what kind of curve does the loop become? If the loop stabilizes, what are the only forces acting on it? (2) unroll the cone.

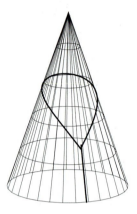

FIGURE 5.10.   Lassoing a cone

EXERCISE 5.4. Find an unrolling map for the cylinder $\mathbf{x}(u, v) = (u, \cos v, \sin v)$ and show that the geodesics on the cylinder go to straight lines in the plane under the isometry I defined by $t = 1$. Hint: for $0 \le t \le 1$, find the line segment joining $(u, \cos v, \sin v)$ to $(u, v, 0)$.

The fact that the unrolling maps I for the cone and cylinder carry geodesics to geodesics is not a phenomenon special to these surfaces. In fact,

Theorem 5.2. *Let $\alpha(t)$ be a geodesic on a surface M and let $I\colon M \to N$ be an isometry. Then, the curve $\beta(t) = I(\alpha(t))$ is a geodesic of N.*

Proof. In case the functions r and s above are the projections, the proof is trivial. On the other hand, if $r = r(u, v)$ and $s = s(u, v)$ are general smooth functions, then the calculations are quite tedious. We shall take a middle road and prove the theorem for the case $r = r(u)$ and $s = s(v)$. This will give the flavor of the calculations involved, but avoid unnecessary complications. With this in mind, take $I(\mathbf{x}(u, v)) = \mathbf{y}(r(u), s(v))$ and $I(\alpha(t)) = I(\mathbf{x}(u(t), v(t))) = \mathbf{y}(r(u(t)), s(v(t)))$. We wish to show that $r(u(t))$ and $s(v(t))$ satisfy the geodesic equations in $\mathbf{y}$'s metric. Our assumption on r and s implies the following relationships between the metrics for $\mathbf{x}$ and $\mathbf{y}$,

$$E_{\mathbf{x}} = E_{\mathbf{y}} r_u^2 \qquad G_{\mathbf{x}} = G_{\mathbf{y}} s_v^2 \qquad E_{\mathbf{y}} = E_{\mathbf{x}} u_r^2 \qquad G_{\mathbf{y}} = G_{\mathbf{x}} v_s^2.$$

Also, the chain rule gives

$$r' = r_u u' \quad s' = s_v v' \quad u' = u_r r' \quad v' = v_s s' \quad r'' = r_{uu} u'^2 + r_u u''.$$

Now we put r and s into the first geodesic expression and substitute the calculations above to get

$$r'' + \frac{E_{\mathbf{y}r}}{2E_{\mathbf{y}}}r'^2 + \frac{E_{\mathbf{y}s}}{E_{\mathbf{y}}}r's' - \frac{G_{\mathbf{y}r}}{2E_{\mathbf{y}}}s'^2$$

$$= r'' + \frac{E_{\mathbf{x}u}u_r^3 + 2E_{\mathbf{x}}u_r u_{rr}}{2E_{\mathbf{x}}u_r^2}r'^2 + \frac{E_{\mathbf{x}v}v_s u_r^2}{E_{\mathbf{x}}u_r^2}r's' - \frac{G_{\mathbf{x}u}v_s^2 u_r}{2E_{\mathbf{x}}u_r^2}s'^2$$

$$= r'' + \frac{E_{\mathbf{x}u}}{2E_{\mathbf{x}}}u_r r'^2 + \frac{u_{rr}}{u_r}r'^2 + \frac{E_{\mathbf{x}v}}{E_{\mathbf{x}}}v_s r's' - \frac{G_{\mathbf{x}u}}{2E_{\mathbf{x}}}\frac{v_s^2 s'^2}{u_r}$$

$$= r'' + \frac{E_{\mathbf{x}u}}{2E_{\mathbf{x}}}u'^2 r_u + \frac{u''}{u_r} - r'' + \frac{E_{\mathbf{x}v}}{E_{\mathbf{x}}}r_u u'v' - \frac{G_{\mathbf{x}u}}{2E_{\mathbf{x}}}\frac{v'^2}{u_r}$$

$$= \frac{1}{u_r}\left[u'' + \frac{E_u}{2E}u'^2 + \frac{E_v}{E}u'v' - \frac{G_u}{2E}v'^2 \right]$$

$$= 0$$

since α satisfies M's geodesic equations. (Note that factoring out $1/u_r$ gave $r_u u_r = 1$ in the numerators of the middle terms.) Therefore, β satisfies the geodesic equations for N. □

EXERCISE 5.5. Show that r and s satisfy the second geodesic equation also.

EXERCISE 5.6. Use the standard polar coordinate patch $\mathbf{x} = (u\cos v, u\sin v)$ for the hyperbolic plane H and define a map $I\colon H \to P$ by

$$I(\mathbf{x}(u,v)) = \left(\frac{4u\sin v}{u^2 + 4u\cos v + 4}, \frac{4 - u^2}{u^2 + 4u\cos v + 4}, 0 \right) = \mathbf{y}(r(u,v), s(u,v))$$

where P is the Poincaré plane parametrized by $\mathbf{y}(r, s) = (r, s)$. Show that I is an isometry. The map I is, in fact, a one-to-one onto map (with smooth inverse) to the upper half plane, so it is a global isometry. The formula above results from the Möbius (or linear fractional) transformation of complex variable theory,

$$T(z) = -i\frac{z - 2}{z + 2},$$

where $z = x + iy$. As we shall see in Chapter 7, complex analysis plays a large role in differential geometry. Here is a sample of the complex approach in terms of something we already know. Consider the hyperbolic plane H with standard polar coordinate u-Clairaut parametrization $\mathbf{x}(u,v) = (u\cos v, u\sin v)$. Since radial lines through the origin are u-parameter curves, they are geodesics. Now, given any point $p = (u_0, v_0)$ in the disk $u = 2$, there exists a linear fractional transformation of the form

$$T(z) = \frac{az + 2\bar{c}}{\frac{c}{2}z + \bar{a}} \qquad |a|^2 - |c|^2 = 1$$

which, in real plane coordinates and with the hyperbolic metric, is an isometry of H and which carries the origin to p. Geodesics through p are then images of radial lines through the origin under T. But it is known that linear fractional transformations take circles and lines to circles and lines. Also, since T is an isometry, and since radial lines through the origin meet $u = 2$ perpendicularly, so must the geodesics through p. Hence, geodesics through p must be circles which meet the boundary circle $u = 2$ at right angles. This is what we calculated earlier.

EXERCISE 5.7. Suppose we parametrize the helicoid by $\mathbf{x}(u, v) = (\sinh r \cos s,$ $\sinh r \sin s, s)$. Smoothly bend the helicoid into the catenoid by defining (from $t = \frac{\pi}{2}$ to $t = \pi$)

$$\mathbf{x}^t(u, v) = (\sin t \sinh u \cos v - \cos t \cosh u \sin v,$$
$$\sin t \sinh u \sin v + \cos t \cosh u \cos v,$$
$$u \cos t + v \sin t).$$

Only at the last step is the bending not one-to-one. Show that, for every $\frac{\pi}{2} \leq t < \pi$, $I(\mathbf{x}(u, v)) = \mathbf{x}^t(u, v)$ is a global isometry and for $t = \pi$, I is a local isometry. Also show that at each stage Gaussian curvature is preserved (since it is determined by the metric) and each intermediate surface is minimal.

Notice that steregraphic projection from the ordinary (i.e. not the stereographic) sphere $S^2 \backslash \{N\}$ to $\mathbb{R}^2$ is not an isometry. Recall that, on tangent vectors, we have

$$\text{St}_*(\mathbf{x}_u) = \left(\frac{-\sin u \cos v}{1 - \sin v}, \frac{\cos u \cos v}{1 - \sin v}, 0 \right)$$
$$\text{St}_*(\mathbf{x}_v) = \left(\frac{\cos u}{1 - \sin v}, \frac{\sin u}{1 - \sin v}, 0 \right).$$

Hence, taking dot products in $\mathbb{R}^3$, we get

$$\text{St}_*(\mathbf{x}_u) \cdot \text{St}_*(\mathbf{x}_u) = \frac{\cos^2 v}{(1 - \sin v)^2}$$
$$= \frac{1}{(1 - \sin v)^2} \mathbf{x}_u \cdot \mathbf{x}_u$$

$$\text{St}_*(\mathbf{x}_v) \cdot \text{St}_*(\mathbf{x}_v) = \frac{1}{(1 - \sin v)^2}$$
$$= \frac{1}{(1 - \sin v)^2} \mathbf{x}_v \cdot \mathbf{x}_v$$

$$\text{St}_*(\mathbf{x}_u) \cdot \text{St}_*(\mathbf{x}_v) = 0.$$

While the factor $1/(1 - \sin v)$ indicates that stretching is taking place and, therefore, St cannot be an isometry, it also shows that the stretching is uniform and angles are preserved. Taking stereographic projection as our basic example, let us weaken the isometry condition by requiring a mapping $I: M \to N$ to satisfy the following proportionality condition. For orthogonal patches $\mathbf{x}$ and $\mathbf{y}$, say that $I(\mathbf{x}(u, v)) = \mathbf{y}(r(u, v), s(u, v))$ is a *conformal* map if $0 = I_*(\mathbf{x}_u) \circ_{\mathbf{y}} I_*(\mathbf{x}_v)$ and

$$E_{\mathbf{x}} = \lambda(u, v)^2 \left[E_{\mathbf{y}} r_u^2 + G_{\mathbf{y}} s_u^2 \right] \qquad G_{\mathbf{x}} = \lambda(u, v)^2 \left[E_{\mathbf{y}} r_v^2 + G_{\mathbf{y}} s_v^2 \right]$$

for a function $\lambda(u, v)$ called the *scaling factor*. Again note that, when $r(u, v) = u$ and $s(u, v) = v$, I is conformal exactly when the metrics are proportional at each point $p \in M$ and its image $I(p) \in N$,

$$E_{\mathbf{x}} = \lambda(u, v)^2 E_{\mathbf{y}} \qquad G_{\mathbf{x}} = \lambda(u, v)^2 E_{\mathbf{y}}.$$

EXERCISE 5.8. Show that if I is conformal, then angles between tangent vectors are preserved. Hint: it is sufficient to show this for $\mathbf{x}_u$ and $\mathbf{x}_v$.

Note that stereographic projection is a conformal mapping from the sphere to the plane with the metrics induced by $\mathbb{R}^3$. This fact will prove to be essential for the complex analytic approach to minimal surfaces in Chapter 7. Furthermore, we shall see in that chapter that the following exercise has a far-reaching generalization.

EXERCISE 5.9. Recall that the Gauss map of a surface $M \colon \mathbf{x}(u, v) \subseteq \mathbb{R}^3$ is denoted by $G \colon M \to S^2$ and defined by $G(\mathbf{x}(u, v)) = U(u, v)$, where U is the unit normal of M. For the helicoid $\mathbf{x}(u, v) = (u \cos v, u \sin v, v)$, the catenoid $\mathbf{x}(u, v) = (\cosh u \cos v, \cosh u \sin v, u)$ and Enneper's surface $\mathbf{x}(u, v) = (u - u^3/3 + uv^2, -v + v^3/3 - vu^2, u^2 - v^2)$, show that the Gauss maps are conformal with respective scaling factors $\lambda(u, v) = 1 + u^2$, $\cosh^2 u$, $(1 + u^2 + v^2)^2/2$.

EXERCISE 5.10. Let $\mathbf{x}(u, v) = (R \cos u \cos v, R \sin u \cos v, R \sin v)$ denote the usual coordinate patch for the sphere of radius R. Define

$$\begin{aligned} I(\mathbf{x}(u, v)) &= \left(u, \ln \tan\left(\frac{v}{2} + \frac{\pi}{4}\right), 0\right) \\ &= \mathbf{y}(r(u, v), s(u, v)) \end{aligned}$$

where $\mathbf{y}(r, s) = (r, s, 0)$ is a patch for a strip in the coordinate plane. Show that I is conformal with scaling factor $\lambda(u, v) = R \cos v$.

Suppose $\alpha = \mathbf{x}(u, v(u))$ is a curve on the sphere which intersects every longitude (i.e. a v-parameter curve) at a constant angle. Show that $v(u) = Cu/R^2$ and show that such a curve has a straight line as its I-image. The map I is called the **Mercator Projection** of a sphere onto a rectangle. In particular, the Mercator projection may be used to construct a map of the Earth. The curve property we have mentioned allowed sailors in earlier times to plot courses on a map as straight lines and then steer the plotted courses by keeping constant compass headings (i.e. making a constant angle with longitudes). Although these courses were not geodesics, the ease in navigation made up for the greater distance. For a nice discussion of the mathematics of map making see [RW].

5.6 Geodesics and MAPLE

It has always been hard for geometers to visualize geodesics. This is due to the fact that solutions of the geodesic equations must generally be obtained numerically and then, since the solutions are in terms of the parameters u and

v, plugged into the surface parametrization. In years past, this process would require a great deal of complicated programming to mix the numerical analysis with 3D-graphics. Today however, a short and simple MAPLE procedure can accomplish the same task. Geodesics may be graphed on their surfaces and may give just the right kind of intuitive information to allow rigorous analysis. For example, the picture Figure 5.6 provides the intuitive insight necessary to solving Exercise 2.10. Of course, it is a simple matter for MAPLE to generate the geodesic equations for a given parametrization. The following procedures do exactly this. First, we need the metric and all the preliminary procedures.

```
> with(plots):
```

```
> dp := proc(X,Y)
>       X[1]*Y[1]+X[2]*Y[2]+X[3]*Y[3];
>       end:
```

```
> nrm := proc(X)
>       sqrt(dp(X,X));
>       end:
> xp := proc(X,Y)
>       local a,b,c;
>       a := X[2]*Y[3]-X[3]*Y[2];
>       b := X[3]*Y[1]-X[1]*Y[3];
>       c := X[1]*Y[2]-X[2]*Y[1];
>       [a,b,c];
>       end:
```

```
> Jacf := proc(X)
>       local Xu,Xv;
>       Xu := [diff(X[1],u),diff(X[2],u),diff(X[3],u)];
>       Xv := [diff(X[1],v),diff(X[2],v),diff(X[3],v)];
>       simplify([Xu,Xv]);
>       end:
```

```
> EFG := proc(X)
>       local E,F,G,Y;
>       Y := Jacf(X);
>       E := dp(Y[1],Y[1]);
>       F := dp(Y[1],Y[2]);
>       G := dp(Y[2],Y[2]);
>       simplify([E,F,G]);
>       end:
```

Once we have the metric coefficients E, F and G, we can write the geodesic equations. As usual, for simplicity, we take orthogonal patches so that $F = 0$.

```
> geoeq:=proc(X)
>       local M,eq1,eq2;
>       M:=EFG(X);
>       eq1:=diff(u(t),t$2)+subs({u=u(t),v=v(t)},
         diff(M[1],u)/(2*M[1]))*diff(u(t),t)^2
         +subs({u=u(t),v=v(t)},diff(M[1],v)/(M[1]))*diff(u(t),t)*diff(v(t),t)
         -subs({u=u(t),v=v(t)},diff(M[3],u)/(2*M[1]))*diff(v(t),t)^2=0;
>       eq2:=diff(v(t),t$2)-subs({u=u(t),v=v(t)},
         diff(M[1],v)/(2*M[3]))*diff(u(t),t)^2
         +subs({u=u(t),v=v(t)},diff(M[3],u)/(M[3]))*diff(u(t),t)*diff(v(t),t)
         + subs({u=u(t),v=v(t)},diff(M[3],v)/(2*M[3]))*diff(v(t),t)^2=0;
>       eq1,eq2;
>       end:
```

Finally, we come to the centerpiece of this section, a MAPLE procedure which plots geodesics on surfaces. This procedure has a bewildering number of inputs, but each serves a function. The first input is, of course, the parametrization of the surface in question. The next four inputs give bounds for the parameters u and v of the surface. The inputs $u0$ and $v0$ give an initial point on the surface for the geodesic while $Du0$ and $Dv0$ give an initial direction. The input n is a *vector* whose first two components bound the geodesic parameter and whose third component makes the picture smoother as it increases (but also makes the rendering slower). The input "gr" is a vector which allows the grid size to be changed from its usual 25 by 25. Again, the larger the grid the better the picture and the slower the rendering. Finally, the inputs theta and phi orient the output picture.

```
> plotgeo:=proc(X,ustart,uend,vstart,vend,u0,v0,Du0,Dv0,n,gr, theta,phi)
>       local sys,desys,dequ,deqv,listp,j,geo,plotX;
>       sys:=geoeq(X);
>       desys:=dsolve({sys,u(0)=u0,v(0)=v0,D(u)(0)=Du0,D(v)(0)=Dv0},
            {u(t),v(t)}, type=numeric, output=listprocedure);
>       dequ:=subs(desys,u(t)); deqv:=subs(desys,v(t));
>       listp:=[seq(subs({u=dequ(j/n[3]),v=deqv(j/n[3])},X), j=n[1]..n[2])];
>       geo:=spacecurve({listp}, color=black,thickness=2):
>       plotX:=plot3d(X,u=ustart..uend,v=vstart..vend,grid= [gr[1],gr[2]]):
>       display({geo,plotX},style=wireframe,scaling=constrained,
            orientation=[theta,phi]);
>       end:
```

Let's consider the torus. The geodesic equations of Example 2.2 are found by the following.

> Tor:=[(5+cos(u))*cos(v),(5+cos(u))*sin(v),sin(u)];

> geoeq(Tor);

$$\frac{d^2u}{dt^2} + (5 + \cos u) \sin u \left(\frac{dv}{dt}\right)^2 = 0 \qquad \frac{d^2v}{dt^2} - \frac{2 \sin u}{5 + \cos u} \frac{du}{dt} \frac{dv}{dt} = 0$$

The geodesic of Figure 5.3 is then given by

> plotgeo(Tor,0,2*Pi,0,2*Pi,Pi/2,0,0,1,[0,240,15],[20,20],0,60);

Notice that we are starting the geodesic at $(u, v) = (\pi/2, 0)$ with direction $(0, 1)$ along the top parallel and we let the geodesic parameter run from 0 to 240 with a smoothness coefficient of $n = 15$. The grid for the torus is 20 by 20 and the picture is viewed from $\theta = 0$ and $\phi = 60$ (in degrees). The picture gives a beautiful illustration of the Clairaut analysis in Example 2.2. Namely, by rotating the picture (in the viewing window) to look from the top down, we can see that the geodesic stays on the outside of the torus.

EXERCISE 6.1. For the Whirling Witch of Agnesi of Exercise 2.9 and Figure 5.5, plot the geodesic referred to by the following.

> Witch:=[2*tan(u),2*cos(u)^2*cos(v),2*cos(u)^2*sin(v)];
> plotgeo(Witch,-1.2,1.2,0,2*Pi,Pi/3,0,0,1,[0,500,15],[20,15],90,60);

Can you see the effect of the Clairaut relation?

EXERCISE 6.2. Plot geodesics for various initial directions and starting points on the sphere, cylinder, cone and paraboloid using

> Sph:=[cos(u)*cos(v),cos(v)*sin(u),sin(v)];

> Cyl:=[cos(v),sin(v),u];

> Co:=[u*cos(v),u*sin(v),u];

> Par:=[u*cos(v),u*sin(v),u^2];

Relate your pictures to examples in the book. In particular, relate the cone and paraboloid to Exercises 2.8 and 2.10 respectively.

EXERCISE 6.3. Plot geodesics for various initial directions and starting points on the catenoid and on the hyperboloid of one sheet.

EXERCISE 6.4. For Exercise 5.2, plot a geodesic with the same initial conditions on the unrolling cone for $t = 0$, $t = 0.25$, $t = 0.5$, $t = 0.75$ and $t = 1$. This illustrates that geodesics on the cone go to straight lines in the plane under the isometric unrolling.

EXERCISE 6.5. For Exercise 5.7, as the helicoid is isometrically bent into the catenoid, plot a geodesic with the same initial conditions for $t = \pi/2$, $t = 2\pi/3$, $t = 5\pi/6$, and $t = \pi$.

Finally, for surfaces not in $\mathbb{R}^3$, but which are plane regions with *conformal metric*, we can still write MAPLE procedures to calculate the geodesic equations. We again need the metric first. The inputs now are a parametrization X and a scaling function g.

```
> EFGconf:=proc(X,g)
>       local E,F,G,Y;
>       Y := Jacf(X);
>       E := dp(Y[1],Y[1])/g^2;
>       F := dp(Y[1],Y[2])/g^2;
>       G := dp(Y[2],Y[2])/g^2;
>       simplify([E,F,G]);
>       end:
```

Now the geodesic equations for the conformal metric are given by the following procedure.

```
> geoeqconf:=proc(X,g)
>       local M,eq1,eq2;
>       M:=EFGconf(X,g);
>       eq1:=diff(u(t),t$2)+subs({u=u(t),v=v(t)},
         diff(M[1],u)/(2*M[1]))*diff(u(t),t)^2
         +subs({u=u(t),v=v(t)},diff(M[1],v)/(M[1]))*diff(u(t),t)*diff(v(t),t)
         -subs({u=u(t),v=v(t)},diff(M[3],u)/(2*M[1]))*diff(v(t),t)^2=0;
>       eq2:=diff(v(t),t$2)-subs({u=u(t),v=v(t)},
         diff(M[1],v)/(2*M[3]))*diff(u(t),t)^2
         +subs({u=u(t),v=v(t)},diff(M[3],u)/(M[3]))*diff(u(t),t)*diff(v(t),t)
         + subs({u=u(t),v=v(t)},diff(M[3],v)/(2*M[3]))*diff(v(t),t)^2=0;
>       eq1,eq2;
>       end:
```

EXERCISE 6.6. Plot the geodesic on the Poincaré plane given in Figure 5.7. In other words, modify the plotgeo procedure to handle this situation.

Chapter 6

HOLONOMY
AND THE
GAUSS-BONNET THEOREM

6.1 INTRODUCTION

Euclid's *parallel postulate* states that through any point off a given line there is a unique parallel line. The following exercise reminds us of how such a postulate may be used to derive essential geometric properties.

EXERCISE 1.1. Use the parallel postulate to prove that the sum of the angles of a triangle is π. Hint: draw a line through a vertex of the triangle parallel to the opposite side.

At one time it was thought that the parallel postulate could be derived from other natural axioms of Euclidean geometry. It was only through the work of Lobachevsky, Gauss and Bolyai in the 1800's that mathematicians realized just how misguided this attempt had been. Perhaps it shouldn't be such a surprise that Gauss would see the necessity of actually *postulating* the parallel postulate. His investigations into the geometry of surfaces led not only to his Theorem Egregium concerning Gauss curvature (which, in fact, had previously been defined by O. Rodrigues), but to an understanding of various fundamental geometric quantities on surfaces. In particular, the "lines" of a surface may be considered to be its geodesics and Gauss could see that geodesics did not always follow the dictates of the parallel postulate.

EXERCISE 1.2. Discuss the parallel postulate's validity for the geometries of the sphere and the hyperbolic plane where lines are geodesics.

As the previous exercise demonstrates, the parallel postulate may fail in two ways. Because we have seen above that the parallel postulate may be used to calculate the sum of angles in a triangle, we might expect that its failure would lead to variability among such calculations. Indeed, we find that there are three possibilities for the sum of the angles of a triangle.

(1) **Euclidean.** In the Euclidean plane , the angles of a triangle have $\sum_{i=1}^{3} \phi_i = \pi$.

(2) **Hyperbolic (Gauss–Lobachevsky–Bolyai).** In the hyperbolic plane, the angles of a triangle have $\sum_{i=1}^{3} \phi_i < \pi$.

(3) **Elliptic (Riemann).** In the sphere, the angles of a triangle have $\sum_{i=1}^{3} \phi_i > \pi$.

In this chapter, we will see how Gauss curvature affects even the most basic quantities of geometry such as angle sums of triangles. Further, we shall see that Gauss curvature makes its presence felt even in the most basic of geometric notions, the notion of when two vectors are parallel.

Before we begin, we must recall how to integrate on a surface. Readers of Chapter 4 may skip the following redundant discussion. First, consider how we compute the surface area of a surface. For example, the surface area of an R-sphere in $4\pi R^2$ and the intuitive reasoning behind this is as follows. Let M be a surface and consider a patch **x**. To approximate the area of the patch we might find the area of each parallelogram

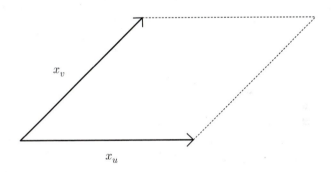

FIGURE 6.1. An infinitesimal piece of area

and then sum them up. Of course, as usual in mathematics, we would then take a limit to make our approximations approach the true area. The limit then gives a continuous sum, the integral. It is important to note that we are *defining* the integral here. For sufficiently complicated "surfaces" outside our restricted smooth definitions, this notion of integration may fail. The area of such a parallelogram may be found by Exercise 1.3.7. Namely, Area = $|\mathbf{x}_u \times \mathbf{x}_v|$. Hence, we have

$$\text{Area of patch } \mathbf{x} = \int_{v_0}^{v_1} \int_{u_0}^{u_1} |\mathbf{x}_u \times \mathbf{x}_v| \, du \, dv.$$

Example 1.1: The R-sphere.
Recall that $|\mathbf{x}_u \times \mathbf{x}_v| = R^2 \cos v$ and $0 \le u < 2\pi$, $-\frac{\pi}{2} \le v \le \frac{\pi}{2}$.

$$
\begin{aligned}
\text{Area of } R\text{-sphere} &= \int_0^{2\pi} \int_{-\frac{\pi}{2}}^{\frac{\pi}{2}} R^2 \cos v \, dv \, du \\
&= \int_0^{2\pi} \Big|_{-\frac{\pi}{2}}^{\frac{\pi}{2}} R^2 \sin v \, du \\
&= \int_0^{2\pi} 2R^2 \, du \\
&= \Big|_0^{2\pi} 2R^2 u \\
&= 4\pi R^2.
\end{aligned}
$$

Note that, by Lagrange's Identity (Exercise 1.3.4), $|\mathbf{x}_u \times \mathbf{x}_v| = \sqrt{EG - F^2}$.

EXERCISE 1.3. Show that the surface area of the torus is $4\pi^2 r R$.

EXERCISE 1.4. In calculus, two of the usual integrals used to calculate surface area are:

(1) **Surfaces of revolution:** $SA = \int_a^b 2\pi f(x) \sqrt{1 + f'(x)^2} \, dx$,

(2) **Graphs of functions** $z = f(x, y)$: $SA = \int_a^b \int_c^d \sqrt{1 + f_x^2 + f_y^2} \, dx \, dy$.

Show that these formulas may be derived from the definition of surface area given above.

Now, how do we integrate a function over a surface? A function simply gives different weights to the points in a region and then "adds up" the weighted area. Hence, we simply multiply the function by $|\mathbf{x}_u \times \mathbf{x}_v|$ and integrate.

Definition 1.2.

$$
\int_M f \overset{\text{def}}{=} \int_{\mathbf{x}} \int f \, |\mathbf{x}_u \times \mathbf{x}_v| \, du \, dv
$$

$$
= \int_{\mathbf{x}} \int f \sqrt{EG - F^2} \, du \, dv.
$$

We shall be interested in the integral of a specific function — the Gaussian curvature K. The integral $\int_M K$ is called the *Total Gaussian Curvature* of M (or of a patch $\mathbf{x}$).

Examples 1.3.

(1) **The R-sphere.** The Gauss curvature is the constant $K = 1/R^2$, so

$$\int_M K = \frac{1}{R^2} \int \int |\mathbf{x}_u \times \mathbf{x}_v| \, du \, dv$$

$$= \frac{1}{R^2} \cdot \text{Area}$$

$$= \frac{1}{R^2} \cdot 4\pi R^2$$

$$= 4\pi.$$

(2) **The Pseudosphere:** The Gauss curvature is constant here as well, $K = -1/c^2$. Hence,

$$\int_M K = -\frac{1}{c^2} \cdot \text{Area} = -\frac{1}{c^2} \cdot 2\pi c^2 = -2\pi.$$

EXERCISE 1.5. Find the surface area of the pseudosphere.

EXERCISE 1.6. Find the total Gaussian curvature of the torus and of the catenoid.

6.2 The Covariant Derivative Revisited

In our initial approach to the geometry of surfaces in $\mathbb{R}^3$ we used the covariant derivative to define the shape operator and the various basic curvature invariants followed from this. Later, we defined Gauss curvature for surfaces not in $\mathbb{R}^3$ by taking a particular formula for the curvature and using it as a definition. In this way we circumvented defining a covariant derivative for such abstract surfaces. In this section, we will see that defining and understanding a notion of covariant derivative in a more abstract context is worthwhile for reasons beyond the definition of curvature alone.

To begin, let's take $\mathbf{x}$ to be a patch with (as usual) $F = 0$ and suppose there is a closed curve (assumed to be unit speed or, at least, constant speed) $\alpha(t) = \mathbf{x}(a(t), b(t))$. Let $\mathcal{E}_1 = \mathbf{x}_u/\sqrt{E}$ and $\mathcal{E}_2 = \mathbf{x}_v/\sqrt{G}$ and note that $\mathcal{E}_1 \cdot \mathcal{E}_1 = 1$ and $\mathcal{E}_2 \cdot \mathcal{E}_2 = 1$ (with $\mathcal{E}_1 \cdot \mathcal{E}_2 = 0$). Thus, $\mathcal{E}_1$ and $\mathcal{E}_2$, which we say are *fields of vectors* along the curve, form an orthonormal basis for the tangent space at any point $\mathbf{x}(u, v)$. For this reason the basis of vector fields $\{\mathcal{E}_1, \mathcal{E}_2\}$ is called a *moving frame* on the patch. (More generally, such a basis may be found even in the case of a nonorthogonal patch.) Further, $\mathcal{E}_1$, $\mathcal{E}_2$ and U form an orthonormal basis for $\mathbb{R}^3$. Now, just from this information and what we know about covariant derivatives, we will analyze how the vector fields $\mathcal{E}_1$ and $\mathcal{E}_2$ change along α.

First, we will consider the most concrete case of a surface in 3-space. This will allow us to see how the covariant derivatives of 3-space and *the surface itself* relate to each other. Indeed, the latter is defined in terms of the former. Then, once we understand what the covariant derivative tells us in this situation, we can eliminate 3-space from the discussion and abstract the qualities of the covariant derivative to other surfaces.

We will use the notation $\nabla^{\mathbb{R}^3}$ for the covariant derivative (or Jacobian matrix) in $\mathbb{R}^3$ and ∇ for either the projection of $\nabla^{\mathbb{R}^3}$ to the tangent plane of the surface or simply for the covariant derivative of M considered without regard to 3-space. Let $Z = (z^1, z^2, z^3)$ be a vector field in $\mathbb{R}^3$ defined by functions $z^i \colon \mathbb{R}^3 \to \mathbb{R}$. Recall that the $\mathbb{R}^3$-covariant derivative $\nabla_{\mathbf{v}}^{\mathbb{R}^3} Z$ is simply a coordinatewise version of the directional derivative (see Chapter 2.2). Then, for $M \subseteq \mathbb{R}^3$, the projection of $\nabla_{\mathbf{v}}^{\mathbb{R}^3} Z$ onto the tangent plane $T_p(M)$ is obtained by subtracting off its normal component,

$$\nabla_{\mathbf{v}} Z = \nabla_{\mathbf{v}}^{\mathbb{R}^3} Z - (\nabla_{\mathbf{v}}^{\mathbb{R}^3} Z \cdot U)\, U.$$

This is then the *definition* of the covariant derivative intrinsic to the surface itself.

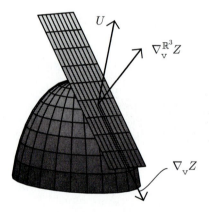

FIGURE 6.2. Definition of $\nabla_{\mathbf{v}} Z$

EXERCISE 2.1. Show that $\nabla_{\alpha'}^{\mathbb{R}^3} \alpha' = \alpha''$. Hence, α is a *geodesic* if $\nabla_{\alpha'} \alpha' = 0$. Hint: Consider $\frac{d\alpha^i}{dt}$ as a function of the 3 variables α^1, α^2 and α^3. This allows the formula for the chain rule to apply to $\alpha'[f] = \frac{d(f \circ \alpha)}{dt}$, where $f = \frac{d\alpha^i}{dt}$. Then see Chapter 2.

Now we write $\nabla_{\alpha'}^{\mathbb{R}^3} \mathcal{E}_1 = \omega_{11} \mathcal{E}_1 + \omega_{21} \mathcal{E}_2 + s_1 U$ and $\nabla_{\alpha'}^{\mathbb{R}^3} \mathcal{E}_2 = \omega_{12} \mathcal{E}_1 + \omega_{22} \mathcal{E}_2 + s_2 U$, where the ω's are functions of the curve α's parameter t. Using the product rule coordinatewise and the definition of the $\mathbb{R}^3$ covariant derivative, we compute

$$0 = \alpha'[\mathcal{E}_1 \cdot \mathcal{E}_1] = 2\nabla_{\alpha'}^{\mathbb{R}^3} \mathcal{E}_1 \cdot \mathcal{E}_1 = \omega_{11}.$$

In general, the formula $\alpha'[V \cdot W] = \nabla_{\alpha'}^{\mathbb{R}^3} V \cdot W + V \cdot \nabla_{\alpha'}^{\mathbb{R}^3} W$ holds for vector fields $V = (v^1, v^2, v^3)$ and $W = (w^1, w^2, w^3)$. This may be seen by the following computation (which is reminiscent of the proof of Lemma 2.4.1):

$$\begin{aligned}
\alpha'[V \cdot W] &= \alpha' \left[\sum v^i w^i \right] \\
&= \sum \alpha'[v^i] w^i + v^i \alpha'[w^i] \\
&= \sum (\nabla_{\alpha'}^{\mathbb{R}^3} V)^i w^i + v^i (\nabla_{\alpha'}^{\mathbb{R}^3} W)^i \\
&= \nabla_{\alpha'}^{\mathbb{R}^3} V \cdot W + V \cdot \nabla_{\alpha'}^{\mathbb{R}^3} W
\end{aligned}$$

Similarly, since $0 = \alpha'[\mathcal{E}_1 \cdot \mathcal{E}_2] = \nabla_{\alpha'}^{\mathbb{R}^3} \mathcal{E}_1 \cdot \mathcal{E}_2 + \mathcal{E}_1 \cdot \nabla_{\alpha'}^{\mathbb{R}^3} \mathcal{E}_2$, we have $\omega_{21} = -\omega_{12}$.

EXERCISE 2.2. Show $s_1 = S(\alpha') \cdot \mathcal{E}_1$ and $s_2 = S(\alpha') \cdot \mathcal{E}_2$ where $S(\alpha')$ is the shape operator applied to α' and s_2 is the U-coefficient of $\nabla_{\alpha'}^{\mathbb{R}^3} \mathcal{E}_2$.

Finally, we obtain

$$\nabla_{\alpha'}^{\mathbb{R}^3} \mathcal{E}_1 = \omega_{21} \mathcal{E}_2 + (S(\alpha') \cdot \mathcal{E}_1) U$$
$$\nabla_{\alpha'}^{\mathbb{R}^3} \mathcal{E}_2 = -\omega_{21} \mathcal{E}_1 + (S(\alpha') \cdot \mathcal{E}_2) U.$$

Therefore, remembering that the unadorned notation ∇ refers to the projection of $\nabla^{\mathbb{R}^3}$ onto the surface's tangent planes, we have

† $$\nabla_{\alpha'} \mathcal{E}_1 = \omega_{21} \mathcal{E}_2 \qquad \nabla_{\alpha'} \mathcal{E}_2 = -\omega_{21} \mathcal{E}_1.$$

This description of the covariant derivative ∇ in terms of the coordinatewise directional derivatives in $\mathbb{R}^3$ (i.e. $\nabla^{\mathbb{R}^3}$) implies properties such as the following.

EXERCISE 2.3. Show

(1) Sum Rule. $\nabla_{f\alpha' + g\beta'} Z = f \nabla_{\alpha'} Z + g \nabla_{\beta'} Z$.

(2) Leibniz Rule. $\nabla_{\alpha'} f Z = \frac{d(f \circ \alpha)}{dt} Z + f \nabla_{\alpha'} Z$.

(3) Commutation Rule. If $\mathbf{x}(u, v)$ is an orthogonal patch, then "mixed second partials are equal"
$$\nabla_{\mathbf{x}_u} \mathbf{x}_v = \nabla_{\mathbf{x}_v} \mathbf{x}_u .$$

(4) Compatibility Rule. For a tangent vector $\mathbf{w}$, $E = \mathbf{x}_u \cdot \mathbf{x}_u$, $F = 0$ (say) and $G = \mathbf{x}_v \cdot \mathbf{x}_v$,
$$\mathbf{w}[E] = 2 \nabla_{\mathbf{w}} \mathbf{x}_u \cdot \mathbf{x}_u \qquad \mathbf{w}[G] = 2 \nabla_{\mathbf{w}} \mathbf{x}_v \cdot \mathbf{x}_v.$$

These are the types of properties which we can abstract from the situation of a surface M in $\mathbb{R}^3$. In particular, the last two rules generalize to give a useful covariant derivative associated to the so-called *Riemannian connection*. We shall need these later when we discuss the influence of Gaussian curvature on surfaces like the hyperbolic plane.

6.3 Parallel Vector Fields and Holonomy

Suppose now that V_0 is a unit tangent vector at $\alpha(t)$. One of the first questions of classical differential geometry is the following. Along the curve α, what tangent vectors should be considered parallel to V_0? After all, the notion of vectors or lines being parallel is basic to geometry, so it should not be so surprising that the question of parallelism arises in differential geometry as well. Because the covariant derivative $\nabla_{\alpha'} V$ tells us how the (tangent) vector field V changes along α, the condition $\nabla_{\alpha'} V = 0$ precisely expresses the desired parallelism of all the vectors comprising V. In other words, a vanishing covariant derivative implies (as it should) that V changes *only* in the normal direction to the surface and so is *unchanging from the viewpoint of residents of the surface*. Therefore, we say that V is a *parallel vector field along* α if $\nabla_{\alpha'} V = 0$.

EXERCISE 3.1. Show that a parallel vector field has constant length. Hint: use the Leibniz rule.

Of course we need to know that parallel vector fields actually exist. In fact, we can *uniquely* determine a parallel vector field along a curve from initial data alone.

Theorem 3.1. *Let M be a surface with covariant derivative ∇ and suppose $\alpha\colon I \to M$ is a curve on M. For any tangent vector V_0 at $\alpha(0)$, there exists a parallel vector field V along α with $V_{\alpha(0)} = V_0$.*

Proof. Without loss of generality take V_0 to have unit length. Because a parallel vector field must have constant length, in fact every vector in the field will have length one as well. Then we may write $V = \cos\theta\, \mathcal{E}_1 + \sin\theta\, \mathcal{E}_2$ where θ is the angle from V to $\mathcal{E}_1$. The desired condition $\nabla_{\alpha'} V = 0$ then becomes (by product and chain rules)

$$0 = -\sin\theta\, \frac{d\theta}{dt}\, \mathcal{E}_1 + \cos\theta\, \nabla_{\alpha'}\, \mathcal{E}_1 + \cos\theta\, \frac{d\theta}{dt}\, \mathcal{E}_2 + \sin\theta\, \nabla_{\alpha'}\, \mathcal{E}_2.$$

Using our previous calculations of the covariant derivatives of $\mathcal{E}_1$ and $\mathcal{E}_2$ along α (formulas †), we obtain

$$0 = -\sin\theta\left[\omega_{21} + \frac{d\theta}{dt}\right]\mathcal{E}_1 + \cos\theta\left[\omega_{21} + \frac{d\theta}{dt}\right]\mathcal{E}_2.$$

Since $\sin\theta$ and $\cos\theta$ cannot be zero simultaneously, we must have $\frac{d\theta}{dt} = -\omega_{21}$ or

$$\theta(t) = \theta(0) - \int \omega_{21}\, dt.$$

This formula then defines θ and, hence, the parallel vector field V. $\qquad\square$

The construction of a parallel vector field V along α from an initial V_0 allows us to speak of the *parallel transport* of V_0 along α. That is, the parallel transport of V_0 along α to a point $\alpha(t)$ is uniquely determined (and defined) to be $V_{\alpha(t)}$. The angle of rotation (which only depends on α and $\theta(0)$), $-\int \omega_{21}\, dt$, is called the *holonomy* along α.

EXERCISE 3.2. Show that a parallel vector field V along a geodesic α makes a constant angle with α'. More generally, suppose V is a parallel vector field along α and W is a vector field along α of constant length. Show that W is parallel if and only if the angle between V and W is constant. Use this to show that the holonomy around a curve does not depend on the parallel vector field along the curve.

EXERCISE 3.3. Prove the following formula for *geodesic curvature* (see Theorem 5.4.1).

$$\kappa_g = \frac{d\theta}{dt} + \omega_{21}$$

Here, θ is the angle between α' and $\mathcal{E}_1$. Hints: (1) write $\alpha' = \cos\theta\, \mathcal{E}_1 + \sin\theta\, \mathcal{E}_2$ just as for V in Theorem 3.1 and carry out the proof of the theorem to get

$$\nabla_{\alpha'}\alpha' = \left[\frac{d\theta}{dt} + \omega_{21} \right] \left[-\sin\theta\, \mathcal{E}_1 + \cos\theta\, \mathcal{E}_2 \right].$$

(2) Note that $\alpha''_{\tan} = \nabla_{\alpha'}\alpha'$ and compare the formula above to the geodesic curvature formula of Theorem 5.4.1. (3) Why must it be true that, up to sign, $T \times U = -\sin\theta\, \mathcal{E}_1 + \cos\theta\, \mathcal{E}_2$?

EXERCISE 3.4. Let $\alpha\colon [a, b] \to M$ be a unit speed curve on a surface. Take a frame along the curve, $\{T, V = T \times U, U\}$, where U is the unit normal of M. Show that the natural ("Frenet") equations for this frame are

$$T' = \kappa_g V + k U$$
$$V' = -\kappa_g T - \tau_g U$$
$$U' = -k T - \tau_g V.$$

Here, k is the normal curvature of T on M, κ_g is the geodesic curvature of α and τ_g is defined by $\tau_g = -U' \cdot V = V' \cdot U$ and is called the *geodesic torsion* of α. Recall that a curve α is a line of curvature of M if T is always an eigenvector of the shape operator of M. Using the natural equations above, show that $\tau_g = 0$ if and only if α is a line of curvature.

EXERCISE 3.5. Let $U \cdot N = \cos\theta$, where N is the principal normal of the unit speed curve α. If τ denotes the usual torsion of α, show that the geodesic torsion obeys

$$\tau_g = -\left(\tau + \frac{d\theta}{dt} \right).$$

Show that, if α is a geodesic, then $\tau_g = -\tau$. Also, show that the converse of this statement does not hold by showing that each term of the above equation vanishes for any latitude circle on the sphere.

Notice what the proof of Theorem 3.1 and Exercise 3.2 above say in the special case of a unit speed u-parameter curve. In this situation, $\alpha' = \mathcal{E}_1$, so we have that either the parameter curve is a geodesic, in which case $\theta = \theta(0)$, or the parameter curve is not a geodesic, in which case the initial vector V_0 is rotated by the holonomy θ (from the viewpoint of 3-space) as it moves along the curve. It is now time for the *prime* example.

Example 3.1. Consider the R-sphere with patch

$$\mathbf{x}(u, v) = (R \cos u \cos v, R \sin u \cos v, R \sin v).$$

The parameter curve tangent vectors are $\mathbf{x}_u = (-R \sin u \cos v, R \cos u \cos v, 0)$ and $\mathbf{x}_v = (-R \cos u \sin v, -R \sin u \sin v, R \cos v)$, while the unit normal is given by $U = (\cos u \cos v, \sin u \cos v, \sin v)$. An orthonormal basis for the tangent plane is then

$$\mathcal{E}_1 = (-\sin u, \cos u, 0) \qquad \mathcal{E}_2 = (-\cos u \sin v, -\sin u \sin v, \cos v).$$

Now take a latitude circle on the sphere at latitude v_0,

$$\alpha(u) = (R \cos u \cos v_0, R \sin u \cos v_0, R \sin v_0).$$

Note that $|\alpha'| = R \cos v_0$ is constant, so the covariant derivatives of $\mathcal{E}_1$ and $\mathcal{E}_2$ with respect to α' truly describe the changes of $\mathcal{E}_1$ and $\mathcal{E}_2$ along α. Further, since the parametrization of α is compatible with that of $\mathcal{E}_1$ and $\mathcal{E}_2$ along α *and* $\alpha' = \mathbf{x}_u$, the $\mathbb{R}^3$-covariant derivatives are simply derivatives with respect to u,

$$\nabla^{\mathbb{R}^3}_{\alpha'(u)} \mathcal{E}_1 = (-\cos u, \ -\sin u, \ 0)$$
$$\nabla^{\mathbb{R}^3}_{\alpha'(u)} \mathcal{E}_2 = (\sin u \sin v_0, \ -\cos u \sin v_0, \ 0).$$

In order to find $\nabla_{\alpha'} \mathcal{E}_1 = \omega_{21} \mathcal{E}_2$, we must decompose $\nabla^{\mathbb{R}^3}_{\alpha'} \mathcal{E}_1 = \omega_{21} \mathcal{E}_2 + s_1 U$ (and similarly for $\nabla^{\mathbb{R}^3}_{\alpha'} \mathcal{E}_2 = -\omega_{21} \mathcal{E}_1 + s_2 U$). From the formulas above, we obtain

$$(-\cos u, -\sin u, 0) = (-\omega_{21} \cos u \sin v_0, -\omega_{21} \sin u \sin v_0, \omega_{21} \cos v_0)$$
$$+ (s_1 \cos u \cos v_0, s_1 \sin u \cos v_0, s_1 \sin v_0)$$

Therefore, we have

$$\omega_{21} = \sin v_0 \qquad s_1 = -\cos v_0.$$

EXERCISE 3.6. For $\nabla_{\alpha'}^{\mathbb{R}^3} \mathcal{E}_2 = -\omega_{21}\,\mathcal{E}_1 + s_2\,U$, show that $s_2 = 0$.

$$\nabla_{\alpha'}\mathcal{E}_1 = \sin v_0\,\mathcal{E}_2 \qquad \nabla_{\alpha'}\mathcal{E}_2 = -\sin v_0\,\mathcal{E}_1.$$

By the proof of the theorem above, the *holonomy* along α is given by

$$-\int_0^{2\pi} \omega_{21}\,du = -\int_0^{2\pi} \sin v_0\,du$$

$$= -2\pi \sin v_0.$$

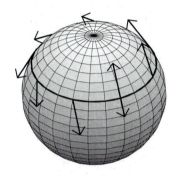

FIGURE 6.3. Holonomy around a latitude circle

EXERCISE 3.7. Calculate the total Gaussian curvature over the portion of the R-sphere above latitude v_0 and compare your answer to the holonomy around the v_0-latitude circle. Up to the addition of a multiple of 2π, what do you notice?

The calculation of holonomy above says that, from our outside viewpoint, parallel tangent vectors rotate as they move along a latitude circle. Of course, as the terminology "parallel" signifies, 2-dimensional residents of the sphere see the vectors as parallel — so, from their viewpoint, not rotating at all.

EXERCISE 3.8. What happens at the Equator and why is the Equator special among the circles of latitude?

EXERCISE 3.9. Give the cone $z^2 = a^2\,(x^2 + y^2)$ a parametrization

$$\mathbf{x}(u, v) = \left(u\,\sin\phi\,\cos\frac{v}{\sin\phi},\; u\,\sin\phi\,\sin\frac{v}{\sin\phi},\; u\,\cos\phi\right),$$

where ϕ is the vertex angle of the cone and $\sin\phi = 1/\sqrt{1 + a^2}$. Take a parallel $\alpha(v) = (u_0\,\sin\phi\,\cos\frac{v}{\sin\phi},\; u_0\,\sin\phi\,\sin\frac{v}{\sin\phi},\; u_0\,\cos\phi)$ with $0 \le v \le 2\pi \sin\phi$ and, by imitating the example above, show that the holonomy around α is $-2\pi \sin\phi$.

EXERCISE 3.10. Explain the result of the previous exercise by setting a sphere inside the cone so that the latitude $v = -\phi$ circle and the cone parallel α coincide. Show that the unit normals of the sphere and the cone (and hence the tangent planes) also coincide along the curve. Therefore, the covariant derivatives of the cone and sphere agree along the curve as well. Hint: for the cone with vertex at the origin and vertex angle ϕ, the center and radius of the sphere should be $(0, 0, u_0/\cos\phi)$ and $u_0 \sin\phi/\cos\phi$ respectively.

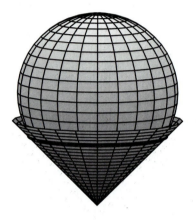

FIGURE 6.4.   Cone on sphere along a latitude circle

6.4 FOUCAULT'S PENDULUM

Nature uses only the longest thread to weave her patterns, so each small piece of fabric reveals the organization of the entire tapestry.

— Richard P. Feynman

In 1851 Jean Foucault (1819 − 1868) built a pendulum consisting of a heavy iron ball on a wire 200 feet long to demonstrate the rotation of the Earth. (Foucault was also responsible for developing means for measuring the speed of light in various media. In particular, he measured the speed of light in air and in water and showed that the speed varied inversely with the index of refraction (i.e. Snell's Law).) Foucault observed that the swing-plane of the pendulum precessed, or rotated, as time went on, eventually returning to its original direction after a period of $T = \frac{24}{\sin v_0}$ hours, where v_0 is the latitude where the *Foucault* pendulum resides. The usual explanation for this phenomenon of precession is in terms of rotating reference frames which produce a horizontal Coriolis "force" displacing the swing-plane (see [Sym] and [Arn]). Here we shall look at the Pendulum in the framework of *geometric phases* or *holonomy* (see [WS], [M] and [Op]).

In order to analyze the Foucault pendulum from the viewpoint of geometry, assume the Earth to be nonrotating and the pendulum to be situated at latitude v_0. Instead of the Earth rotating to move the pendulum, we move the pendulum once around the latitude circle in 24 hours at constant speed on this stationary Earth. This is clearly equivalent to the standard situation in terms of the ultimate movement of the pendulum. The long cable of the pendulum and the slow progression around the latitude circle have two consequences (which are given by the usual physics arguments).

First, the long cable provides a relatively small swing for the pendulum which is then approximately flat. Hence, we may consider each swing as a tangent vector to the sphere. By orienting these vectors consistently, we obtain a vector field of *pendulum swing plane directions* V. At each moment of time t there is such a swing direction vector $V(t)$ and all these vectors may be placed along the latitude circle $\alpha(u)$ by associating a given moment of time t with the unique point describing the pendulum's movement along $\alpha(u)$. Hence, we write $V(u)$ for the swing plane vector field.

Secondly, because we move around the latitude circle slowly, the consequent centripetal force on the pendulum is negligible. (In fact, it is approximately $\frac{1}{290}$ of the downward force mg.) This says that we may take the only force F felt by the pendulum to be gravity in the normal direction U. But such a normal force does not affect the vertical swing plane of the pendulum tangentially, so the swing plane appears unchanging to a 2-dimensional resident of the sphere. That is, *projected to the tangent plane*,

$$\nabla_{\alpha'} V = \text{proj} \, \frac{dV(u)}{du} = 0,$$

where the covariant derivative again reduces to the ordinary derivative due to our special parametrization. By our earlier discussion, we then have

Proposition 4.1. *The vector field V associated to the Foucault pendulum is parallel along a latitude circle.*

Of course, as we transport the Foucault pendulum once around the latitude circle α, *holonomy* rotates the parallel vector field V by $-2\pi \sin v_0$ radians. In particular, the angular speed of this vector rotation is then $\omega = \frac{2\pi \sin v_0 \text{ rads}}{24 \text{ hours}}$. The equivalence of our geometric situation with the physical one then gives

Theorem 4.2. *The period of the Foucault pendulum's precession is*

$$\frac{2\pi \text{ rads}}{\omega} = \frac{24}{\sin v_0} \text{ hours}.$$

Of course, this is precisely the period obtained in physics. Here, however, the precession of the swing plane of the Foucault pendulum results from the holonomy along α induced by the curvature of the Earth. Further, since we view the whole pendulum apparatus as stationary relative to the Earth, what can explain the observed precession of the swing plane? As Foucault argued, we must have

Corollary 4.3. *The Earth rotates along its latitude circles.*

EXERCISE 4.1. Suppose a Foucault pendulum is transported around a latitude circle on a torus. (You should still assume the only force is normal to the torus.) Compute the holonomy and explain whether this experiment alone can tell you whether we live on a sphere or torus.

6.5 THE ANGLE EXCESS THEOREM

Now let us return to geometry itself. Consider a unit speed simple closed curve β parametrized by $0 \leq s \leq 2\pi$. Let B denote the region "inside" β and assume that B is simply connected. That is, we assume that β may be continuously shrunk to a point in B. Of course we may use the same notation and formulas for β as for α above. *Note, however, that we do not assume that our surface is in 3-space. We abstract the properties of the covariant derivative from our discussion above (see Exercise 2.3) and simply note that such a thing actually does exist.* For example, we write $\nabla_{\beta'}\mathcal{E}_1 = \omega_{21}\mathcal{E}_2$ and θ denotes the angle between β' and $\mathcal{E}_1$. Because $\mathcal{E}_1$ and $\mathcal{E}_2$ are orthonormal, we may identify ω_{21} in two ways: first,

$$\nabla_{\beta'}\mathcal{E}_1 \cdot \mathcal{E}_2 = \omega_{21}$$

$$= \kappa_g - \frac{d\theta}{ds}$$

by our formula for geodesic curvature (of β). Secondly, the lefthandside may be computed using the definitions of $\mathcal{E}_1$ and $\mathcal{E}_2$ as $\mathbf{x}_u/\sqrt{E}$ and $\mathbf{x}_v/\sqrt{G}$ respectively. Note that $\beta' = \mathbf{x}_u u' + \mathbf{x}_v v'$, so we have

$$\nabla_{\beta'}\mathcal{E}_1 \cdot \mathcal{E}_2 = \left[\left(\frac{\mathbf{x}_u}{\sqrt{E}} \right)_u \frac{du}{ds} + \left(\frac{\mathbf{x}_u}{\sqrt{E}} \right)_v \frac{dv}{ds} \right] \cdot \mathcal{E}_2$$

$$= \left[\frac{\mathbf{x}_{uu}\sqrt{E} - \mathbf{x}_u \frac{1}{2}E^{-1/2}E_u}{E} \frac{du}{ds} + \frac{\mathbf{x}_{uv}\sqrt{E} - \mathbf{x}_u \frac{1}{2}E^{-1/2}E_v}{E} \frac{dv}{ds} \right] \cdot \mathcal{E}_2$$

$$= \left(\left[\frac{\mathbf{x}_{uu}}{\sqrt{E}} - \frac{\mathbf{x}_u E_u}{2E^{3/2}} \right] \frac{du}{ds} + \left[\frac{\mathbf{x}_{uv}}{\sqrt{E}} - \frac{\mathbf{x}_u E_v}{2E^{3/2}} \right] \frac{dv}{ds} \right) \cdot \frac{\mathbf{x}_v}{\sqrt{G}}.$$

Using our usual formulas for $\mathbf{x}_{uu}$ and $\mathbf{x}_{uv}$ (see Chapter 3.4), we obtain

$$\mathbf{x}_{uu} \cdot \mathbf{x}_v = -\frac{E_v}{2} \text{ and } \mathbf{x}_{uv} \cdot \mathbf{x}_v = \frac{G_u}{2}.$$

Hence, these calculations and the assumed orthogonality of $\mathbf{x}_u$ and $\mathbf{x}_v$ imply

$$\omega_{21} = \nabla_{\beta'}\frac{\mathbf{x}_u}{\sqrt{E}} \cdot \frac{\mathbf{x}_v}{\sqrt{G}} = \frac{-E_v}{2\sqrt{EG}}\frac{du}{ds} + \frac{G_u}{2\sqrt{EG}}\frac{dv}{ds}.$$

EXERCISE 5.1. Verify the formula above using the expressions for $\mathbf{x}_{uu}$ and $\mathbf{x}_{uv}$ derived in Chapter 3 (and ignoring U components).

EXERCISE 5.2. The general formula for ω_{21} derived above allows the determination of holonomy in a variety of situations. For instance, consider the Poincaré plane $P = \{(x, y) \in \mathbb{R}^2 \mid y > 0\}$ with the patch $\mathbf{x}(u, v) = (u, v)$ and conformal metric,

$$\mathbf{w}_1 \circ \mathbf{w}_2 = \frac{\mathbf{w}_1 \cdot \mathbf{w}_2}{v^2} \quad \text{where} \quad \mathbf{w}_1, \mathbf{w}_2 \in T_p(P) \quad \text{and} \quad p = (u, v).$$

Show that the holonomy along the horizontal line $v = 1$, from $(0, 1)$ to $(a, 1)$ is equal to $-a$. This means that a tangent vector at $(0, 1)$ is rotated by $-a$ radians as it moves along $v = 1$ to $(a, 1)$. Hint: Show that, in general for P, $\omega_{21} = 1/v_0$ along a u-parameter curve with $v = v_0$.

For the following we must recall Green's Theorem (see Chapter 4). Let C be a simple closed curve which may be continuously shrunken to a point in its interior. We say such a curve is *shrinkable*. The line integral $\int_C P\, dx + Q\, dy$ may be computed as an area integral

$$\oint_C P\, dx + Q\, dy = \int_R \int \left(\frac{\partial Q}{\partial x} - \frac{\partial P}{\partial y} \right) dx\, dy$$

where R is the region inside C. Now, by the computation of ω_{21} above and the identification $\omega_{21} = \kappa_g - \frac{d\theta}{ds}$, we obtain

$$\frac{d\theta}{ds} - \kappa_g = \frac{E_v}{2\sqrt{EG}} \frac{du}{ds} - \frac{G_u}{2\sqrt{EG}} \frac{dv}{ds}.$$

We then integrate both sides with respect to s to get

$$\int \frac{d\theta}{ds}\, ds - \int \kappa_g\, ds = \int \frac{E_v}{2\sqrt{EG}} \frac{du}{ds} - \frac{G_u}{2\sqrt{EG}} \frac{dv}{ds}\, ds$$

$$= \int_\beta \frac{E_v}{2\sqrt{EG}}\, du - \frac{G_u}{2\sqrt{EG}}\, dv$$

$$= \int_B \int -\frac{1}{2} \left[\frac{\partial}{\partial u} \left(\frac{G_u}{\sqrt{EG}} \right) + \frac{\partial}{\partial v} \left(\frac{E_v}{\sqrt{EG}} \right) \right] du\, dv$$

by Green's Theorem. Then, by Theorem 3.4.1, we have

$$= \int_B \int K\sqrt{EG}\, du\, dv$$

$$= \int_B K$$

by the definition of integration on a surface (or patch) and the fact that, since $F = 0$, $|\mathbf{x}_u \times \mathbf{x}_v| = \sqrt{EG - F^2} = \sqrt{EG}$.

Theorem 5.1. *For a shrinkable simple closed curve β, the holonomy around β may be identified with the total Gaussian curvature evaluated on the region B inside β. Further, the total change in θ around β is given by*

$$\theta(2\pi) - \theta(0) = \int \kappa_g \, ds + \int_B K.$$

Corollary 5.2. *If β is a geodesic, then*

$$\theta(2\pi) - \theta(0) = \int_B K.$$

Now, by our definition of θ as the angle between β' and $\mathcal{E}_1$, we see that, for a closed curve β, the total change in θ must be an integral multiple of 2π.

EXERCISE 5.3. Explain this statement.

Hence, we have $2\pi \cdot A = \int \kappa_g \, ds + \int_B K$, where A is an integer. Now, to see what A is, intuitively we might think of what happens as we shrink β to a point. Since the shrinking process is continuous and the total turning is an *integral* multiple of 2π, the total turning must remain constant throughout. Near a point, we may think of the curve as a tiny circle in the plane and $\mathcal{E}_1$ as a fixed vector ($\mathbf{e}_1 = (1, 0)$ say). In this case it is clear that the tangent vector rotates exactly once as it is transported around the circle; that is, the total turning is 2π. By continuity of the process, we must, therefore, have $A = 1$. To make this argument precise requires a version of the *Hopf Umlaufsatz* (see [Hsi] for example).

Finally, we come to the main goal enunciated at the start of this chapter — an understanding of a fundamental concept in traditional geometry from the differential point of view. Consider a triangle (Figure 6.5) which is made up of three curves in a surface M and which is shrinkable. We say the angles i_j are *interior angles* and their supplements $\pi - i_j$ are *exterior angles* of the triangle. Suppose we move around the triangle in the direction shown and consider the turning of the tangent vector(s) to the curve(s). The same intuitive argument above shows that we obtain a total turning of 2π. There are two contributions to turning: (1) the contribution obtained as above via the geodesic curvature formula, $\int \kappa_g(\triangle) + \int_\triangle K$ and (2) the contributions obtained at the corners of the triangle, $(\pi - i_1) + (\pi - i_2) + (\pi - i_3)$. Hence, we have

$$2\pi = \int \kappa_g(\triangle) + \int_\triangle K + (\pi - i_1) + (\pi - i_2) + (\pi - i_3).$$

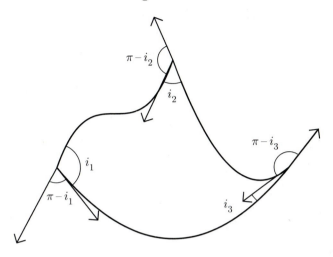

FIGURE 6.5.   A triangle in M

Theorem 5.3. *If a triangle in a surface is shrinkable to a point, then*

$$2\pi = \int \kappa_g(\triangle) + \int_{\triangle} K + \sum_{j=1}^{3}(\pi - i_j).$$

Moving the angle sum to the other side and multiplying by -1 produces

$$-\int \kappa_g(\triangle) - \int_{\triangle} K = -2\pi + (\pi - i_1) + (\pi - i_2) + (\pi - i_3)$$
$$= \pi - (i_1 + i_2 + i_3).$$

Corollary 5.4. *If a triangle in a surface is shrinkable to a point and is made up of geodesic segments, then the sum of the interior angles of the triangle differs from π by (+ or $-$) the total Gaussian curvature.*

Corollary 5.5. *If the surface M has constant Gaussian curvature K, then* $\pi - (i_1 + i_2 + i_3) = -K \cdot \text{Area}(\triangle)$.

Examples 5.1.

  (1) *The R-sphere.* $K = \frac{1}{R^2}$, so

$$\int_{\triangle} K = \frac{1}{R^2} \int_{\triangle}$$
$$= \frac{\text{Area of } \triangle}{R^2}.$$

Hence, $\sum i_j - \pi = \frac{\text{Area of } \triangle}{R^2}$. Therefore, on a sphere,

$$\sum i_j = \pi + \frac{\text{Area } \triangle}{R^2} > \pi$$

which verifies the traditional "parallel postulate" approach.

(2) *The Hyperbolic (or Poincaré) Plane.*

EXERCISE 5.4. Show that the sum of the angles of a triangle in H or P is strictly less than π.

Thus, curvature affects even the most fundamental of geometric quantities, the sum of the angles of a triangle. In this sense, from the differential point of view, Gaussian curvature is the basic structural element of geometry. In the next section we shall consider a global version of the Theorem above — the famous *Gauss-Bonnet Theorem*.

6.6 THE GAUSS-BONNET THEOREM

It is a fact that every surface M may be *triangulated*. That is, the surface may be completely covered by shrinkable triangles in the surface which meet only along edges or at vertices. Furthermore, an orientation on a surface induces orientations on each of the triangles so that the edge orientations are opposite when considered in adjacent triangles (see Figure 6.6).

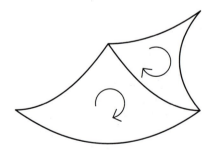

FIGURE 6.6.   Oriented triangles meeting along an edge

Suppose M has a triangulation with V total vertices, E total edges and F total triangles. Theorem 5.3 applies to each triangle, so we can add up both sides of the formula over all triangles to get

$$\sum_{\triangle} \int \kappa_g(\triangle) + \sum_{\triangle} \int_{\triangle} K + \sum_{\triangle} \sum_{j=1}^{3} (\pi - i_j) = \sum_{\triangle} 2\pi.$$

Now, M may have *boundary curves* (which we will assume are smooth) just as for a cylinder of finite height. We denote this boundary by ∂M. In this case, we may split up the sets of vertices and edges according to whether they lie on the boundary or in the interior of M. Let $V = V_I + V_B$ and $E = E_I + E_B$, where the subscripts I and B refer to the *interior* and *boundary* respectively. Then

$$\int_{\partial M} \kappa_g + \int_M K + 3\pi\, F - 2\pi\, V_I - \pi\, V_B = 2\pi\, F$$

since the sum of the interior angles at a particular vertex in V_I, for all triangles intersecting that vertex, is 2π and the sum of the interior angles at a particular vertex in V_B, for all triangles intersecting that vertex, is the angle of the tangent line to the boundary at that vertex, π. Furthermore, each triangle has 3 edges surrounding it and two triangles share a common edge, *except for edges in E_B* which meet only one triangle. This leads to the equality $3F = 2E_I + E_B$. Replacing $3F\pi$ by $(2E_I + E_B)\pi$, we then have

$$\int_{\partial M} \kappa_g + \int_M K + 2\pi\, E_I + \pi\, E_B - 2\pi\, V_I - \pi\, V_B = 2\pi\, F$$

$$\int_{\partial M} \kappa_g + \int_M K + 2\pi\, E_I + \pi\, E_B + \pi\, E_B - \pi\, V_B - 2\pi\, V_I - \pi\, V_B = 2\pi\, F$$

since $V_B = E_B$. Then

$$\int_{\partial M} \kappa_g + \int_M K + 2\pi\, (E_I + E_B) - 2\pi\, (V_I + V_B) = 2\pi\, F$$

$$\int_{\partial M} \kappa_g + \int_M K + 2\pi\, E - 2\pi\, V = 2\pi\, F$$

$$\int_{\partial M} \kappa_g + \int_M K = 2\pi\, (V - E + F)$$

Therefore, if we define $\chi(M) = V - E + F$ to be the *Euler Characteristic* of M, then we have

Gauss-Bonnet Theorem 6.1. *If M is a compact oriented surface with boundary ∂M made up of a finite number of smooth closed curves, then*

$$\int_{\partial M} \kappa_g + \int_M K = 2\pi\, \chi(M).$$

Corollary 6.2. *If M is a compact oriented surface without boundary, then*

$$\int_M K = 2\pi\, \chi(M).$$

Of course, the one point we have neglected up to this point is that the quantity $\chi(M) = V - E + F$ may not be characteristic of M, but may depend on which triangulation is chosen. That this is not the case is a result which goes back to Euler. In fact, the result is more general. A *graph* G consists of finite sets of vertices $V = \{v_k\}$ and edges $E = \{e_{ij}\}$ where e_{ij} denotes the unique edge joining v_i and v_j. The graph is *connected* if there is an edge path joining any two vertices. We shall only consider connected graphs. A graph G is *embedded* in a surface M if there is a map $f: G \to M$ such that the images of edges never intersect except at vertex images. An embedding is a *2-cell embedding* if the surface polygons (or faces) determined by the embedding are simply connected (and, hence, look like deformed 2-disks).

EXERCISE 6.1. Embed a small graph in the plane and calculate $V - E + F$. Be sure to include the face at infinity, the face which surrounds the entire graph. Do the same for 2-cell embeddings on the sphere, torus and cylinder. Be sure your embeddings are 2-cell.

Theorem 6.3 (Euler). *Any embedding of a graph in the plane gives* $V - E + F = 2$.

Proof. The proof is by induction on the number of edges in the graph. The reader can easily check that the result is true for graphs with 1 or 2 edges. Let G be a graph with E edges. There are two cases. The graph G may be a *tree*; a graph enclosing no faces. Then $E = V - 1$ and $F = 1$ (the face at infinity), so $V - E + F = V - (V - 1) + 1 = 2$ as desired. If G is not a tree, then G has a cycle, a path of edges which form a closed curve in the plane. This cycle has an inside and outside by the Jordan curve Theorem, so there is an edge e which bounds two faces. Remove this edge from G. Note the new graph $\bar{G}$ is still connected since e was on a cycle and the number of faces has been reduced to $F - 1$. By the inductive hypothesis on $\bar{G}$ (since it has $E - 1$ edges), $V - (E - 1) + (F - 1) = 2$. But simplifying the lefthand side gives $V - E + F = 2$ which applies to the original graph G. $\qquad\square$

EXERCISE 6.2. Using the proof above as a model, show that the Euler characteristics $\chi = V - E + F$ for the sphere, torus, disk and cylinder are, respectively, 2, 0, 1 and 0. For the disk (i.e. circle union its interior), use the theorem for the plane and remove the face at infinity. Explain why the sphere and plane both give $V - E + F = 2$ for 2-cell embeddings. Hint: stereographic projection.

If we disregard geometry and only consider shape up to continuous deformation (for example), then any compact oriented surface (without boundary) is known to assume the form of a *sphere with handles*. A sphere with g handles, S_g, is obtained by cutting $2g$ disks out of the sphere and gluing in g cylinders along the boundary circles. Of course, the sphere is simply S_0 and the torus is S_1.

EXERCISE 6.3. Show that any 2-cell embedding of a graph on S_g gives $V - E + F = 2 - 2g$. Hence, the Euler characteristic of a compact oriented surface without boundary is always of the form $2 - 2g$.

EXERCISE 6.4. We have seen in Chapter 5 that a torus may be given a metric so that the Gauss curvature $K = 0$ at every point. Show, using Gauss-Bonnet, that the torus can never be given a metric with $K \leq 0$ globally and $K < 0$ at some point. This indicates how a nongeometric (i.e. but deformation) invariant such as the Euler characteristic can constrain the geometry of a surface. Further, show that any surface with positive Gauss curvature is deformable to the sphere. In fact, such a surface must be "topologically equivalent" to the sphere.

Consider what the Gauss-Bonnet Theorem is really saying. For instance, a sphere may be stretched and twisted (without ripping) to produce various geometrically distinct surfaces such as the "pretzel" of Figure 6.7. But, while the Gauss curvatures of these surfaces are clearly not constant, because these surfaces are deformed spheres, the Gauss-Bonnet Theorem says that the *total* Gauss curvatures are always constrained to be 4π. Somehow, the deformation shape of the surface — which is reflected in the global invariant χ — acts to determine a global *geometric* invariant, the total Gaussian curvature. The idea of constraining geometry by such deformation invariants has led to many major results in Mathematics. In this way, the Gauss-Bonnet Theorem is the ancestor of much of the spirit of modern differential geometry and algebraic topology. (See [Go] for an interesting perspective on Gauss-Bonnet and the Euler characteristic.)

FIGURE 6.7.   A pretzel

There are other ways in which Gauss-Bonnet constrains the geometry of a surface. The following example indicates how geodesics may be affected.

Example 6.1: The hyperboloid of one sheet.
Consider $M : x^2 + y^2 - z^2 = 1$, the hyperboloid of one sheet. Recall that M is a surface of revolution parametrized by

$$\mathbf{x}(u, v) = (\cosh u \cos v, \ \cosh u \sin v, \ \sinh u).$$

The central (v-parameter) circle $\alpha = \mathbf{x}(0, v) = (\cos v, \sin v, 0)$ is a (closed) geodesic since $h(u) = \cosh u$ and $h'(0) = \sinh(0) = 0$. In fact, using the Gauss-Bonnet Theorem, we can show that this circle is *the only closed geodesic on M.* To see this, suppose that β is another closed geodesic on M.

There are two possibilities; either α and β are disjoint or they intersect. Suppose they are disjoint. Then α and β are the smooth boundary curves of a portion of the hyperboloid $\overline{M}$ which can be deformed to a cylinder and which, therefore, by the exercises above, has Euler characteristic zero. Now, as part of the hyperboloid, $\overline{M}$ has $K < 0$ at each point. Hence, $\int_{\overline{M}} K < 0$ contradicts the Gauss-Bonnet formula $\int_{\overline{M}} K = 2\pi\chi(\overline{M}) = 0$. (Here we have used the fact that α and β are geodesics to obtain $\int_{\partial\overline{M}} k_g = 0$.)

Now suppose that α and β intersect in (necessarily) two points making interior angles ϕ_i, both of which are less than π.

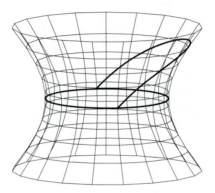

FIGURE 6.8.   A possible closed geodesic?

EXERCISE 6.5. Justify this last statement, $\phi_i < \pi$. Hint: use the uniqueness of geodesics through a point with given initial tangent vector.

The region R (see Figure 6.8) bounded by α and β with interior angles ϕ_i is simply connected and, so, is deformable to a disk. Indeed, we may consider the region to be a triangle with one vertex on β (say) with angle π and the other two vertices to be the points of intersection with interior angles ϕ_i. Again, since R is part of the hyperboloid, $\int_R K < 0$. The local Gauss-Bonnet Theorem then says

$$\int_R K + (\pi - \pi) + (\pi - \phi_1) + (\pi - \phi_2) = 2\pi$$

$$\int_R K = \phi_1 + \phi_2$$

$$\int_R K > 0$$

which is a contradiction. Hence, there are no other closed geodesics.

EXERCISE 6.6. Show that, on a surface with $K \leq 0$, no closed geodesic bounds a simply connected region (i.e. a region deformable to a disk).

6.7 Geodesic Polar Coordinates

Finally, let us look at an alternative framework for the angle-sum theorem above. Although we have mentioned *geodesic polar coordinates* previously in Chapter 5, we have not described their construction. We shall do this now and see how they allow a conceptually different approach to the angle-sum theorem as well as yet another example of the influence of curvature on global geometry. A full (and rigorous) discussion of geodesic polar coordinates is rather technical and we will not go into it here. A nice (and relatively elementary) discussion may be found in [O'N1] and much of what we do below mimics this. Many of the details below are left to the reader.

Fix a point $p \in M$ and let $\mathbf{x}(u, v)$ denote a mapping from a portion of the plane obtained by using v as the angle of a unit tangent vector $\mathbf{w}$ at p from a fixed unit tangent vector $\mathbf{e}_1$ and u as the "distance" along the unique unit speed geodesic α with $\alpha(0) = p$, $\alpha'(0) = \mathbf{w}$. More specifically, let $\mathbf{e}_1$ and $\mathbf{e}_2$ be perpendicular unit tangent vectors at p and write $\mathbf{w} = \cos v\, \mathbf{e}_1 + \sin v\, \mathbf{e}_2$. Then $\mathbf{x}(u, v) = \alpha_{\mathbf{w}}(u)$, where the subscript is to remind us that α is determined by $\mathbf{w}$, which, in turn, is determined by the parameter v. Of course we must restrict u so that $\mathbf{x}(u, v)$ is defined, so $\mathbf{x}(u, v)$ maps onto a (generally small) neighborhood of p. The key result, which follows from the inverse function theorem and which we shall not prove, is

Theorem 7.1. *The mapping* $\mathbf{x}(u, v)$ *is a patch for* $u > 0$. *That is,* $\mathbf{x}(u, v)$ *is a smooth map with smooth inverse from an open set in the plane onto an open neighborhood of* $p \in M$ *and is one-to-one for* $u > 0$.

Note the exclusion of p from the parametrization. This exclusion follows for the same reason that ordinary polar coordinates exclude the origin in the plane. As we have mentioned, such a patch is called a *geodesic polar coordinate patch*.

FIGURE 6.9. A geodesic polar patch

EXERCISE 7.1. Using the fact that straight lines are geodesics in the plane (with the usual Euclidean metric), show that geodesic polar coordinates about the origin in the plane are, in fact, ordinary polar coordinates.

EXERCISE 7.2. Let $\mathbf{e}_1 = (1,0)$, $\mathbf{e}_2 = (0,1)$ and *switch the roles of u and v* in the usual patch for the unit sphere. Show that this patch provides geodesic polar coordinates around the North pole $N = \mathbf{x}(\pi/2, v)$. Here, note that it isn't necessary to begin the geodesic parametrization at $t = 0$. How much of the sphere does this patch cover?

EXERCISE 7.3. The usual polar patch for the hyperbolic plane $\mathbf{x}(u, v) = (u \cos v, u \sin v)$, does not give unit speed radial lines through $(0,0)$. Show that the arclength of these lines is given by $s(u) = 2 \operatorname{arctanh}(u/2)$ and that a unit speed reparametrization is then given by

$$\beta(s) = \left(2 \tanh \frac{s}{2} \cos v_0, \ 2 \tanh \frac{s}{2} \sin v_0\right)$$

for fixed v_0. Remember to use the hyperbolic metric here! Now reparametrize the patch $\mathbf{x}$ and show that you get geodesic polar coordinates around zero.

The key to using geodesic polar coordinates is the following

Gauss's Lemma 7.2. *The patch $\mathbf{x}(u, v)$ has $E = 1$, $F = 0$ and $G > 0$.*

Proof. $E = \mathbf{x}_u \cdot \mathbf{x}_u = 1$ because u-parameter curves $\alpha_{\mathbf{w}_0}(u)$ are unit speed. Hence, $\mathbf{x}_v[E] = E_v = 0$. The product rule, however, gives

$$
\begin{aligned}
0 = E_v \\
= \nabla_{\mathbf{x}_v}(\mathbf{x}_u \cdot \mathbf{x}_u) \\
= 2 \nabla_{\mathbf{x}_v}\mathbf{x}_u \cdot \mathbf{x}_u \\
= 2 \nabla_{\mathbf{x}_u}\mathbf{x}_v \cdot \mathbf{x}_u
\end{aligned}
$$

by the commutation rule, Exercise 2.3. Hence,

$$0 = 2 \nabla_{\mathbf{x}_u}\mathbf{x}_v \cdot \mathbf{x}_u + 2 \mathbf{x}_v \cdot \nabla_{\mathbf{x}_u}\mathbf{x}_u$$

since u-parameter curves are geodesics with $\nabla_{\mathbf{x}_u}\mathbf{x}_u = 0$. So,

$$
\begin{aligned}
0 = 2 \nabla_{\mathbf{x}_u}(\mathbf{x}_v \cdot \mathbf{x}_u) \\
= 2 F_u.
\end{aligned}
$$

Therefore, F only depends on v. But now consider the v-parameter curve $\mathbf{x}(u_0, v)$ and note that, at $u = 0$, $\mathbf{x}(0, v) = p$ is constant. Therefore, at p, $\mathbf{x}_v(0, v) = 0$ for every v and (for fixed v)

$$\lim_{u_0 \to 0} F = \lim_{u_0 \to 0} \mathbf{x}_u(u_0, v) \cdot \mathbf{x}_v(u_0, v) = 0.$$

Since F doesn't depend on u, $F = 0$ along this $v =$constant curve. But the same argument applies for every v, so we must have $F = 0$ identically in the patch. Thus, the u-parameter curves are radial geodesics (from p) and the v-parameter curves are the orthogonal trajectories to these geodesics. Finally, because $\mathbf{x}$ is a patch, $G = EG - F^2 > 0$.   □

Remark 7.1. Note that the proof above contains the fact that $\sqrt{G}(0, v) = |\mathbf{x}_v(0, v)| = 0$. We will use this below as an initial condition for a very special differential equation.

EXERCISE 7.4. Use the argument of Chapter 5 to show that, for any point q in the geodesic polar patch, the radial geodesic from p (i.e. the u-parameter curve) gives the shortest arclength of any curve joining p to q. Further, from the argument, infer that no other curve gives this arclength.

Since the point p (called the *pole*) is a trouble spot for the patch, we must be careful in making inferences about the geometry near it. In order to understand what happens there, as is our custom now, we first construct a frame from the geodesic polar patch,

$$\mathcal{E}_1 = \mathbf{x}_u \qquad \mathcal{E}_2 = \frac{\mathbf{x}_v}{\sqrt{G}}.$$

The u-parameter curves are geodesics, so $\mathcal{E}_1$ is parallel along them. The vector field $\mathcal{E}_2$, however, has constant length 1 and maintains a constant angle of $\pi/2$ with $\mathcal{E}_1$, so (by Exercise 3.2) is parallel along radial geodesics as well. That is, $\nabla_{\mathbf{x}_u} \mathcal{E}_2 = 0$. Let

$$\mathcal{E}_1(0) = \cos v_0\, \mathbf{e}_1 + \sin v_0\, \mathbf{e}_2 \qquad \mathcal{E}_2(0) = -\sin v_0\, \mathbf{e}_1 + \cos v_0\, \mathbf{e}_2$$

where $\mathbf{e}_1$, $\mathbf{e}_2$ are fixed unit tangent vectors at p. Let $\alpha(u) = \mathbf{x}(u, v_0)$ and consider the following covariant derivative along this u-parameter curve,

$$\nabla_{\mathbf{x}_u} \mathbf{x}_v = \nabla_{\mathbf{x}_u}(\sqrt{G}\, \mathcal{E}_2)$$
$$= (\sqrt{G})_u\, \mathcal{E}_2 + \sqrt{G}\, \nabla_{\mathbf{x}_u} \mathcal{E}_2$$

But $\mathcal{E}_2$ is parallel along α, so $\nabla_{\mathbf{x}_u} \mathcal{E}_2 = 0$. Thus,

$$\nabla_{\mathbf{x}_v} \mathbf{x}_u = \nabla_{\mathbf{x}_u} \mathbf{x}_v$$
$$= (\sqrt{G})_u\, \mathcal{E}_2$$

In particular, as we approach p,

$$\mathbf{x}_{uv}(0, v_0) = \lim_{u \to 0} (\sqrt{G})_u(u, v_0))\mathcal{E}_2(0).$$

We can also compute this quantity another way using $\mathbf{x}_u = \mathcal{E}_1$. We have $\mathcal{E}_1(0) = \cos v_0\, \mathbf{e}_1 + \sin v_0\, \mathbf{e}_2 = \mathbf{x}_u(0, v_0)$ and we may write $\mathbf{x}_u(0, v) = \cos v\, \mathbf{e}_1 + \sin v\, \mathbf{e}_2$. Then we compute $\nabla_{\mathbf{x}_v} \mathbf{x}_u(0, v) = -\sin v\, \mathbf{e}_1 + \cos v\, \mathbf{e}_2$ and $\nabla_{\mathbf{x}_v} \mathbf{x}_u(0, v_0) = -\sin v_0\, \mathbf{e}_1 + \cos v_0\, \mathbf{e}_2 = \mathcal{E}_2(0)$. Therefore,

$$\lim_{u \to 0} (\sqrt{G})_u(u, v_0)) \mathcal{E}_2(0) = \mathcal{E}_2(0)$$

and

$$\lim_{u \to 0} (\sqrt{G})_u(u, v_0)) = 1.$$

Notice what this calculation says about the metric *near* p. We knew previously, of course, that $E = 1$ and $F = 0$ everywhere on the patch, so the fact that G is close to u^2 near p says that the metric around p is close to being Euclidean (in polar coordinates)— at least to first order derivatives. This indicates also that curvature should be given by a second derivative of $\sqrt{G}$. This is verified in the important exercise below.

EXERCISE 7.5. Show that $K = -(1/\sqrt{G})(\sqrt{G})_{uu}$ by writing out the chain-rule formula for $(\sqrt{G})_{uu}$ and comparing the result to that obtained by plugging $E = 1$ into Theorem 3.4.1. A little algebra produces the *Jacobi Equation* along a radial geodesic, $(\sqrt{G})_{uu} + K \sqrt{G} = 0$.

In order to understand the angle-excess theorem from the viewpoint of geodesic polar coordinates, we must deal with nonradial geodesics as well as radial ones to form a triangle. With this in mind, let γ be a unit speed geodesic which intersects a radial geodesic $\alpha(u) = \mathbf{x}(u, v_0)$. We may write $\gamma(t) = \mathbf{x}(u(t), v(t))$ and note that $\gamma' = \mathbf{x}_u \frac{du}{dt} + \mathbf{x}_v \frac{dv}{dt}$ with $(du/dt)^2 + G(dv/dt)^2 = 1$ since γ is unit speed. Also, the angle ϕ at which γ and α intersect is found by taking the inner product (denoted $\circ$) of their tangent vectors,

$$\cos \phi = \mathbf{x}_u \circ \gamma' = \frac{du}{dt}.$$

Hence, we obtain $G(dv/dt)^2 = 1 - (du/dt)^2 = \sin^2 \phi$ and so $\sin \phi = \sqrt{G}\, dv/dt$.

Now we differentiate $\cos \phi = \mathbf{x}_u \circ \gamma'$ along γ to get

$$-\sin \phi \, \frac{d\phi}{dt} = \nabla_{\gamma'} \mathbf{x}_u \circ \gamma' + \mathbf{x}_u \circ \nabla_{\gamma'} \gamma'$$

$$= \nabla_{\gamma'} \mathbf{x}_u \circ \gamma'$$

$$= \nabla_{\mathbf{x}_u \frac{du}{dt} + \mathbf{x}_v \frac{dv}{dt}} \mathbf{x}_u \circ \gamma'$$

$$= \frac{du}{dt} \nabla_{\mathbf{x}_u} \mathbf{x}_u \circ \gamma' + \frac{dv}{dt} \nabla_{\mathbf{x}_v} \mathbf{x}_u \circ \gamma'$$

$$= \frac{dv}{dt} \nabla_{\mathbf{x}_v} \mathbf{x}_u \circ \gamma'$$

$$= \frac{dv}{dt} (\sqrt{G})_u \, \mathcal{E}_2 \circ \gamma'$$

$$= (\frac{dv}{dt})^2 (\sqrt{G})_u \, \sqrt{G}.$$

Then, replacing $\sin \phi$ with $\sqrt{G} \frac{dv}{dt}$ produces

$$\frac{d\phi}{dt} = -(\sqrt{G})_u \frac{dv}{dt}$$

or, in differential form,

$$d\phi = -(\sqrt{G})_u \, dv.$$

The calculation above relates the angle of intersection of geodesics to the rate of change of $\sqrt{G}$ (along a radial geodesic). We can, therefore, apply this to a triangle made up of two radial geodesic geodesics and a nonradial geodesic which intersects them both to give a different proof of the angle-excess theorem.

EXERCISE 7.6. (Following [Cox]) Suppose a geodesic triangle in M is made up of two radial geodesics $\alpha(u) = \mathbf{x}(u, v_0)$, $\beta(u) = \mathbf{x}(u, v_1)$ and a unit speed geodesic $\gamma(t) = \mathbf{x}(u(t), v(t))$ which intersects α and β as shown in Figure 6.10.

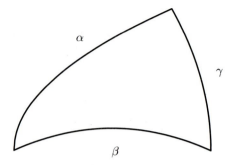

FIGURE 6.10.   A geodesic triangle

Prove that

$$\int_\triangle K = A + B + C - \pi.$$

Hints: (1) Use geodesic polar coordinates. (2) Start with the formula for total Gauss curvature and justify the calculation below,

$$\int_\triangle K = \int\int K\sqrt{EG - F^2}\, du\, dv$$

$$= \int\int K\sqrt{G}\, du\, dv$$

$$= \int\int_0^u -(\sqrt{G})_{uu}\, du\, dv$$

$$= \int 1 - (\sqrt{G})_u\, dv$$

$$= \int_{v_0}^{v_1} dv + \int_{\pi - B}^{C} d\phi$$

$$= A + B + C - \pi.$$

Now, from the start, we constructed a geodesic polar coordinate patch *only* in a neighborhood of a point $p \in M$. It may (and does) happen that, if we try to extend the patch beyond this neighborhood — even to its boundary — then we lose the qualities of a patch. Namely, we may lose one-to-oneness or we may find that there is a point $\mathbf{x}(u_0, v)$ with $G(u_0, v) = 0$ so that regularity of the patch fails. This is a situation we can analyze and through which, once again, we will see the powerful effect of curvature on geometry.

Let $\mathbf{x}(u, v)$ be a geodesic polar patch on M with pole p. Say that $\tilde{p} = \mathbf{x}(u_0, v)$ is *conjugate* to p along the (radial) geodesic determined by v *if* $G(u_0, v) = 0$. Of course, we can take geodesic polar coordinates about any point $p \in M$, so we can get away from taking a specific parametrization when we need to and simply consider conjugate points along arbitrary geodesics. The following theorem is the key to understanding the relationships among geodesics, conjugate points and shortest arclengths. The same argument as in Exercise 7.6 above (as well as Chapter 5) essentially proves the first part. The second part is more technical and harder, but plausible from the discussion above. The third part is the Hopf-Rinow theorem which we mentioned in Chapter 5. (In the following, we say that a curve β is *close to* α if the distance from $\alpha(t)$ to $\beta(t)$ never exceeds a certain small fixed amount.)

Theorem 7.3.

(1) *If α is a geodesic joining $p \in M$ to $q \in M$ and there are no conjugate points to p along α between p and q, then α gives the shortest arclength of any curve which is close to α and which joins p and q.*

(2) *If $\tilde{p} = \alpha(t_0)$ is conjugate to $p = \alpha(0)$ along the geodesic α, then α cannot give the shortest arclength of any curve (even close to α) joining p to any $q = \alpha(t_1)$ for all $t_1 > t_0$.*

(3) *If M is geodesically complete, then any two points of M may be joined by a geodesic which has the shortest arclength of any curve between the two points.*

Example 7.2: The Sphere.
The parametrization of the unit sphere obtained by switching the roles of u and v in the usual geographic parametrization gives $G = \cos^2 u$. The North pole, $u = \pi/2$, gives $G = 0$ and the next such zero occurs at the conjugate point $u = -\pi/2$, the South pole S. Of course a longitude (i.e. u-parameter curve) is a geodesic which gives the shortest arclength of any curve joining N and S, but once we go past S on the great circle to a point q, then the geodesic giving the shortest arclength from N to q is the longitude on the other side of the sphere.

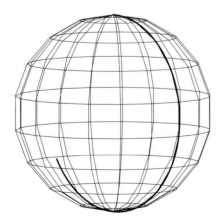

FIGURE 6.11.   A nonminimum-length geodesic

Conjugate points may be determined from curvature in the following way.

We have seen in Exercise 7.5 that $\sqrt{G}$ along a geodesic obeys the following *Jacobi* differential equation with initial conditions:

$$(\sqrt{G})_{uu} + K\sqrt{G} = 0$$

subject to

$$\sqrt{G}(0, v) = 0$$
$$(\sqrt{G})_u(0, v) = 1$$

for all v. But the solution to such an equation is uniquely determined by the usual uniqueness theorems of differential equations theory. Therefore, in order to find a zero of G (i.e. a conjugate point of p) along a geodesic, it is sufficient to solve the differential equation abstractly and then find a zero of the abstract solution. Specifically, solve the differential equation (which we also call *the Jacobi equation*)

$$f'' + K f = 0$$

subject to

$$f(0) = 0$$
$$f'(0) = 1$$

for a function f *and then find* u_0 with $f(u_0) = 0$.

Example 7.3: The Sphere.
The unit sphere, with the parametrization of Exercise 7.2 (and Example 7.2), has $K = 1$ and initial conditions $f(\pi/2) = 0$, $f'(\pi/2) = 1$. The general solution of $f'' + f = 0$ is given by $f(u) = A\cos u + B\sin u$ and the initial conditions give $B = 0$ and $A = -1$. Then, $f(u) = -\cos u$ and the next zero for f after $u = \pi/2$ is $u_0 = -\pi/2$ — the South pole. Note that, because radial geodesics from the North pole N are unit speed, the arclength to the South pole is

$$L(\alpha) = \left| \int_{\pi/2}^{-\pi/2} |\alpha'| \, dt \right|$$

$$= \left| \int_{\pi/2}^{-\pi/2} 1 \, dt \right|$$

$$= \pi.$$

EXERCISE 7.7.

(1) For a plane, where and when do geodesics give the shortest arclength? Find conjugate points by solving the Jacobi equation with $K = 0$.

(2) For the hyperbolic plane, where and when do geodesics give the shortest arclength? Find conjugate points by solving the Jacobi equation with $K = -1$.

(3) Generalize the first two parts to the following: if a surface M has $K \leq 0$, then there are no conjugate points along any geodesic. Hence, any geodesic gives the shortest arclength for all curves near it. Hints: show that $f(u) \geq u$ by using the initial conditions $f(0) = 0$, $f'(0) = 1$ and the Jacobi equation to derive the consequence $f'(u) \geq 1$. Hence, $f(u) \neq 0$, so no conjugate point can occur.

Finally, we come to the main result of this section. In order to give a proof, we first need a basic comparison result from differential equations. (Also, we use the fact that, just as in the sphere example above, because geodesics are unit speed, the parameter u along the geodesic is precisely the arclength along the geodesic.)

Theorem 7.4 (Sturm-Liouville). *Let $y''(t) + g(t)\, y(t) = 0$ be a second order linear differential equation with initial conditions $y(0) = 0$ and $y'(0) \neq 0$. Suppose, for all t, that $a^2 \leq g(t) \leq b^2$ and let $t_0 > 0$ be the first positive solution to $y(t) = 0$. Then*

$$\frac{\pi}{b} \leq t_0 \leq \frac{\pi}{a}.$$

We can apply the Sturm-Liouville comparison theorem to the Jacobi equation in the situation where Gauss curvature is positive and bounded away from zero. (In order to ensure the proper domains for geodesics, we again refer to the notion of "completeness" mentioned in Chapter 5.)

Lemma 7.5. *Suppose M is complete and $K \geq a^2 > 0$. Then any geodesic $\alpha \colon [0, \infty) \to M$ has a conjugate point somewhere in the interval $(0, \pi/a)$.*

Proof. We apply the Sturm-Liouville theorem to the Jacobi equation $f'' + K f = 0$ to see that the first conjugate point (i.e. first zero of f) occurs for $u_0 \leq \pi/a$. □

We can globalize this local result by asking how big M itself can be under the curvature constraint $K \geq a^2 > 0$. From Theorem 7.3 above, if M is complete, then any two points $p, q \in M$ are joined by a shortest length geodesic. Thus, it makes sense to define the *diameter* of M to be the length of the longest shortest-length geodesic joining any two points of M. In fact, this length may not be attained, so we must take a supremum of such lengths. Formally, we define the distance of p from q, $d(p, q)$, to be the arclength of the shortest geodesic joining p and q. Then

$$\mathrm{Diam}(M) \overset{\mathrm{def}}{=} \sup\{d(p, q) | p \text{ and } q \in M\}.$$

Note that a finite diameter for $M \subseteq \mathbb{R}^3$ implies that M is bounded. Recall that a closed bounded subset of $\mathbb{R}^3$ is called *compact*. The notion of compactness is, in fact, more general, but if it is unfamiliar, then the reader should simply consider the $\mathbb{R}^3$ situation. Lemma 7.5 then gives

Theorem 7.6 (Bonnet). *Suppose M is complete and $K \geq a^2 > 0$. Then $Diam(M) \leq \pi/a$ and, consequently, M is compact.*

Proof. Let $p, q \in M$ be arbitrary points. Since M is complete, there is a shortest arclength geodesic joining them. Lemma 7.5 implies that the first conjugate point occurs along this geodesic at $u_0 \leq \pi/a$. Since the geodesic is shortest length and we know this cannot be the case after the first conjugate point, then the arclength of the geodesic must be $\leq \pi/a$. Because this is true *for all* $p, q \in M$, then $Diam(M) \leq \pi/a$. Hence, M is bounded and, because it is complete, it is therefore compact. $\square$

Once again we see that hypotheses on Gauss curvature have powerful consequences. We saw in Chapter 3 that compactness implies a local result; namely, that the curvature is positive for at least one point of a surface. Now we have seen that the global hypothesis of positive curvature (bounded away from zero) throughout the surface implies compactness of the surface. The following exercises point out the limits of the hypotheses and conclusions.

EXERCISE 7.8. The hyperboloid of two sheets $x^2 + y^2 - z^2 = -1$ and the paraboloid $z = x^2 + y^2$ have nonnegative Gauss curvature and are complete since they are closed in $\mathbb{R}^3$. They are not bounded however, so they are not compact. What goes wrong here?

EXERCISE 7.9. Is the converse of Bonnet's theorem true. That is, if M is compact (say, closed and bounded in $\mathbb{R}^3$) with $Diam(M) \leq \pi/a$, then is it true that $K \geq a^2 > 0$? Hint: estimate the diameter of a torus and compare its Gauss curvature.

EXERCISE 7.10. In Example 7.3 and Exercise 7.7 above, the Jacobi equation $f'' + Kf = 0$ was solved for $K = 1$, $K = 0$ and $K = -1$. Use these solutions to interpret how geodesics starting from the same point come together or grow apart. To do this, note that a solution (by uniqueness) must be $\sqrt{G}$ for a particular geodesic polar patch and that $\sqrt{G} = |\mathbf{x}_v|$ measures the instantaneous rate of change of a radial geodesic in the v-direction. In Chapter 5 we considered geodesics as paths of particles constrained to lie on surfaces under no external forces. The interpretation above shows that, in particular, equilibrium positions of such particles on surfaces of negative curvature exhibit exponential instability (see [Arn, Appendix 1 IJK]).

Chapter 7

MINIMAL SURFACES
AND
COMPLEX VARIABLES

At the end of Chapter 4 we hinted that there was a close relationship between the subjects of minimal surfaces and complex analysis. In this chapter we describe this relationship and use it to "represent" minimal surfaces in a very convenient way. Basic references for this chapter are [Oss1], [Ni], [DHKW] and [FT]. We shall begin by recalling the basics of complex variable theory.

We denote $\sqrt{-1}$ by i and the field of *complex numbers* by $\mathbb{C} = \{z = x + iy | x, y \in \mathbb{R}\}$. A function $f \colon \mathbb{C} \to \mathbb{C}$ is *continuous* at z_0 if $\lim_{z \to z_0} f(z) = f(z_0)$. If this is true for all z_0 in some open set D, then f is continuous in D. The function f is *complex differentiable* at $z_0 \in \mathbb{C}$ if

$$\lim_{z \to z_0} \frac{f(z) - f(z_0)}{z - z_0}$$

exists. In this case, the limit is denoted $f'(z_0)$. If the limit exists for all $z_0 \in D$, D open, then we say that f is *analytic* or *holomorphic* in D. Notice that, although this definition resembles the usual single variable calculus definition of derivative, it is much more subtle since z may approach z_0 from any direction along any kind of path. We can use this subtlety to our advantage, however. Because the range of f is $\mathbb{C}$, we can write $f(z) = f(x+iy) = \phi(x,y) + i\psi(x,y)$ where ϕ and ψ are real-valued functions of the two real variables x and y. The function ϕ is the *real part* of f while ψ is f's *imaginary part*. Now let's assume the limit above exists and compute it in the special case of $z = iy \to z_0 = iy_0$.

$$\lim_{y \to y_0} \frac{\phi(x_0, y) + i\psi(x_0, y) - [\phi(x_0, y_0) + i\psi(x_0, y_0)]}{i(y - y_0)}$$

$$= \lim_{y \to y_0} \frac{\phi(x_0, y) - \phi(x_0, y_0) + i\psi(x_0, y) - i\psi(x_0, y_0)}{i(y - y_0)}$$

$$= \lim_{y \to y_0} \frac{\phi(x_0, y) - \phi(x_0, y_0)}{i(y - y_0)} + \frac{i\psi(x_0, y) - i\psi(x_0, y_0)}{i(y - y_0)}$$

$$= \frac{1}{i} \frac{\partial \phi}{\partial y} + \frac{\partial \psi}{\partial y}$$

$$= \frac{\partial \psi}{\partial y} - i \frac{\partial \phi}{\partial y}.$$

EXERCISE 1.1. Show that, when $z = x \to z_0 = x_0$, then the limit is $\frac{\partial \phi}{\partial x} + i \frac{\partial \psi}{\partial x}$.

Of course, if f is complex differentiable at z_0, then both of these limits are equal to $f'(z_0)$ and, hence, to each other. Therefore, we have

$$\frac{\partial \phi}{\partial x} = \frac{\partial \psi}{\partial y} \qquad \frac{\partial \phi}{\partial y} = -\frac{\partial \psi}{\partial x}.$$

These are the *Cauchy-Riemann equations*. In fact, the analysis above may be enhanced to show that f is holomorphic on D if and only if $\frac{\partial \phi}{\partial x}, \frac{\partial \phi}{\partial y}, \frac{\partial \psi}{\partial x}, \frac{\partial \psi}{\partial y}$ exist and are continuous on D *and* the Cauchy-Riemann equations hold. We note here, without proof, that if f is holomorphic, so are all of its derivatives $f', f'', \ldots$.

EXERCISE 1.2. Show that $f(z) = z^2$ is holomorphic and compute $f'(z)$ from the limit directly and from the Cauchy-Riemann equations.

EXERCISE 1.3. The *complex conjugate* of $z = x + iy$ is $\bar{z} = x - iy$. Show that $f(z) = \bar{z}$ is not holomorphic.

EXERCISE 1.4. The *modulus* of $z = x + iy$ is $|z| = \sqrt{x^2 + y^2} = \sqrt{z\bar{z}}$. If $f(z) = u + iv$ and $\bar{f}(z) = u - iv$, then we write $|f| = \sqrt{u^2 + v^2} = \sqrt{f\bar{f}}$. Show that a holomorphic function f with $|f|$ constant is itself constant.

Suppose f is holomorphic. The Cauchy-Riemann equations give

$$\frac{\partial^2 \phi}{\partial x^2} + \frac{\partial^2 \phi}{\partial y^2} = \frac{\partial}{\partial x} \frac{\partial \psi}{\partial y} - \frac{\partial}{\partial y} \frac{\partial \psi}{\partial x}$$

$$= \frac{\partial^2 \psi}{\partial x \partial y} - \frac{\partial^2 \psi}{\partial y \partial x}$$

$$= 0$$

since mixed partials are equal. Thus, ϕ and (similarly) ψ satisfy the Laplace equation $\Delta \eta = 0$ (where $\Delta = \frac{\partial^2}{\partial x^2} + \frac{\partial^2}{\partial y^2}$) and, so, are harmonic functions. Conversely, if ϕ is a twice differentiable harmonic function of x, y, then on some open set there is another harmonic function ψ such that $f = \phi + i\psi$ is holomorphic. Harmonic functions ϕ and ψ which give such an f are said to be *harmonic conjugates*. From now on, for convenience, we shall use the notation ϕ_x for the partial derivative with respect to x etc..

EXERCISE 1.5. Let $\phi = x^2 - y^2$ and find its harmonic conjugate ψ as follows. From Cauchy-Riemann, $\phi_x = \psi_y$, so integrate ϕ_x with respect to y to get ψ up to a function of x alone. Now use $\phi_y = -\psi_x$ to determine this function. Are you surprised at your result?

Integration of complex functions is accomplished through the usual line integrals of vector calculus. Suppose $f = \phi + i\psi$ is continuous and $\gamma(t) : [a, b] \to \mathbb{C}$ is a curve. Then we define the integral of f along γ to be

$$\int_\gamma f = \int_a^b f(\gamma(t))\gamma'(t)\,dt.$$

EXERCISE 1.6. Show that $\int_\gamma f = \int_\gamma \phi(x, y)\,dx - \psi(x, y)\,dy + i\int_\gamma \phi(x, y)\,dy + \psi(x, y)\,dx$ where the integrals on the righthand side are real variable line integrals.

EXERCISE 1.7. Suppose f is holomorphic with continuous derivative on and inside a closed curve γ. Use Green's theorem on each integral in Exercise 1.6 to show that $\int_\gamma f = 0$. This is a weak version of Cauchy's theorem. Show that this implies that integrals of holomorphic functions only depend on the endpoints and not on the paths chosen over which to integrate.

The most important thing for *us* to remember is that there is a Fundamental Theorem of Calculus for complex integrals. Namely, if f is holomorphic, then

$$\int_\gamma f' = f(b) - f(a).$$

Therefore, many of the formulas from ordinary calculus carry over into complex analysis.

With a view to the future, when we shall consider a patch $\mathbf{x}(u, v)$ with complex coordinates, we write $z = u + iv$, $\bar{z} = u - iv$ and introduce the following notation for complex partial differentiation:

$$\frac{\partial}{\partial z} = \frac{1}{2}\left(\frac{\partial}{\partial u} - i\frac{\partial}{\partial v}\right) \qquad \frac{\partial}{\partial \bar{z}} = \frac{1}{2}\left(\frac{\partial}{\partial u} + i\frac{\partial}{\partial v}\right).$$

One advantage of this notation is that it provides an easy test for f to be holomorphic.

EXERCISE 1.8. Show that f is holomorphic if and only if $\frac{\partial f}{\partial \bar{z}} = 0$.

EXERCISE 1.9. Show that

$$\Delta f \overset{\text{def}}{=} f_{uu} + f_{vv} = 4 \left(\frac{\partial}{\partial z} \left(\frac{\partial f}{\partial \bar{z}} \right) \right).$$

7.2 Isothermal Coordinates

As we saw in Chapter 4, the key to introducing complex analysis into minimal surface theory is the existence of isothermal coordinates on a minimal surface. Recall that a patch $\mathbf{x}(u, v)$ is *isothermal* if $E = \mathbf{x}_u \cdot \mathbf{x}_u = \mathbf{x}_v \cdot \mathbf{x}_v = G$ and $F = 0$. In fact, isothermal coordinates exist on all surfaces, but the proof is much harder than the one given below (due to Osserman [Oss1]) for minimal surfaces.

Theorem 2.1. *Isothermal coordinates exist on any minimal surface $M \subseteq \mathbb{R}^3$.*

Proof. Fix a point $m \in M$. Choose a coordinate system for $\mathbb{R}^3$ so that m is the origin, the tangent plane to M, $T_m M$, is the xy-plane, and near m, M is the graph of a function $z = f(x, y)$. Furthermore, the quotient and chain rules give

$$\left(\frac{1 + f_x^2}{w} \right)_y - \left(\frac{f_x f_y}{w} \right)_x = -\frac{f_y}{w} \left[f_{xx}(1 + f_y^2) - 2 f_x f_y f_{xy} + f_{yy}(1 + f_x^2) \right]$$

$$\left(\frac{1 + f_y^2}{w} \right)_x - \left(\frac{f_x f_y}{w} \right)_y = -\frac{f_x}{w} \left[f_{xx}(1 + f_y^2) - 2 f_x f_y f_{xy} + f_{yy}(1 + f_x^2) \right]$$

where $w = \sqrt{1 + f_x^2 + f_y^2}$. As is traditional (and convenient), let $p = f_x$, $q = f_y$. Because M is minimal, f satisfies the minimal surface equation

$$f_{xx}(1 + f_y^2) - 2 f_x f_y f_{xy} + f_{yy}(1 + f_x^2)$$

so we have $(\frac{1+p^2}{w})_y - (\frac{pq}{w})_x = 0$ and $(\frac{1+q^2}{w})_x - (\frac{pq}{w})_y = 0$. Define two vector fields in the xy-plane by

$$V = \left(\frac{1 + p^2}{w}, \frac{pq}{w} \right) \qquad \text{and} \qquad W = \left(\frac{pq}{w}, \frac{1 + q^2}{w} \right)$$

and apply Green's Theorem to any closed curve C contained in a simply connected region $\mathcal{R}$ to obtain

$$\int_C V = \int \int_{\mathcal{R}} \left(\frac{pq}{w} \right)_x - \left(\frac{1 + p^2}{w} \right)_y dx \, dy = 0$$

$$\int_C W = \int \int_{\mathcal{R}} \left(\frac{1+q^2}{w}\right)_x - \left(\frac{pq}{w}\right)_y \, dx \, dy = 0.$$

Since the line integrals are zero for all closed curves in $\mathcal{R}$, V and W must have potential functions (see [MT]). That is, there exist μ and ρ with $\text{grad}(\mu) = V$ and $\text{grad}(\rho) = W$. (Here we use "grad" to stand for the gradient to avoid confusion with the covariant derivative.) Considered coordinatewise, these equations imply $\mu_x = \frac{1+p^2}{w}, \mu_y = \frac{pq}{w}$ and $\rho_x = \frac{pq}{w}, \rho_y = \frac{1+q^2}{w}$. Define a mapping $T \colon \mathcal{R} \to \mathbb{R}^2$ by

$$T(x,y) = (x + \mu(x,y), y + \rho(x,y)).$$

The Jacobian matrix of this mapping is then

$$J(T) = \begin{bmatrix} 1 + \mu_x & \mu_y \\ \rho_x & 1 + \rho_y \end{bmatrix} = \begin{bmatrix} 1 + \frac{1+p^2}{w} & \frac{pq}{w} \\ \frac{pq}{w} & 1 + \frac{1+q^2}{w} \end{bmatrix}$$

and we calculate the determinant to be $\det J(T) = \frac{(1+w)^2}{w} > 0$. The Inverse Function Theorem then says that, near $m = (0,0)$, there is a smooth inverse function $T^{-1}(u,v) = (x,y)$ with $J(T^{-1}) = J(T)^{-1} =$

$$\frac{1}{\det J(T)} \begin{bmatrix} 1 + \frac{1+q^2}{w} & -\frac{pq}{w} \\ -\frac{pq}{w} & 1 + \frac{1+p^2}{w} \end{bmatrix} = \frac{1}{(1+w)^2} \begin{bmatrix} w + 1 + q^2 & -pq \\ -pq & w + 1 + p^2 \end{bmatrix}.$$

Of course, the last matrix is just

$$\begin{bmatrix} x_u & x_v \\ y_u & y_v \end{bmatrix}$$

by the definition of the Jacobian. We will put these calculations to use in showing that the following patch (in the uv coordinates described above)

$$\mathbf{x}(u,v) \overset{\text{def}}{=} (x(u,v), y(u,v), f(x(u,v), y(u,v)))$$

is isothermal. First we calculate

$$\mathbf{x}_u = \left(\frac{w+1+q^2}{(1+w)^2}, \frac{-pq}{(1+w)^2}, p\left(\frac{w+1+q^2}{(1+w)^2}\right) + q\left(\frac{-pq}{(1+w)^2}\right)\right)$$

and

$$
\begin{aligned}
E &= \mathbf{x}_u \cdot \mathbf{x}_u \\
&= \frac{1}{(1+w)^4} \Big[(w+1+q^2)^2 + p^2 q^2 + p^2(w+1+q^2)^2 \\
&\qquad\qquad\qquad - 2p^2 q^2 (w+1+q^2) + p^2 q^4 \Big] \\
&= \frac{1}{(1+w)^4} \Big[(1+w)^2 (1+q^2+p^2) \Big] \\
&= \frac{w^2}{(1+w)^2}.
\end{aligned}
$$

EXERCISE 2.1. Show that $\mathbf{x}_v = \left(\frac{-pq}{(1+w)^2}, \frac{w+1+p^2}{(1+w)^2}, p\left(\frac{-pq}{(1+w)^2}\right) + q\left(\frac{w+1+p^2}{(1+w)^2}\right) \right)$ and $G = \mathbf{x}_v \cdot \mathbf{x}_v = E$ and $F = 0$.

Hence, the patch $\mathbf{x}(u,v)$ is isothermal. □

EXERCISE 2.2. If $M \colon \mathbf{x}(u,v)$ is a surface with isothermal parametrization, show that the formula for mean curvature reduces to $H = \frac{l+n}{2E}$. Hence, for M minimal, $l = -n$.

7.3 THE WEIERSTRASS-ENNEPER REPRESENTATIONS

From now on, unless stated otherwise, we shall take M to be a minimal surface described by an isothermal patch $\mathbf{x}(u,v)$. We let $z = u + iv$ denote the corresponding complex coordinate and recall $\frac{\partial}{\partial z} = \frac{1}{2}\left(\frac{\partial}{\partial u} - i\frac{\partial}{\partial v}\right)$. Since $u = \frac{z+\bar{z}}{2}$ and $v = \frac{-i(z-\bar{z})}{2}$, we may write

$$
\mathbf{x}(z,\bar{z}) = \big(x^1(z,\bar{z}), x^2(z,\bar{z}), x^3(z,\bar{z}) \big).
$$

We regard $x^i(z,\bar{z})$ as a complex valued function which happens to take real values and we have $\frac{\partial x^i}{\partial z} = \frac{1}{2}(x_u^i - ix_v^i)$. Define

$$
\phi \overset{\text{def}}{=} \frac{\partial \mathbf{x}}{\partial z} = (x_z^1, x_z^2, x_z^3).
$$

Let's examine ϕ a bit more closely. We shall use the following notation: $(\phi)^2 = (x_z^1)^2 + (x_z^2)^2 + (x_z^3)^2$ and $|\phi|^2 = |x_z^1|^2 + |x_z^2|^2 + |x_z^3|^2$, where $|z| = \sqrt{u^2 + v^2}$ is the modulus of z. First, note that $(x_z^i)^2 = \frac{1}{4}((x_u^i)^2 - (x_v^i)^2 - 2ix_u^i x_v^i)$.

Therefore,

$$(\phi)^2 = \frac{1}{4}\left(\sum_{j=1}^{3}(x_u^i)^2 - \sum_{j=1}^{3}(x_v^i)^2 - 2i\sum_{j=1}^{3}x_u^i x_v^i\right)$$

$$= \frac{1}{4}(|\mathbf{x}_u|^2 - |\mathbf{x}_v|^2 - 2i\mathbf{x}_u \cdot \mathbf{x}_v)$$

$$= \frac{1}{4}(E - G - 2iF)$$

$$= 0$$

since $\mathbf{x}(u, v)$ is isothermal. By comparing real and imaginary parts, we see that the converse is true as well. Namely, if $(\phi)^2 = 0$, then the patch must be isothermal.

EXERCISE 3.1. Show that $|\phi|^2 = \dfrac{E}{2} \neq 0$.

Finally, $\frac{\partial \phi}{\partial \bar{z}} = \frac{\partial}{\partial \bar{z}}\left(\frac{\partial \mathbf{x}}{\partial z}\right) = \frac{1}{4}\Delta\mathbf{x} = 0$ since $\mathbf{x}$ is isothermal. Therefore, each $\phi^i = \frac{\partial \mathbf{x}}{\partial z}$ is holomorphic. Conversely, the same calculation shows that, if each ϕ^i is holomorphic, then each x^i is harmonic and, therefore, M minimal. All of these observations together give

Theorem 3.1. *Suppose M is a surface with patch $\mathbf{x}$. Let $\phi = \frac{\partial \mathbf{x}}{\partial z}$ and suppose $(\phi)^2 = 0$ (i.e. $\mathbf{x}$ is isothermal). Then M is minimal if and only if each ϕ^i is holomorphic.*

In case each ϕ^i is holomorphic we say that ϕ is holomorphic. The result above says that any minimal surface may be described near each of its points by a triple of holomorphic functions $\phi = (\phi^1, \phi^2, \phi^3)$ with $(\phi)^2 = 0$. Indeed, in this case we may construct an isothermal patch for a minimal surface by taking

Corollary 3.2. $x^i(z, \bar{z}) = c_i + 2\,\mathrm{Re}\int \phi^i\,dz.$

Proof. Because $z = u + iv$, we may write $dz = du + idv$. (This is "differential" shorthand for $\frac{dz}{dt} = \frac{du}{dt} + i\frac{dv}{dt}$.) Then

$$\phi^i\,dz = \frac{1}{2}[(x_u^i - ix_v^i)(du + idv)] = \frac{1}{2}[x_u^i\,du + x_v^i\,dv + i(x_u^i\,dv - x_v^i\,du)],$$

$$\bar{\phi}^i\,d\bar{z} = \frac{1}{2}[(x_u^i + ix_v^i)(du - idv)] = \frac{1}{2}[x_u^i\,du + x_v^i\,dv - i(x_u^i\,dv - x_v^i\,du)].$$

We then have $dx^i = \frac{\partial x^i}{\partial z}dz + \frac{\partial x^i}{\partial \bar{z}}d\bar{z} = \phi^i\,dz + \bar{\phi}^i\,d\bar{z} = 2\mathrm{Re}\,\phi\,dz$ and we can now integrate to get x^i. $\qquad\square$

So, in a real sense, the problem of constructing minimal surfaces reduces to finding $\phi = (\phi^1, \phi^2, \phi^3)$ with $(\phi)^2 = 0$. A nice way of constructing such a ϕ

is to take a holomorphic function f and a *meromorphic* function g (with fg^2 holomorphic) and form

$$\phi^1 = \frac{1}{2}f(1-g^2) \quad \phi^2 = \frac{i}{2}f(1+g^2) \quad \phi^3 = fg.$$

EXERCISE 3.2. Show that this ϕ satisfies $(\phi)^2 = 0$.

EXERCISE 3.3. Show that $g = \frac{\phi^3}{\phi^1 - i\phi^2}$.

A function g is *meromorphic* if all its singularities are poles. That is, around each singularity z_0 there is a Laurent expansion (generalizing the Taylor expansion) of the form $g(z) = \frac{b_n}{(z-z_0)^n} + \cdots + \frac{b_1}{(z-z_0)} + \sum_{j=0}^{\infty} a_j(z-z_0)^j$ for some finite n with coefficients determined by g. For us, the most important examples of meromorphic functions are rational functions $g(z) = \frac{\mathcal{P}(z)}{\mathcal{Q}(z)}$ for polynomials $\mathcal{P}$, $\mathcal{Q}$. Therefore, we obtain

Theorem 3.3: Weierstrass-Enneper Representation I. *If f is holomorphic on a domain D, g is meromorphic on D and fg^2 is holomorphic on D, then a minimal surface is defined by $\mathbf{x}(z, \bar{z}) = \big(x^1(z, \bar{z}), x^2(z, \bar{z}), x^3(z, \bar{z})\big)$, where*

$$x^1(z, \bar{z}) = \operatorname{Re} \int f(1-g^2)\, dz$$

$$x^2(z, \bar{z}) = \operatorname{Re} \int i\, f(1+g^2)\, dz$$

$$x^3(z, \bar{z}) = \operatorname{Re} 2\int fg\, dz.$$

Before we look at some examples, it is worthwhile noting another form of the Weierstrass-Enneper Representation. Suppose g is holomorphic and has an inverse function g^{-1} in a domain D which is holomorphic as well. Then we can consider g as a new complex variable $\tau = g$ with $d\tau = g'dz$. Define $F(\tau) = f/g'$ and obtain $F(\tau)d\tau = f\, dz$. Therefore, if we replace g by τ and $f\, dz$ by $F(\tau)d\tau$, we get

Theorem 3.4: Weierstrass-Enneper Representation II. *For any holomorphic function $F(\tau)$, a minimal surface is defined by the parametrization $\mathbf{x}(z, \bar{z}) = \big(x^1(z, \bar{z}), x^2(z, \bar{z}), x^3(z, \bar{z})\big)$, where*

$$x^1(z, \bar{z}) = \operatorname{Re} \int (1-\tau^2)F(\tau)\, d\tau$$

$$x^2(z, \bar{z}) = \operatorname{Re} \int i\, (1+\tau^2)F(\tau)\, d\tau$$

$$x^3(z, \bar{z}) = \operatorname{Re} 2\int \tau F(\tau)\, d\tau.$$

Note the corresponding

$$\phi = \left(\frac{1}{2}(1 - \tau^2)F(\tau), \ \frac{i}{2}(1 + \tau^2)F(\tau), \ \tau F(\tau) \right).$$

This representation tells us that *any* holomorphic function $F(\tau)$ defines a minimal surface! Of course, we can't expect every function to give complex integrals which may be evaluated to nice formulas. Nevertheless, we shall see that we may calculate much information about a minimal surface directly from its representations. In order to look at some of our standard minimal surfaces from the representation viewpoint, we need to recall some of the basic functions of complex analysis. For this, write $z = u + iv$ and define

$$e^z = e^u(\cos v + i \sin v)$$

and

$$\log(z) = \ln \sqrt{u^2 + v^2} + i \arctan\left(\frac{v}{u}\right).$$

We haven't been precise here about branches of the log, but this technicality will not concern us. Using the definition of e^z, we may define

$$\sin z = \frac{e^{iz} - e^{-iz}}{2i} \qquad \sinh z = \frac{e^z - e^{-z}}{2}$$

$$\cos z = \frac{e^{iz} + e^{-iz}}{2} \qquad \cosh z = \frac{e^z + e^{-z}}{2}.$$

One reason these definitions are chosen is that they extend the usual real functions of the same name. For example, if $z = u$, then the definition of e^z gives $\sin z$ as the real function $\sin u$. Similarly, for $z = u$, $\sinh z = \sinh u$. While these complex functions have exactly the same differentiation, integration, sum and difference rules as their real cousins, it is often useful to expand the complex functions into their real and imaginary parts. To accomplish this, we replace z by $u + iv$ and use the definition of e^z above.

EXERCISE 3.4. Derive the following formulas:

(1) $\sin z = \sin u \cosh v + i \cos u \sinh v.$
(2) $\cos z = \cos u \cosh v + i \sin u \sinh v.$
(3) $\sinh z = \sinh u \cos v + i \cosh u \sin v.$
(4) $\cosh z = \cosh u \cos v + i \sinh u \sin v.$

Example 3.1: The Catenoid. Let $F(\tau) = \frac{1}{2\tau^2}$. Then, using the substitution $\tau = e^z$ below, we obtain

$$x^1 = \text{Re} \int (1 - \tau^2) \frac{1}{2\tau^2} \, d\tau \qquad\qquad x^2 = \text{Re}\, i \int (1 + \tau^2) \frac{1}{2\tau^2} \, d\tau$$

$$= \text{Re} \int \frac{1}{2\tau^2} - \frac{1}{2} \, d\tau \qquad\qquad = \text{Re} \int i \left[\frac{1}{2\tau^2} + \frac{1}{2} \right] d\tau$$

$$= -\text{Re} \left[\frac{1}{2\tau} + \frac{\tau}{2} \right] \qquad\qquad = -\text{Re}\, i \left[\frac{1}{2\tau} - \frac{\tau}{2} \right]$$

$$= -\text{Re} \frac{e^{-z} + e^z}{2} \qquad\qquad = \text{Re}\, i \frac{-e^{-z} + e^z}{2}$$

$$= -\text{Re} \cosh z \qquad\qquad = \text{Re}\, i \sinh z$$

$$= -2 \cosh u \cos v. \qquad\qquad = -2 \cosh u \sin v.$$

$$x^3 = \text{Re}\, 2 \int \tau \frac{1}{2\tau^2} \, d\tau$$

$$= \text{Re} \int \frac{1}{\tau} \, d\tau$$

$$= \text{Re} \ln \tau$$

$$= \text{Re}\, z$$

$$= u.$$

We obtain (without the negative signs) a parametrization $\mathbf{x}(u, v) = (\cosh u \cos v, \cosh u \sin v, u)$ which we recognize as a catenoid.

EXERCISE 3.5. Let $F(\tau) = \frac{i}{2\tau^2}$. Show that the representation associated to $F(\tau)$ is a helicoid.

EXERCISE 3.6. Show that the catenoid and helicoid, respectively, also have representations of the form $(f, g) = (-\frac{e^{-z}}{2}, -e^z)$ and $(f, g) = (-i \frac{e^{-z}}{2}, -e^z)$.

EXERCISE 3.7. Show that the helicoid has another representation $(f, g) = (-\frac{i}{2}, \frac{1}{z})$. This representation seems simpler than the first two, but there is a small problem. Namely, the integrals of the representation are not *path independent*. They are then said to have *real periods*. Show that if we carry out the integration for x^3 (of the representation above) about the entire unit circle, we obtain a real period equal to 2π. To fix this ambiguity, we take a smaller domain $\mathbb{C}/\{0 \cup \mathbb{R}^-\}$. We therefore obtain one loop of the helicoid from this representation. Other loops may be similarly obtained and fitted together to create the whole surface. Thus, the simplicity of the representation $(f, g) = (-\frac{i}{2}, \frac{1}{z})$ hides the subtlety underlying complex integration.

EXERCISE 3.8. Let $F(\tau) = i(\frac{1}{\tau} - \frac{1}{\tau^3})$. Show that the associated representation is Catalan's surface $\mathbf{x}(u, v) = (u - \sin u \cosh v, \ 1 - \cos u \cosh v, \ 4 \sin \frac{u}{2} \sinh \frac{v}{2})$. Hint: after integrating, replace τ by $e^{-iz/2}$ and use the expansion of $\sin z$.

EXERCISE 3.9. Let $F(\tau) = 1 - \frac{1}{\tau^4}$. Show that the associated representation is Henneberg's surface

$$\mathbf{x}(u, v) = (2\sinh u \cos v - \frac{2}{3}\sinh 3u \cos 3v,$$

$$2\sinh u \sin v + \frac{2}{3}\sinh 3u \sin 3v,$$

$$2\cosh 2u \cos 2v).$$

Let $v = \frac{\pi}{2}$ and show that you obtain Neil's parabola $(z - 2)^3 = 9x^2$. Hint: for the first part, after integrating, replace τ by e^z, use the expansions of $\sin z$ and $\cos z$ and replace y by $-y$.

EXERCISE 3.10. Let $F(\tau) = 1$ (or, equivalently, $(f, g) = (1, z)$. Show that the Weierstrass-Enneper representation gives you Enneper's surface. If $(f, g) = (1, z^n)$, then an nth-order Enneper surface is obtained. Calculate the $\mathbf{x}(u, v)$ for the 2nd-order Enneper surface. Hint: MAPLE might be useful here.

EXERCISE 3.11. Let $(f, g) = (z^2, \frac{1}{z^2})$. Calculate the $\mathbf{x}(u, v)$ given by the Weierstrass-Enneper representation. This surface is called *Richmond's surface*.

EXERCISE 3.12. Let $F(\tau) = \frac{2}{1 - \tau^4}$. Show that the Weierstrass-Enneper representation gives Scherk's surface $z = \ln(\frac{\cos y}{\cos x})$. Hints: (1) for x^1, use partial fractions to get $\int \frac{2}{1+\tau^2} d\tau = i\log(1 - i\tau) - i\log(1 + i\tau)$, (2) since the log's are multiplied by i and you want the real part, the definition of $\log(z)$ gives $x^1 = \arctan(\frac{u}{1-v}) - \arctan(\frac{-u}{1+v})$ (i.e. $1 + i\tau = 1 - v + iu$ and $1 - i\tau = 1 + v - iu$), (3) the formula for the tangent of a difference of two angles then gives $x^1 = \arctan(\frac{2u}{1-(u^2+v^2)})$, (4) similarly, $x^2 = \arctan(\frac{-2v}{1-(u^2+v^2)})$, (5) also, $x^3 = \text{Re}(\log(\tau^2 + 1) - \log(\tau^2 - 1)) = \frac{1}{2}\ln\left(\frac{(u^2-v^2+1)^2+4u^2v^2}{(u^2-v^2-1)^2+4u^2v^2}\right)$, (6) by drawing appropriate right triangles, write $\cos x^2$ and $\cos x^3$ in terms of u and v and show $x^3 = \ln(\frac{\cos x^2}{\cos x^1})$. Note that the integrals have real periods so that, just as for the helicoid, we can take one piece at a time to build up the whole surface.

While we have seen that the Weierstrass-Enneper representations give us our standard array of minimal surfaces, their true value becomes apparent when we realize that we can analyze many facets of minimal surfaces directly from the representing functions (f, g) and $F(\tau)$. In particular, this applies even to surfaces whose Weierstrass-Enneper integrals may not be explicitly computable. As an example of this, we shall compute the Gaussian curvature K of a minimal surface in terms of $F(\tau)$. First note that isothermal parameters give

$$
\begin{aligned}
K &= -\frac{1}{2\sqrt{EG}}\left(\frac{\partial}{\partial v}\left(\frac{E_v}{\sqrt{EG}}\right) + \frac{\partial}{\partial u}\left(\frac{G_u}{\sqrt{EG}}\right)\right) \\
&= -\frac{1}{2E}\left(\frac{\partial}{\partial v}\left(\frac{E_v}{E}\right) + \frac{\partial}{\partial u}\left(\frac{E_u}{E}\right)\right) \\
&= -\frac{1}{2E}\left(\frac{\partial^2}{\partial v^2}\ln E + \frac{\partial^2}{\partial u^2}\ln E\right) \\
&= -\frac{1}{2E}\Delta(\ln E),
\end{aligned}
$$

where Δ is the Laplace operator $\frac{\partial^2}{\partial u^2} + \frac{\partial^2}{\partial v^2}$. By Exercise 3.1, we know that $E = 2|\phi|^2$, so we compute E for $\phi = (\frac{1}{2}(1 - \tau^2)F(\tau), \frac{i}{2}(1 + \tau^2)F(\tau), \tau F(\tau))$. We have

$$
\begin{aligned}
E &= 2\left[|\frac{1}{2}(1 - \tau^2)F(\tau)|^2 + |\frac{i}{2}(1 + \tau^2)F(\tau)|^2 + |\tau F(\tau)|^2\right] \\
&= \frac{1}{2}|F|^2\left[|\tau^2 - 1|^2 + |\tau^2 + 1|^2 + 4|\tau|^2\right].
\end{aligned}
$$

Now, $\tau^2 = u^2 - v^2 + 2iuv$, so $|\tau^2 - 1|^2 = (u^2 - v^2 - 1)^2 + 4u^2v^2$. Similarly, $|\tau^2 + 1|^2 = (u^2 - v^2 + 1)^2 + 4u^2v^2$ and $4|\tau|^2 = 4(u^2 + v^2)$. Then

$$
\begin{aligned}
E &= \frac{1}{2}|F|^2 2\left[(u^2 - v^2)^2 + 1 + 4u^2v^2 + 2u^2 + 2v^2\right]. \\
&= |F|^2\left[u^4 + 2u^2v^2 + v^4 + 1 + 2u^2 + 2v^2\right] \\
&= |F|^2\left[1 + u^2 + v^2\right]^2.
\end{aligned}
$$

EXERCISE 3.13. Show directly from the definition of ϕ in terms of the representation (f, g) that $E = |f|^2\left[1 + |g|^2\right]^2$.

Now, $\ln E = \ln|F|^2 + 2\ln(1 + u^2 + v^2)$ and it is an easy task to prove the following.

EXERCISE 3.14. Show that $\Delta(2\ln(1 + u^2 + v^2)) = \dfrac{8}{(1 + u^2 + v^2)^2}$.

We must now calculate $\Delta(\ln|F|^2) = \Delta(\ln F\bar{F}) = \Delta(\ln F + \ln\bar{F})$. Previously, we have seen that $\Delta = 4\partial^2/\partial\bar{z}\partial z$. Further, because F is holomorphic, $\bar{F}$ cannot be, so $\partial\bar{F}/\partial z = 0$ and, consequently, $\partial\ln\bar{F}/\partial z = 0$. We are left with $\Delta(\ln F) = 4\partial^2\ln F/\partial\bar{z}\partial z = 4\partial(F'/F)/\partial\bar{z} = 0$ since F, F' and, hence, F'/F are holomorphic. Thus, $\Delta(\ln|F|^2) = 0$ and $\Delta(\ln E) = 8/(1 + u^2 + v^2)^2$.

Theorem 3.5. *The Gaussian curvature of the minimal surface determined by the Weierstrass-Enneper representation II is*

$$K = \frac{-4}{|F|^2(1 + u^2 + v^2)^4}.$$

Proof. From the calculations above,

$$K = -\frac{1}{2E}\Delta(\ln E)$$

$$= \frac{-8}{2|F|^2(1 + u^2 + v^2)^4}$$

$$= \frac{-4}{|F|^2(1 + u^2 + v^2)^4}.$$

$\square$

EXERCISE 3.15. Use the identification $F = \frac{f}{g'}$ to derive the formula

$$K = \frac{-4|g'|^2}{|f|^2(1 + |g|^2)^4}$$ in terms of the Weierstrass-Enneper representation I.

EXERCISE 3.16. Explain the apparent discrepancy between the two formulas for K. That is, the first formula never allows $K = 0$ while the second has $K = 0$ at points where $g' = 0$. Hint: what was our assumption about g which allowed the transformation from Weierstrass-Enneper I to Weierstrass-Enneper II.

EXERCISE 3.17. Recall that a point is umbilic if the two principal curvatures at the point are equal. Show that umbilic points on a minimal surface with representation (f, g) correspond to the zeros of g'. Hence, umbilic points on a minimal surface are *flat*.

EXERCISE 3.18. A minimal surface described by (f, g) or $F(\tau)$ has an *associated family* of minimal surfaces given by (respectively) $(e^{it}f, g)$ or $e^{it}F(\tau)$. Two surfaces of the family described by t_0 and t_1 are said to be *adjoint* if $t_1 - t_0 = \frac{\pi}{2}$. Show that all the surfaces of an associated family are locally isometric. Hints: (1) since we have isothermal coordinates, it is enough to show that E remains the same no matter what t is taken and (2) $|e^{it}| = 1$ for all t.

EXERCISE 3.19. The catenoid has a representation $(f, g) = (-\frac{e^{-z}}{2}, -e^z)$. Write the $\mathbf{x}(u, v)$ for its associated family and its adjoint surface.

EXERCISE 3.20. Find the adjoint surface to Henneberg's surface with $F(\tau) = -i(1 - \frac{1}{\tau^4})$. Set $u = 0$ and show that the resulting v-parameter curve is the astroid $x^{2/3} + y^{2/3} = (8/3)^{2/3}$. This astroid is a geodesic.

So far, we have treated the Weierstrass-Enneper representation more from an algebraic and analytic point of view than from a geometric one. But, after all, it is the geometry of minimal surfaces which interests us, so it is reasonable at this point to connect the discussion above with our usual differential geometric constructs. For this, recall that the Gauss map of a surface $M: \mathbf{x}(u, v)$ is a mapping from the surface to the unit sphere S^2, denoted $G: M \to S^2$ and given by $G(p) = U_p$, where U_p is the unit normal to M at p. In terms of the parametrization, we may write $G(\mathbf{x}(u, v)) = U(u, v)$ and, for a small piece of M, think of $U(u, v)$ as a parametrization of the sphere S^2. Recall also that the induced linear transformation of tangent planes is given, for the basis $\{\mathbf{x}_u, \mathbf{x}_v\}$, by $G_*(\mathbf{x}_u) = U_u = -S(\mathbf{x}_u)$ and $G_*(\mathbf{x}_v) = U_v = -S(\mathbf{x}_v)$.

Proposition 3.6. *Let $M: \mathbf{x}(u, v)$ be a minimal surface parametrized by isothermal coordinates. Then the Gauss map of M is a conformal map.*

Proof. Recall from Chapter 5 that, in order to show G to be conformal, we need only show $|G_*(\mathbf{x}_u)| = \rho(u, v)|\mathbf{x}_u|$, $|G_*(\mathbf{x}_v)| = \rho(u, v)|\mathbf{x}_v|$ and $G_*(\mathbf{x}_u) \cdot G_*(\mathbf{x}_v) = \rho^2(u, v)\, \mathbf{x}_u \cdot \mathbf{x}_v$. Since isothermal coordinates have $E = G$ and $F = 0$, we get

$$G_*(\mathbf{x}_u) = U_u = -\frac{l}{E}\mathbf{x}_u - \frac{m}{E}\mathbf{x}_v \quad , \quad G_*(\mathbf{x}_v) = U_v = -\frac{m}{E}\mathbf{x}_u - \frac{n}{E}\mathbf{x}_v.$$

where we have used the formulas for U_u and U_v developed in Chapter 5. Taking dot products gives

$$|U_u|^2 = \frac{1}{E}\left[l^2 + m^2\right], \qquad |U_v|^2 = \frac{1}{E}\left[m^2 + n^2\right]$$

$$U_u \cdot U_v = \frac{m}{E}\left[l + n\right].$$

But, by Exercise 2.2, $l = -n$, so we obtain

$$|U_u|^2 = \frac{1}{E}\left[l^2 + m^2\right] = |U_v|^2 \quad \text{and} \quad U_u \cdot U_v = 0.$$

Since $|\mathbf{x}_u| = \sqrt{E} = |\mathbf{x}_v|$ and $\mathbf{x}_u \cdot \mathbf{x}_v = 0$, we see that the Gauss map G is conformal with conformality factor $\sqrt{l^2 + m^2}/E$. $\square$

EXERCISE 3.21. Show that the conformality factor $\sqrt{l^2 + m^2}/E$ is equal to $\sqrt{|K|}$, where K is the Gaussian curvature.

EXERCISE 3.22. Suppose $M: \mathbf{x}(u, v)$ is a surface whose Gauss map $G: M \to S^2$ is conformal. Show that either M is (part of) a sphere or M is a minimal surface. For simplicity, assume that the patch $\mathbf{x}(u, v)$ is orthogonal (but not necessarily isothermal).

EXERCISE 3.23. Show that a nonplanar minimal surface $M : \mathbf{x}(u, v)$ cannot have constant Gaussian curvature. Hint: (1) if so, then define $\widetilde{M} : \mathbf{y}(u, v) = \sqrt{|K|}\,\mathbf{x}(u, v)$ and show that $\widetilde{M}$ has constant Gauss curvature -1 as well as the same unit normal as M. (2) Show the Gauss map of $\widetilde{M}$ is a local isometry. (3) Why is this a contradiction?

Recall that *stereographic projection from the North pole* is denoted by St: $S^2/\{N\} \to \mathbb{R}^2$ and defined by

$$\mathrm{St}(\cos u \cos v, \sin u \cos v, \sin v) = \left(\frac{\cos u \cos v}{1 - \sin v}, \frac{\sin u \cos v}{1 - \sin v}, 0 \right).$$

We have seen previously in Chapter 5 that stereographic projection is conformal with conformality factor $1/(1 - \sin v)$. In cartesian coordinates, stereographic projection is simply $\mathrm{St}(x, y, z) = (x/(1-z), y/(1-z), 0)$. We may identify $\mathbb{R}^2$ with $\mathbb{C}$ and extend St to a one-to-one onto mapping St: $S^2 \to \mathbb{C} \cup \{\infty\}$ with the North pole mapping to ∞. With these identifications, we have

Theorem 3.7. *Let $M : \mathbf{x}(u, v)$ be a minimal surface in isothermal coordinates with Weierstrass-Enneper representation (f, g). Then the Gauss map of M, $G : M \to \mathbb{C} \cup \{\infty\}$, may be identified with the meromorphic function g.*

Proof. The proof is a long calculation which brings together all of our complex analytic ingredients. Recall that $\phi = \frac{\partial \mathbf{x}}{\partial z}$, $\bar{\phi} = \frac{\partial \mathbf{x}}{\partial \bar{z}}$ and

$$\phi^1 = \frac{1}{2}f(1 - g^2) \quad \phi^2 = \frac{i}{2}f(1 + g^2) \quad \phi^3 = fg.$$

We will describe the Gauss map in terms of ϕ^1, ϕ^2 and ϕ^3. First, we write $\mathbf{x}_u \times \mathbf{x}_v = ((\mathbf{x}_u \times \mathbf{x}_v)^1, (\mathbf{x}_u \times \mathbf{x}_v)^2, (\mathbf{x}_u \times \mathbf{x}_v)^3) = (x_u^2 x_v^3 - x_u^3 x_v^2, x_u^3 x_v^1 - x_u^1 x_v^3, x_u^1 x_v^2 - x_u^2 x_v^1)$. Let's consider the first component $(\mathbf{x}_u \times \mathbf{x}_v)^1 = x_u^2 x_v^3 - x_u^3 x_v^2$. We have

$$x_u^2 x_v^3 - x_u^3 x_v^2 = \mathrm{Im}[(x_u^2 - ix_v^2)(x_u^3 + ix_v^3)]$$
$$= \mathrm{Im}[2(\partial x^2/\partial z) \cdot 2(\partial x^3/\partial \bar{z})]$$
$$= 4\,\mathrm{Im}(\phi^2 \bar{\phi}^3).$$

Similarly, $(\mathbf{x}_u \times \mathbf{x}_v)^2 = 4\,\mathrm{Im}(\phi^3 \bar{\phi}^1)$ and $(\mathbf{x}_u \times \mathbf{x}_v)^3 = 4\,\mathrm{Im}(\phi^1 \bar{\phi}^2)$. Hence,

$$\mathbf{x}_u \times \mathbf{x}_v = 4\,\mathrm{Im}(\phi^2 \bar{\phi}^3, \phi^3 \bar{\phi}^1, \phi^1 \bar{\phi}^2) = 2(\phi \times \bar{\phi}),$$

where the last equality follows from $z - \bar{z} = 2\,\mathrm{Im}\,z$. Now, since $\mathbf{x}(u, v)$ is isothermal, $|\mathbf{x}_u \times \mathbf{x}_v| = |\mathbf{x}_u| \cdot |\mathbf{x}_v| = |\mathbf{x}_u|^2 = E = 2|\phi|^2$. Therefore,

$$U = \frac{\mathbf{x}_u \times \mathbf{x}_v}{|\mathbf{x}_u \times \mathbf{x}_v|} = \frac{2(\phi \times \bar{\phi})}{2|\phi|^2} = \frac{\phi \times \bar{\phi}}{|\phi|^2}.$$

We now compute the Gauss map $G\colon M \to \mathbb{C} \cup \{\infty\}$.

$$G(\mathbf{x}(u,v)) = \mathrm{St}(U(u,v))$$

$$= \mathrm{St}\left(\frac{\phi \times \bar{\phi}}{|\phi|^2}\right)$$

$$= \mathrm{St}\left(\frac{2\,\mathrm{Im}(\phi^2\bar{\phi}^3, \phi^3\bar{\phi}^1, \phi^1\bar{\phi}^2)}{|\phi|^2}\right)$$

$$= \left(\frac{2\,\mathrm{Im}(\phi^2\bar{\phi}^3)}{|\phi|^2 - 2\,\mathrm{Im}(\phi^1\bar{\phi}^2)}, \frac{2\,\mathrm{Im}(\phi^3\bar{\phi}^1)}{|\phi|^2 - 2\,\mathrm{Im}(\phi^1\bar{\phi}^2)}, 0\right).$$

The last equality follows since

$$\frac{x}{1-z} = \frac{2\,\mathrm{Im}(\phi^2\bar{\phi}^3)}{|\phi|^2} \cdot \frac{1}{1 - \frac{2\,\mathrm{Im}(\phi^1\bar{\phi}^2)}{|\phi|^2}}$$

$$= \frac{2\,\mathrm{Im}(\phi^2\bar{\phi}^3)}{|\phi|^2} \cdot \frac{|\phi|^2}{|\phi|^2 - 2\,\mathrm{Im}(\phi^1\bar{\phi}^2)}$$

$$= \frac{2\,\mathrm{Im}(\phi^2\bar{\phi}^3)}{|\phi|^2 - 2\,\mathrm{Im}(\phi^1\bar{\phi}^2)}$$

and similarly for $y/(1-z)$. Identifying $(x,y) \in \mathbb{R}^2$ with $x + iy \in \mathbb{C}$ allows us to write

$$G(\mathbf{x}(u,v)) = \frac{2\,\mathrm{Im}(\phi^2\bar{\phi}^3) + 2i\,\mathrm{Im}(\phi^3\bar{\phi}^1)}{|\phi|^2 - 2\,\mathrm{Im}(\phi^1\bar{\phi}^2)}.$$

Now, let's consider the numerator $\mathcal{N}$ of this fraction.

$$\mathcal{N} = 2\,\mathrm{Im}(\phi^2\bar{\phi}^3) + 2i\,\mathrm{Im}(\phi^3\bar{\phi}^1)$$

$$= \frac{1}{i}[\phi^2\bar{\phi}^3 - \bar{\phi}^2\phi^3 + i\,\phi^3\bar{\phi}^1 - i\,\bar{\phi}^3\phi^1]$$

$$= \phi^3(\bar{\phi}^1 + i\,\bar{\phi}^2) - \bar{\phi}^3(\phi^1 + i\,\phi^2).$$

Also, $0 = (\phi)^2 = (\phi^1)^2 + (\phi^2)^2 + (\phi^3)^2 = (\phi^1 - i\,\phi^2)(\phi^1 + i\,\phi^2) + (\phi^3)^2$, so

$$\phi^1 + i\,\phi^2 = \frac{-(\phi^3)^2}{\phi^1 - i\,\phi^2}.$$

Then we have,

$$\mathcal{N} = \phi^3(\bar{\phi}^1 + i\,\bar{\phi}^2) + \bar{\phi}^3\frac{(\phi^3)^2}{\phi^1 - i\,\phi^2}$$

$$= \frac{\phi^3[(\phi^1 - i\,\phi^2)(\bar{\phi}^1 + i\,\bar{\phi}^2) + |\phi^3|^2]}{\phi^1 - i\,\phi^2}$$

$$= \frac{\phi^3}{\phi^1 - i\,\phi^2}\left[|\phi^1|^2 + |\phi^2|^2 + |\phi^3|^2 + i(\bar{\phi}^2\phi^1 - \phi^2\bar{\phi}^1)\right]$$

$$= \frac{\phi^3}{\phi^1 - i\,\phi^2}\left[|\phi|^2 - 2\,\mathrm{Im}(\phi^1\bar{\phi}^2)\right].$$

Hence, the second factor of the numerator $\mathcal{N}$ cancels the denominator of $G(\mathbf{x}(u, v))$ and we end up with

$$G(\mathbf{x}(u, v)) = \frac{\phi^3}{\phi^1 - i\,\phi^2}.$$

By Exercise 3.3, we know that $g = \frac{\phi^3}{\phi^1 - i\,\phi^2}$ as well, so we are done. □

Using the Weierstrass-Enneper representation II, we see that the Gauss map may be identified with the complex variable τ as well.

EXERCISE 3.24. A point $p = \mathbf{x}(u_0, v_0)$ is a *pole* of the Gauss map $G\colon M \to \mathbb{C} \cup \infty$ if $G(p) = \infty$. By the definition of stereographic projection, p is a pole of G if and only if $U(p) = (0, 0, 1)$, the North pole. Verify this by showing that $U(p) = (0, 0, 1)$ if and only if the denominator of the following expression for G vanishes:

$$G(\mathbf{x}(u, v)) = \frac{2\operatorname{Im}(\phi^2 \bar{\phi}^3) + 2i\operatorname{Im}(\phi^3 \bar{\phi}^1)}{|\phi|^2 - 2\operatorname{Im}(\phi^1 \bar{\phi}^2)}.$$

EXERCISE 3.25. Prove **Bernstein's Theorem**: If a minimal surface $M\colon z = f(x, y)$ is defined on the whole xy-plane, then M is a plane. Hints: (1) In the proof of the existence of isothermal coordinates, if the parameter domain is the whole plane, then the potential functions μ and ρ can be extended over the plane as well. The mapping T then becomes a diffeomorphism (i.e. a smooth one-to-one onto map with smooth inverse) between the xy- and uv-planes. Therefore, we may assume that M has parameter domain the whole uv-plane where u and v are isothermal coordinates. (2) The normals for M are contained in a hemisphere. Rotate the sphere to get them in the lower hemisphere. (3) Think of the uv-plane as the complex plane $\mathbb{C}$ and look at the composition $\mathbb{C} \to M \xrightarrow{G} S^2/\{N\} \to \mathbb{C}$. Why is this map holomorphic? Think of g. (4) Liouville's Theorem in complex analysis says that a complex function defined on the entire complex plane which is both bounded and holomorphic is constant.

7.4 BJÖRLING'S PROBLEM

Often, we wish to create minimal surfaces having preassigned properties. For example, in Plateau's problem, we desire a minimal surface which spans a given boundary curve. In this case, preassigning the values of the minimal surface on the boundary puts a considerable constraint on construction of the surface. There are other natural constraints, however, which are easier to handle, but no less interesting. Suppose we wish to construct a minimal surface which contains a prescribed path of a free particle constrained to move on it. Since free particles move along geodesics (see Chapter 8), we are asking for a minimal surface containing a given curve as a geodesic. Can we solve this problem? The answer turns out to be yes. In fact, although we shall give the proof of the solution below for the geodesic problem, it is possible to formulate the problem and solution for any vector field along the given curve which is orthogonal to the curve (see Exercise 4.1 below). First of all, let's be precise about the question.

Björling's Problem 4.1: (Special Case). *Given a real analytic curve* $\alpha\colon I \to \mathbb{R}^3$ *(with* $\alpha'(t) = 0$ *only at isolated points), construct a minimal surface* $M\colon \mathbf{x}(u,v)$ *having* u-*parameter curve* $\alpha(u) = \mathbf{x}(u,0)$ *such that the normal* N *of* α *is the unit normal of* M, $N(u) = U(u,0)$, *for* $u \in I$.

As usual, we may assume α is parametrized by arclength when it is convenient. Recall that a curve in a surface is a geodesic if its acceleration is always parallel to the unit normal of the surface. Since the direction of the acceleration is precisely the direction of N, we see that the condition of Björling's problem is that the given curve be a geodesic in the surface. To solve Björling's problem, we must delve deeper into complex analysis then we have so far. For example, the curve α is required to be real analytic (i.e. have a Taylor series which converges on I) because we can convert such curves automatically to their complex counterparts by replacing the real variable in the Taylor series of α by a complex variable z. The resulting complex Taylor series *defines* a holomorphic complex curve $\alpha(z)\colon D \to \mathbb{C}^3$ on a domain $D \subseteq \mathbb{C}$ containing the interval I. We say that $\alpha(z)$ is a *holomorphic extension* of $\alpha(t)$. Similarly, $N(t)$ may be extended to $N(z)$. The existence of a Taylor series for holomorphic functions has a powerful consequence which we shall use below. For a proof, see [MH] for example.

Identity Theorem 4.1: (Principle of Analytic Continuation). *If* f *and* g *are two holomorphic functions on a connected, open region* $D \subseteq \mathbb{C}$ *and* $f(z_i) = g(z_i)$ *for some convergent sequence* $z_1, z_2, \ldots, z_n, \ldots \to \bar{z}$ *in* D, *then* $f = g$ *on all of* D.

This beautiful and surprising result will be the key to our solution of Björling's problem. So, given $\alpha(t)$ and $N(t)$ as above, let $\alpha(z)$ and $N(z)$ be the respective holomorphic extensions on a simply connected domain D containing I. We then have

Theorem 4.2: (Solution to Björling's Problem). *There is exactly one solution to Björling's problem and it is given by*

$$\mathbf{x}(u,v) = \operatorname{Re}\left[\alpha(z) - i \int_{u_0}^{z} N(w) \times \alpha'(w)\, dw\right],$$

where $u_0 \in I$ *is fixed (and* w *is a complex dummy variable).*

Proof. We start by proving the solution is unique if it exists. Suppose $\mathbf{x}(u,v)$ is a solution to Björling's problem. That is, $\mathbf{x}(u,v)$ is a minimal surface (without loss of generality, in isothermal coordinates) having $\mathbf{x}(u,0) = \alpha(u)$ and

$U(u, 0) = N(u)$. Since $\mathbf{x}(u, v)$ is isothermal, each coordinate function $x^i(u, v)$ is harmonic. Let $y^i(u, v)$ denote the corresponding harmonic conjugate (with $y^i(u_0, 0) = 0$ for definiteness) and note that $x^i(z) + iy^i(z)$ is holomorphic. We can create a holomorphic curve $\beta \colon D \to \mathbb{C}^3$ by defining, for $z = u + iv \in D$,

$$\beta(z) = \mathbf{x}(u, v) + i\, \mathbf{y}(u, v).$$

Now, $\beta'(z) = \mathbf{x}_u + i\, \mathbf{y}_u = \mathbf{x}_u - i\, \mathbf{x}_v$ by usual complex differentiation and the Cauchy-Riemann equations. But $\mathbf{x}(u, v)$ is in isothermal coordinates, so $|\mathbf{x}_u \times \mathbf{x}_v| = |\mathbf{x}_u| \cdot \mathbf{x}_v| = \sqrt{E} \cdot \sqrt{G} = E$ and, hence, $\mathbf{x}_u \times \mathbf{x}_v = E\, U$. This implies that $\mathbf{x}_v = U \times \mathbf{x}_u$ and we obtain

$$\beta' = \mathbf{x}_u - i\, U \times \mathbf{x}_u.$$

If we restrict to $(u, 0) \in I$, then we have $\beta'(u) = \alpha'(u) - i\, N(u) \times \alpha'(u)$ since $\mathbf{x}_u(u, o) = \alpha'(u)$ and $U(u, 0) = N(u)$. We integrate this equation in the usual real variables sense to get $\beta(u) = \alpha(u) - i \int_{u_0}^{u} N(t) \times \alpha'(t)\, dt$ for all $u \in I$. (Here we have also used the initial conditions $y^i(u, 0) = 0$ to determine the constant of integration.) But now we see that, on the whole interval I, β agrees with the holomorphic curve $\gamma(z) = \alpha(z) - i \int N(w) \times \alpha'(w)\, dw$. The Identity Theorem then says that β and γ agree on all of D. Then, since the real part of β is $\mathbf{x}(u, v)$ by definition, we get the required form

$$\mathbf{x}(u, v) = \mathrm{Re} \left[\alpha(z) - i \int_{u_0}^{z} N(w) \times \alpha'(w)\, dw \right].$$

To prove that a solution to Björling's problem exists, we define a holomorphic curve by $\beta(z) = \left[\alpha(z) - i \int_{u_0}^{z} N(w) \times \alpha'(w)\, dw \right]$ on a domain D where the power series converge. Note that, since $N(z)$ and $\alpha'(z)$ are real for $z \in I$, $\mathrm{Re}\, \beta'(z) = \alpha'(z)$ and $\mathrm{Im}\, \beta'(z) = -N(z) \times \alpha'(z)$. Let us denote β' by ϕ to match up with our earlier work. For $z \in I$, the usual rules of vector calculus apply to give $\alpha' \cdot (N \times \alpha') = 0$ and $|N \times \alpha'| = |N| \cdot |\alpha'| = |\alpha'|$. Then, for $z \in I$ we have

$$\begin{aligned}
(\phi)^2 &= \alpha'(z) \cdot \alpha'(z) - 2i\alpha'(z) \cdot (N(z) \times \alpha'(z)) \\
&\quad - (N(z) \times \alpha'(z)) \cdot (N(z) \times \alpha'(z)) \\
&= |\alpha'|^2 - 0 - |N \times \alpha'|^2 \\
&= 0.
\end{aligned}$$

Again, the Identity Theorem implies that the holomorphic function $(\phi)^2$ must be zero on all of D. But this is exactly the situation which led to the Weierstrass-Enneper representations. Hence, we know by Theorem 3.1, that the real part of $\int \phi \, dw = \beta$ is a minimal surface in isothermal coordinates. That is, we have a minimal surface defined by

$$\mathbf{x}(u,v) = \operatorname{Re}\left[\alpha(z) - i\int_{u_0}^{z} N(w) \times \alpha'(w)\, dw\right].$$

We must still check the conditions of Björling. Since $N(u)$ and $\alpha'(u)$ are real for $u \in I$, $\mathbf{x}(u,0) = \operatorname{Re}\beta(u) = \alpha(u)$ and the first condition is satisfied. Also, for $u \in I$, we may compute $\beta'(u)$ two ways, from the definition of $\beta(z)$ and from the harmonic conjugate process used in the proof of uniqueness. We get

$$\alpha'(u) - i\,(N(u) \times \alpha'(u)) = \beta'(u) = \mathbf{x}_u(u,0) - i\,\mathbf{x}_v(u,0).$$

Equating real and imaginary parts gives $\mathbf{x}_u(u,0) = \alpha'(u)$ and $\mathbf{x}_v(u,0) = N(u){\times}\alpha'(u)$. But recall that isothermal coordinates entail $\mathbf{x}_v(u,0) = U(u,0){\times}\mathbf{x}_u(u,0) = U(u,0) \times \alpha'(u)$ and, consequently, $N(u) = U(u,0)$. Thus, the second condition of Björling is satisfied and we are done. $\qquad\square$

EXERCISE 4.1. Show that the solution of Björling's problem presented above carries through without change for any real-analytic unit vector field along α, $\widetilde{N}$, with $\widetilde{N}(t) \cdot \alpha'(t) = 0$. Recall that a *vector field* $\widetilde{N}$ along $\alpha \colon I \to \mathbb{R}^3$ is simply a smooth map $\widetilde{N} \colon I \to \mathbb{R}^3$ defined on the same domain as α. We think of the vectors $\widetilde{N}(t)$ as originating from the points in three-space belonging to the image of α.

EXERCISE 4.2. For the situation where, as in the special case of Björling's problem, $\widetilde{N} = N$, the unit normal of α, show that the solution to Björling's problem may be written

$$\mathbf{x}(u,v) = \operatorname{Re}\left[\alpha(z) + i\int_{u_0}^{z} B|\alpha'(w)|\, dw\right],$$

where B is the binormal of α. Hint: $N = B \times T = B \times \frac{\alpha'}{|\alpha'|}$.

We can use the full solution to Björling's problem provided by Exercise 4.1 in the particular case where we choose the vector field $-N(t)$.

Corollary 4.3. *Let* $\mathbf{x}(u,v)$ *be a solution to Björling's problem with curve* $\alpha(t)$ *and vector field* $N(t)$. *Then the solution* $\tilde{\mathbf{x}}(u,v)$ *to Björling's problem with curve* $\alpha(t)$ *and vector field* $-N(t)$ *is given by*

$$\tilde{\mathbf{x}}(u,v) = \operatorname{Re}\left[\alpha(z) + i\int_{u_0}^{z} N(w) \times \alpha'(w)\, dw\right]$$
$$= \mathbf{x}(u,-v).$$

Proof. Flip the domain D about the u-axis to get a domain $\tilde{D}$ and define $\tilde{\mathbf{x}}(u,v) = \mathbf{x}(u,-v)$. Clearly, this surface is minimal and $\tilde{U}(u,v) = -U(u,-v)$.

Hence, $\tilde{\mathbf{x}}(u, v)$ solves Björling's problem with vector field $-N(t)$. By the uniqueness of such a solution, however, we also have

$$\tilde{\mathbf{x}}(u, v) = \text{Re}\left[\alpha(z) + i \int_{u_0}^{z} N(w) \times \alpha'(w)\, dw\right]$$

and we are done.   □

Remark 4.2. For $u \in I$ (i.e. $v = 0$), $N(u)$, $\alpha'(u)$ and $z = u$ are real, so $\tilde{\mathbf{x}}(u, 0) = \alpha(u) = \mathbf{x}(u, 0)$. Since D and $\tilde{D}$ are open and contain I, they must overlap in an open set. The Identity Theorem applies to show that, in fact, $\mathbf{x}(u, v)$ and $\tilde{\mathbf{x}}(u, v)$ are the same surface on the overlap. This means that we have extended $\mathbf{x}(u, v)$ past its original domain D by piecing it together with $\tilde{\mathbf{x}}(u, v)$ in $\tilde{D}$. This is an example of why the Identity Theorem is also called The Principle of Analytic Continuation. This is also an example of the minimal surface analogue to Schwarz's Reflection Principle in complex analysis. Minimal surfaces, in general, may possess symmetry about lines or planes under appropriate conditions. Indeed, we have

Theorem 4.4: (The Schwarz Reflection Principle).

(1) *A minimal surface is symmetric about any straight line contained in the surface.*

(2) *A minimal surface is symmetric about any plane which intersects the surface orthogonally.*

The notions of symmetry above are the usual ones with respect to lines and planes. Namely, *line symmetry* means that, given a point p and line l on the surface, there is a point q on the surface lying the same distance away from l as p and the line in $\mathbb{R}^3$ joining p and q meets l at a right angle. Similarly, *plane symmetry* means that given a point p and a plane P meeting the surface at a right angle, there is a point q on the surface lying the same distance from P as p and the line joining p and q meets P at a right angle. Theorem 4.4 follows from Lemma 4.5 and Exercise 4.5, as well as the fact that surfaces may be rigidly moved in $\mathbb{R}^3$ without changing their geometric character.

Example 4.3. Consider Scherk's surface $z = \frac{1}{a} \ln(\cos ay / \cos ax)$ on the square $-\pi/2 < ax < \pi/2$, $-\pi/2 < ay < \pi/2$. As we approach any of the vertices of the square, we approach an indeterminate form $\frac{0}{0}$ which, due to the nature of 2-variable limits, can approach any real number. That is, extending Scherk's surface to the vertices of the square requires erecting a vertical line over each vertex. But then we may apply symmetry about a line to continue Scherk's surface to a diagonally adjacent square such as $\pi/2 < ax < 3\pi/2$, $\pi/2 < ay < 3\pi/2$. Continuing in this manner produces the entire Scherk's surface.

EXERCISE 4.3. Show that Scherk's surface $z = \frac{1}{a}\ln(\cos ay/\cos ax)$ on the square $-\pi/2 < ax < \pi/2$, $-\pi/2 < ay < \pi/2$ intersects the xy-plane in the lines $y = \pm x$. Also, show that these lines are lines of symmetry for the surface.

EXERCISE 4.4. Verify the first statement of Theorem 4.4 for a ruling $(0,0,v_0) + u(\cos v_0, \sin v_0, 0)$ of the helicoid $\mathbf{x}(u,v) = (u\cos v, u\sin v, v)$.

Lemma 4.5. *Suppose $M: \mathbf{x}(u,v)$ is a minimal surface and $\mathbf{x}(u,0)$ is contained in the xy-plane. If $\mathbf{x}(u,v)$ meets the xy-plane orthogonally along $\mathbf{x}(u,0)$, then $x^1(u,-v) = x^1(u,v)$, $x^2(u,-v) = x^2(u,v)$ and $x^3(u,-v) = -x^3(u,v)$.*

Proof. Denote $\mathbf{x}(u,0)$ by $\alpha(u)$ and let the vector field $\widetilde{N}(u)$ along α be defined to be the unit normal of the minimal surface $U(u,0)$. Since α is in the xy-plane, it may be written $\alpha(u) = (\alpha^1(u), \alpha^2(u), 0)$. Also, since the surface meets the xy-plane orthogonally, $\widetilde{N}(u) = U(u,0) = (\widetilde{N}^1(u), \widetilde{N}^2(u), 0)$. With a view to applying the solution to Björling's problem, we calculate $\widetilde{N}(u) \times \alpha'(u) = (0, 0, \widetilde{N}^1(u)(\alpha^2)'(u) - \widetilde{N}^2(u)(\alpha^1)'(u))$. The Identity Theorem implies that the holomorphic extensions of these quantities must have exactly the same formulas, but with variable z instead of u. Now we may apply the uniqueness of the solution to Björling's problem. Since the first and second coordinates of $\widetilde{N}(z) \times \alpha'(z)$ are zero, we have $x^1(u,v) = \operatorname{Re}\alpha^1(z)$ and $x^2(u,v) = \operatorname{Re}\alpha^2(z)$. Using Corollary 4.3 with α and $-\widetilde{N}$, however, we see that $x^1(u,-v) = \operatorname{Re}\alpha^1(z)$ and $x^2(u,-v) = \operatorname{Re}\alpha^2(z)$ as well. Hence, the first two relations hold. Since the third coordinate of α is zero, $x^3(u,v) = -i\int \widetilde{N} \times \alpha'\, dw$. Again using Corollary 4.3 with α and $-\widetilde{N}$, we see that $x^3(u,-v) = +i\int \widetilde{N} \times \alpha'\, dw = -x^3(u,v)$. $\qquad\square$

EXERCISE 4.5. Suppose $M: \mathbf{x}(u,v)$ is a minimal surface and $\mathbf{x}(u,0)$ lies on the x-axis. Show that $x^1(u,-v) = x^1(u,v)$, $x^2(u,-v) = -x^2(u,v)$ and $x^3(u,-v) = -x^3(u,v)$. Hint: $\alpha(u) = (\alpha^1(u), 0, 0)$ and $\widetilde{N}(u) = (0, \widetilde{N}^2(u), \widetilde{N}^3(u))$.

EXERCISE 4.6. Suppose α is a plane curve with unit normal N. Show that the minimal surface which solves Björling's problem for this data intersects the plane containing α orthogonally. Thus, α is a geodesic because we choose α's normal N as our vector field and is a line of curvature by Exercise 5.1.6.

EXERCISE 4.7. If $\alpha(u) = (\beta(u), 0, \gamma(u))$ with unit normal vector field $N(u)$, show that the solution to Björling's problem is given by

$$\mathbf{x}(u,v) = \left(\operatorname{Re}\beta(z),\ \operatorname{Im}\int \sqrt{\beta'^2(w) + \gamma'^2(w)}\, dw,\ \operatorname{Re}\gamma(z) \right).$$

Example 4.4: Henneberg's Surface. Recall that (one form of) Henneberg's surface is given by $\mathbf{x}(u,v) = (x^1(u,v), x^2(u,v), x^3(u,v))$, where $x^1(u,v) = -1 + \cosh 2u \cos 2v$, $x^2(u,v) = \sinh u \sin v - \frac{1}{3}\sinh 3u \sin 3v$ and $x^3(u,v) = -\sinh u \cos v + \frac{1}{3}\sinh 3u \cos 3v$. We have already seen that this rather complicated parametrization comes, algebraically, from choosing the right generating function $F(\tau)$ in the Weierstrass-Enneper representation. Here we wish to show that a natural geometric condition on a minimal surface also gives rise to Henneberg's surface. Let $\alpha(u) = (\cosh 2u - 1, 0, -\sinh u + \frac{1}{3}\sinh 3u)$. This is a parametrization for Neil's parabola $2x^3 = 9z^2$.

EXERCISE 4.8. Check this statement. Hint: Show the identities $\cosh 2u = 1 + 2\sinh^2 u$ and $\frac{1}{3}\sinh 3u - \sinh u = \frac{4}{3}\sinh^3 u$.

We want α to be a geodesic in our minimal surface M, so we choose α's unit normal $N(u)$ as our vector field. Note that α also will be a line of curvature in M since it is a plane curve. Then, we have

$$x^1(u,v) = \mathrm{Re}\,(\cosh 2z - 1)$$
$$= \mathrm{Re}\left(-1 + \frac{e^{2z} + e^{-2z}}{2}\right)$$
$$= -1 + \frac{e^{2u}\cos 2v + e^{-2u}\cos(-2v)}{2}$$
$$= -1 + \cosh 2u \cos 2v$$

and, similarly, $x^3(u,v) = \mathrm{Re}(-\sinh z + \frac{1}{3}\sinh 3z) = -\sinh u \cos v + \frac{1}{3}\sinh 3u \cos 3v$. For $x^2(u,v)$, we compute $\alpha'(z) = (2\sinh 2z, 0, -\cosh z + \cosh 3z)$. By Exercise 4.7, we must calculate $\mathrm{Im}\int \sqrt{\beta'^2(w) + \gamma'^2(w)}\,dw =$

$$\mathrm{Im}\int \sqrt{4\cosh^2 2w - 4 + \cosh^2 w - 2\cosh w \cosh 3w + \cosh^2 3w}\,dw.$$

EXERCISE 4.9. Show that the integral above reduces to

$$\int \sinh 3w + \sinh w\,dw = \frac{1}{3}\cosh 3z + \cosh z.$$

The following identities may be helpful (but use of MAPLE is highly recommended): $\cosh^2 2z = 4\cosh^4 z - 4\cosh^2 z + 1$, $\cosh 3z = 4\cosh^3 z - 3\cosh z$, $\cosh z \cosh 3z = 4\cosh^4 z - 3\cosh^2 z$, $\cosh^2 3z = 16\cosh^6 z - 24\cosh^4 z + 9\cosh^2 z$ and $\sinh 3z = 3\sinh z + 4\sinh^3 z$.

Hence, $\text{Im} \int \sqrt{\beta'^2(w) + \gamma'^2(w)} \, dw$

$$= \text{Im} \left(\frac{1}{3} \cosh 3z + \cosh z \right)$$

$$= \text{Im} \left(\frac{1}{3} \cosh 3u \cos 3v + \frac{i}{3} \sinh 3u \sin 3v + \cosh u \cos v + i \sinh u \sin v \right)$$

$$= \frac{1}{3} \sinh 3u \sin 3v + \sinh u \sin v.$$

Thus, $x^2(u, v) = \frac{1}{3} \sinh 3u \sin 3v + \sinh u \sin v$.

EXERCISE 4.10. Take the strip in the domain of u and v defined by $u = 2r - 1 + s$, $v = \pi(r - \frac{1}{4})$ with $0 \leq r \leq 1$, $-.1 \leq s \leq .1$. Graph the surface obtained by applying the Henneberg patch $\mathbf{x}(u, v)$ to this strip. What surface is this? What does this say about Henneberg's surface?

EXERCISE 4.11. Let $\alpha(u) = (1 - \cos u, 0, u - \sin u)$ be a cycloid. Show that the minimal surface containing α as a geodesic (and a line of curvature) is Catalan's surface

$$\mathbf{x}(u, v) = (1 - \cos(u) \cosh(v), \ 4 \sin \frac{u}{2} \sinh \frac{v}{2}, \ u - \sin(u) \cosh(v)).$$

EXERCISE 4.12. Give another proof of the fact that the only minimal surface of revolution is a catenoid. Hint: take, for convenience, $M \colon \mathbf{x}(u, v) = (h(u) \cos v, u, h(u) \sin v)$ and, first, show that if M is minimal (i.e. $H = 0$), there is some $u = u_0$ with $h'(u_0) = 0$. Then apply the solution to Björling's problem to the parallel circle about u_0 with vector field the unit normal of the circle.

7.5 MINIMAL SURFACES WHICH ARE NOT AREA MINIMIZING

It has long been known that minimal surfaces do not always minimize area. We have already seen in Exercise 4.3.4 that this may be so for certain catenoids. In order to understand what is really happening in such examples however, it would be very nice to have a general method for analyzing them. It is to this that we now turn. Our discussion is an expanded version of that in [R]. Again, we begin with a minimal surface $M \colon \mathbf{x}(u, v)$ bounded by a Jordan curve C and we take a variation $\mathbf{y}^t(u, v) = \mathbf{x}(u, v) + tV(u, v)$ where $V(u, v) = \rho(u, v) \, U(u, v)$ is a normal vector field on M of varying length $\rho(u, v)$ with $\rho(C) = 0$. (Here we write $\rho(C) = 0$ to mean that ρ vanishes on the curve in the uv-plane which is carried to C by $\mathbf{x}(u, v)$.) Note that we are taking only a very special type of variation here, so we can only expect to derive necessary conditions for area minimization. We calculate $\mathbf{y}_u^t = \mathbf{x}_u + tV_u$, $\mathbf{y}_v^t = \mathbf{x}_v + tV_v$

and $\mathbf{y}_u^t \times \mathbf{y}_v^t = \mathbf{x}_u \times \mathbf{x}_v + t[\mathbf{x}_u \times V_v + V_u \times \mathbf{x}_v] + t^2 V_u \times V_v$. In the following, we use the notation:

$$\maltese = (\mathbf{x}_u \times \mathbf{x}_v) \cdot (\mathbf{x}_u \times V_v + V_u \times \mathbf{x}_v)$$
$$\maltese\!\!\maltese = 2(\mathbf{x}_u \times \mathbf{x}_v) \cdot (V_u \times V_v) + (\mathbf{x}_u \times V_v) \cdot (\mathbf{x}_u \times V_v)$$
$$+ 2(\mathbf{x}_u \times V_v) \cdot (V_u \times \mathbf{x}_v) + (V_u \times \mathbf{x}_v) \cdot (V_u \times \mathbf{x}_v)$$

$$\mathcal{S} = \sqrt{|\mathbf{x}_u \times \mathbf{x}_v|^2 + 2t\maltese + t^2\maltese\!\!\maltese + O(t^3)}.$$

where $O(t^3)$ denotes terms involving powers of t greater than or equal to three. With this notation, we see that the surface area $A(t) = \int \int |\mathbf{y}_u^t \times \mathbf{y}_v^t| \, du \, dv$ is given by

$$A(t) = \int \int \mathcal{S} \, du \, dv$$

Now, we are assuming M is minimal, so we know $A'(0) = 0$. Further, we have

Lemma 5.1. *If M is minimal, then $\maltese = 0$.*

Proof. First note that

$$\maltese = (\mathbf{x}_u \times \mathbf{x}_v) \cdot (\mathbf{x}_u \times V_v) + (\mathbf{x}_u \times \mathbf{x}_v) \cdot (V_u \times \mathbf{x}_v)$$
$$= (\mathbf{x}_u \cdot \mathbf{x}_u)(\mathbf{x}_v \cdot V_v) - (\mathbf{x}_u \cdot V_v)(\mathbf{x}_v \cdot \mathbf{x}_u)$$
$$+ (\mathbf{x}_u \cdot V_u)(\mathbf{x}_v \cdot \mathbf{x}_v) - (\mathbf{x}_u \cdot \mathbf{x}_v)(\mathbf{x}_v \cdot V_u)$$

by Lagrange's identity. Now, $V_u = \rho_u U + \rho U_u$ and $V_v = \rho_v U + \rho U_v$, so we have

$$\mathbf{x}_u \cdot V_u = \mathbf{x}_u \cdot (\rho_u U + \rho U_u)$$
$$= \rho \mathbf{x}_u \cdot U_u \qquad \text{since } \mathbf{x}_u \cdot U = 0$$
$$= \rho \mathbf{x}_u \cdot (-S(\mathbf{x}_u))$$
$$= -\rho l \qquad \text{by definition of } l.$$

Similarly, $\mathbf{x}_u \cdot V_v = -\rho m$, $\mathbf{x}_v \cdot V_u = -\rho m$ and $\mathbf{x}_v \cdot V_v = -\rho n$. Plugging these quantities into $\maltese$ gives

$$\maltese = -\rho E n + \rho F m - \rho G l + \rho F m$$
$$= -\rho(E n + G l - 2F m)$$
$$= 0$$

since $E n + G l - 2F m$ is the numerator of mean curvature and M is minimal.
$\square$

Hence, $\mathcal{S} = \sqrt{|\mathbf{x}_u \times \mathbf{x}_v|^2 + t^2 \maltese\maltese + O(t^3)}$ and, upon taking two derivatives of area, we have

$$A'(t) = \int\int \frac{t\maltese\maltese + O(t^2)}{\mathcal{S}}\, du\, dv \quad \text{and} \quad A''(t) = \int\int \frac{\maltese\maltese\mathcal{S} - t\maltese\maltese\mathcal{S}'}{\mathcal{S}^2}\, du\, dv.$$

Note that $\mathcal{S}' = \frac{t\maltese\maltese + O(t^2)}{\mathcal{S}}$ by Lemma 5.1, so that

$$\mathcal{S}|_{t=0} = |\mathbf{x}_u \times \mathbf{x}_v| \quad \text{and} \quad \mathcal{S}'|_{t=0} = 0.$$

Hence, we have the expression for $A''(0)$,

$$A''(0) = \int\int \frac{\maltese\maltese}{|\mathbf{x}_u \times \mathbf{x}_v|}\, du\, dv.$$

Now let's rewrite $\maltese\maltese$ using the Lagrange identity and the notation $e = U_u \cdot U_u$, $f = U_u \cdot U_v$, $g = U_v \cdot U_v$. We have

$$\maltese\maltese = 2(\mathbf{x}_u \times \mathbf{x}_v) \cdot (V_u \times V_v) + (\mathbf{x}_u \times V_v) \cdot (\mathbf{x}_u \times V_v)$$

$$+ 2(\mathbf{x}_u \times V_v) \cdot (V_u \times \mathbf{x}_v) + (V_u \times \mathbf{x}_v) \cdot (V_u \times \mathbf{x}_v)$$

$$= 2[(\mathbf{x}_u \cdot V_u)(\mathbf{x}_v \cdot V_v) - (\mathbf{x}_u \cdot V_v)(\mathbf{x}_v \cdot V_u)]$$

$$+ (\mathbf{x}_u \cdot \mathbf{x}_u)(V_v \cdot V_v) - (\mathbf{x}_u \cdot V_v)(V_v \cdot \mathbf{x}_u) + 2[(\mathbf{x}_u \cdot V_u)(V_v \cdot \mathbf{x}_v)$$

$$- (\mathbf{x}_u \cdot \mathbf{x}_v)(V_v \cdot V_u)] + (V_u \cdot V_u)(\mathbf{x}_v \cdot \mathbf{x}_v) - (V_u \cdot \mathbf{x}_v)(\mathbf{x}_v \cdot V_u)$$

$$= 2[(-\rho l)(-\rho n) - (-\rho m)(-\rho m)] + E(\rho_v^2 + \rho^2 g) - \rho^2 m^2$$

$$+ 2[(-\rho l)(-\rho n) - F(\rho_u \rho_v + \rho^2 f)] + G(\rho_u^2 + \rho^2 e) - \rho^2 m^2$$

$$= 4\rho^2[ln - m^2] + \rho^2[Eg + Ge - 2Ff] + E\rho_v^2 + G\rho_u^2 - 2F\rho_u\rho_v.$$

Plugging this, together with $|\mathbf{x}_u \times \mathbf{x}_v| = \sqrt{EG - F^2}$, into $A''(0)$ gives $A''(0) =$

$$\int\int \frac{[4\rho^2[ln - m^2] + \rho^2[Eg + Ge - 2Ff] + E\rho_v^2 + G\rho_u^2 - 2F\rho_u\rho_v]}{\sqrt{EG - F^2}}\, du\, dv.$$

This is a general expression for the *second variation* of a minimal surface. As in ordinary calculus, if $t = 0$ provides a minimum for the function A, then we must have $A''(0) \geq 0$. Of course, the expression above is still much too complicated to analyze, so now we use what we know about minimal surfaces to simplify it. In particular, we may first suppose that the patch $\mathbf{x}(u, v)$ is in isothermal parameters. Hence, $E = G$, $F = 0$, $l = -n$ (since

$H = 0$), $K = -(l^2 + m^2)/E^2$, $|U_u| = |U_v| = \sqrt{l^2 + m^2}/\sqrt{E}$ and $e = U_u \cdot U_u = (l^2 + m^2)/E = U_v \cdot U_v = g$. Putting these expressions into $A''(0)$ gives

$$A''(0) = \int\int \frac{1}{E}\left[4\rho^2(-l^2 - m^2) + 2\rho^2 E \frac{l^2 + m^2}{E} + E(\rho_u^2 + \rho_v^2)\right] du\, dv$$

$$= \int\int \frac{1}{E}\left[-2\rho^2(l^2 + m^2) + E\rho_u^2 + E\rho_v^2\right] du\, dv$$

$$= \int\int -2E\rho^2 \frac{l^2 + m^2}{E^2} + \rho_u^2 + \rho_v^2\, du\, dv$$

$$= \int\int 2E\rho^2 K + \rho_u^2 + \rho_v^2\, du\, dv.$$

Further, suppose the isothermal coordinates come from a Weierstrass-Enneper representation II. We have calculated previously that

$$K = \frac{-4}{|F|^2(1 + u^2 + v^2)^4} \qquad \text{and} \qquad E = |F|^2(1 + u^2 + v^2)^2.$$

Substituting these into $A''(0)$ gives

$$A''(0) = \int\int 2\rho^2 |F|^2 (1 + u^2 + v^2)^2 \frac{-4}{|F|^2(1 + u^2 + v^2)^4} + \rho_u^2 + \rho_v^2\, du\, dv$$

$$= \int\int \frac{-8\rho^2}{(1 + u^2 + v^2)^2} + \rho_u^2 + \rho_v^2\, du\, dv.$$

Although we have not written it explicitly, the integration to calculate $A''(0)$ is carried out over a region $\mathcal{R}$ in the uv-parameter plane. Note that the last expression for $A''(0)$ does not depend on the Weierstrass-Enneper representation at all. It only depends on the region $\mathcal{R}$ and the choice of a function ρ on that region. Hence, if we can find a function ρ (with $\rho(C) = 0$) defined on $\mathcal{R}$ such that $A''(0) < 0$, then the minimal surface M cannot have minimum area among surfaces spanning C. This brings us to

Theorem 5.2: (Schwarz). *Let M be a minimal surface spanning a curve C. If the closed unit disk $D = \{(u, v)|u^2 + v^2 \leq 1\}$ is contained in the interior of $\mathcal{R}$, then a function ρ exists for which $A''(0) < 0$. Hence, M does not have minimum area among surfaces spanning C.*

Before we give Schwarz's proof of this result, let's interpret it geometrically. As soon as we described M in terms of the Weierstrass-Enneper representation II, we identified the parameters u and v with the real and imaginary parts of the complex variable τ. Of course, τ is itself identified with the function g of the Weierstrass-Enneper representation I which, in turn, we have seen is the Gauss map followed by stereographic projection. Therefore, since stereographic projection from the North pole projects the lower hemisphere of S^2 onto the unit disk D, $\mathcal{R}$ contains D in its interior precisely when the image of the Gauss map of M contains the lower hemisphere of S^2 in its interior. Of course, there is nothing special about stereographic projection from the North pole, so in fact we have the following geometrical version of Schwarz's Theorem.

Theorem 5.3. *Let M be a minimal surface spanning a curve C. If the image of the Gauss map of M contains a hemisphere of S^2 in its interior, then M does not have minimum area among surfaces spanning C.*

Proof of Theorem 5.2. Let $\mathcal{D} = \{(u, v, r) \mid u^2 + v^2 \leq r^2\}$ be the domain bounded by the cone $r = \sqrt{u^2 + v^2}$. Define a function on $\mathcal{D}$ by

$$\rho(u, v, r) = \frac{u^2 + v^2 - r^2}{u^2 + v^2 + r^2}$$

and consider

$$\mathcal{A}(r) \overset{\text{def}}{=} \int \int_{D(r)} \frac{-8\rho^2}{(1 + u^2 + v^2)^2} + \rho_u^2 + \rho_v^2 \, du \, dv$$

where $D(r) = \{(u, v) \mid u^2 + v^2 < r^2\}$ is the open r-disk. Of course, this is $A''(0)$ when we are in $\mathcal{R}$, so we wish to show that the choice of ρ above leads to $\mathcal{A}(r) < 0$ for certain values of r. First, let's split the integral into two pieces and look at the integral of the last two terms of the integrand. If we let $P = -\rho\rho_v$, $Q = \rho\rho_u$ and apply Green's Theorem, we obtain

$$\int_{u^2+v^2=r^2} -\rho\rho_v \, du + \rho\rho_u \, dv = \int \int_{D(r)} \rho_u^2 + \rho_v^2 \, du \, dv$$

$$+ \int \int_{D(r)} \rho(\rho_{uu} + \rho_{vv}) \, du \, dv.$$

Now, the left hand side is zero because $\rho(u, v, r) = 0$ whenever $u^2 + v^2 = r^2$, so

$$\int \int_{D(r)} \rho_u^2 + \rho_v^2 \, du \, dv = - \int \int_{D(r)} \rho \, \Delta\rho \, du \, dv.$$

where $\Delta\rho = \rho_{uu} + \rho_{vv}$ is the Laplacian of ρ. Hence,

$$\mathcal{A}(r) = \int \int_{D(r)} -\rho \left[\frac{8\rho^2}{(1 + u^2 + v^2)^2} + \Delta\rho \right] du \, dv.$$

EXERCISE 5.1. Show by direct calculation, for $\rho = \rho(u, v, 1) = \frac{u^2+v^2-1}{u^2+v^2+1}$, that

$$\frac{8\rho^2}{(1 + u^2 + v^2)^2} + \Delta\rho = 0.$$

By the exercise, we see that $\mathcal{A}(1) = 0$. Now, $r = 1$ corresponds to the unit disk, so we would like $\mathcal{A}(r) < 0$ for r's slightly larger than 1. To show this is true, we need to look at $\mathcal{A}'(1)$. For convenience, let us change variables by taking $s = \frac{u}{r}$ and $t = \frac{v}{r}$ with $du = rds$, $dv = rdt$. Then

$$\mathcal{A}(r) = -\int\int \rho \left[\frac{8\rho^2}{(1+r^2(s^2+t^2)^2} + \Delta\rho \right] r^2 \, ds\, dt,$$

$$\rho(s,t,r) = \frac{r^2s^2 + r^2t^2 - r^2}{r^2s^2 + r^2t^2 + r^2} = \frac{s^2+t^2-1}{s^2+t^2+1} \stackrel{\text{def}}{=} \rho(s,t).$$

Therefore, in st-coordinates, ρ does not depend on r. Also, $\rho_u = \rho_s/r$, $\rho_{uu} = \rho_{ss}/r^2$ by the chain rule and similarly for v. Hence, $\Delta_{u,v}\rho = \Delta_{s,t}\,\rho/r^2$ and we notice that $\Delta_{s,t}\,\rho$ does not depend on r. Now we can replace the uv-Laplace operator in the integral above to get

$$\mathcal{A}(r) = -\int\int \rho \left[\frac{8r^2\rho^2}{(1+r^2(s^2+t^2)^2} + \Delta_{s,t}\,\rho \right] ds\, dt.$$

The fact that ρ and $\Delta_{s,t}\,\rho$ are independent of r allows us to easily take the derivative of $\mathcal{A}(r)$ with respect to r. We obtain $\mathcal{A}'(r) =$

$$-\int\int \frac{16\rho^2 r(1+r^2(s^2+t^2))^2 - 32\rho^2 r^3(1+r^2(s^2+t^2))(s^2+t^2)}{(1+r^2(s^2+t^2))^4} \, ds\, dt.$$

At $r = 1$, $s^2 + t^2 < 1$, so replacing ρ by its definition in terms of s and t gives

$$\mathcal{A}'(1) = -\int\int \frac{16\rho^2(1+s^2+t^2) - 32\rho^2(s^2+t^2)}{(1+s^2+t^2)^3} \, ds\, dt$$

$$= -\int\int_{s^2+t^2<1} \frac{16\rho^2(1+s^2+t^2-2s^2-2t^2)}{(1+s^2+t^2)^3} \, ds\, dt$$

$$= -\int\int_{s^2+t^2<1} \frac{16(s^2+t^2-1)^2(1-s^2-t^2)}{(1+s^2+t^2)^5} \, ds\, dt$$

$$= -\int\int_{s^2+t^2<1} \frac{-16(s^2+t^2-1)^2(s^2+t^2-1)}{(1+s^2+t^2)^5} \, ds\, dt$$

$$= 16 \int\int_{s^2+t^2<1} \frac{(s^2+t^2-1)^3}{(1+s^2+t^2)^5} \, ds\, dt.$$

The numerator of the integrand is always negative since $s^2 + t^2 < 1$, so $\mathcal{A}'(1) < 0$. Since this means that $\mathcal{A}(r)$ is decreasing at $r = 1$ and we have seen previously that $\mathcal{A}(1) = 0$, it must be the case that $\mathcal{A}(r) < 0$ for $1 < r < \bar{r}$ (some $\bar{r}$).

Now suppose that the unit disk D is contained in the interior of the parameter domain $\mathcal{R}$. Then there is an r such that $1 < r < \bar{r}$ and $\{(u, v) | u^2 + v^2 \leq r^2\} \subseteq \mathcal{R}$. Define

$$\rho(u, v) = \begin{cases} \rho(u, v, r) & \text{for } u^2 + v^2 \leq r^2 \\ 0 & \text{for } u^2 + v^2 > r^2. \end{cases}$$

Note that $\rho|_{\partial\mathcal{R}} = 0$ and, by the discussion above, $A''(0) < 0$. Hence, the minimal surface given by the Weierstrass-Enneper representation which spans C is not area minimizing. $\square$

Remark 5.1. To be perfectly rigorous, we should note that the partial derivatives ρ_u, ρ_v are not continuous on the boundary circle $\{(u, v) | u^2 + v^2 = r^2\}$, but may be "rounded off" suitably there while keeping $A''(0) < 0$.

We have seen in Exercise 2.1.9 that Enneper's surface $\mathbf{x}(u, v) = (u - u^3/3 + uv^2, -v + v^3/3 - vu^2, u^2 - v^2)$ has no self-intersections for $u^2 + v^2 < 3$. Further, Exercise 2.3.7 showed that the Gauss map of Enneper's surface (restricted to the disk $u^2 + v^2 < 3$) covers more than a hemisphere of S^2. By our discussion above, we infer that Enneper's surface does not minimize area among all surfaces spanning the curve C given by applying the patch $\mathbf{x}$ to the parameter circle $u^2 + v^2 = R^2$, where $1 < R < \sqrt{3}$. Of course, by the theorem of Douglas and Radó, there exists a least area (and hence minimal) surface spanning C. Therefore, there are at least two minimal surfaces spanning C. In fact, the question of uniqueness for the solution to Plateau's problem is a delicate one. Here is a positive result along this line which we mention without proof.

Theorem 5.4: (Ruchert). *For $0 < r \leq 1$, Enneper's surface is the unique solution to Plateau's problem for the curve given by applying the Enneper patch to a parameter circle of radius r.*

EXERCISE 5.2. Show explicitly that Enneper's surface is not least area among surfaces spanning C (as above). Hints: (1) put $|\mathbf{x}_u \times \mathbf{x}_v| = (1 + u^2 + v^2)^2$ into polar coordinates with $u = r \cos\theta$, $v = r \sin\theta$, (2) compute the surface area of Enneper's surface inside C

$$\text{Area} = \int_0^{2\pi} \int_0^R (r^2 + 1)^2 r \, dr \, d\theta$$

where the r-factor comes from the Jacobian determinant. To get an explicit number, take $R = 1.5$ say. (3) In polar coordinates, the curve C has the form C: $(R\cos\theta - \frac{1}{3}R^3 \cos 3\theta, -R\sin\theta - \frac{1}{3}R^3 \sin 3\theta, R^2 \cos 2\theta)$. Define a (nonminimal) cylinder with directrix C by

$$y(s, \theta) = \underbrace{(R \cos \theta - \frac{1}{3} R^3 \cos 3\theta,}_{x(\theta)} \; \underbrace{s - R \sin \theta - \frac{1}{3} R^3 \sin 3\theta,}_{y(\theta)} \; \underbrace{R^2 \cos 2\theta)}_{z(\theta)}$$

where $0 \leq \theta \leq \pi$ and $0 \leq s \leq 2| - R \sin \theta - \frac{1}{3} R^3 \sin 3\theta|$. Numerically (e.g. MAPLE) compute the surface area of this cylinder inside C (with $R = 1.5$)

$$\text{Area} = \int_0^\pi 2|y(\theta)| \sqrt{x'(\theta)^2 + z'(\theta)^2} \, d\theta.$$

(4) Compare with Enneper's surface.

Although this exercise and the result of Douglas and Radó show that there are at least two minimal surfaces spanning the curve C, in fact we only know one such surface explicitly — Enneper's surface. The Douglas-Radó theorem is an existence theorem without an explicit formula for the least area surface spanning a curve. Indeed, at present there seems to be *no* example of two or more *explicit* minimal surfaces spanning a given Jordan curve. Of course, for non-Jordan curves we have the catenoid-disks example.

7.6 MINIMAL SURFACES AND MAPLE

MAPLE may be used to plot minimal surfaces, of course, using the "plot3d" command, but there are more uses which I will discuss here. First, the Weierstrass-Enneper representation may be put into a procedure which has input a holomorphic function (in some domain) and output an isothermal parametrization of a minimal surface. In fact, I will give two procedures, one in rectangular coordinates and the other in polar coordinates. Sometimes one is more convenient than the other.

```
> with(plots):

> Weier := proc(F)
>       local Z1,Z2,X1,X2,X3,Z3,X;
>       Z1 := int(F*(1-z^2),z);
>       Z2 := int(I*F*(1+z^2),z);
>       Z3 := int(2*F*z,z);
>       X1 :=evalc(Re(subs(z = u+I*v,Z1)));
>       X2 := evalc(Re(subs(z = u+I*v,Z2)));
>       X3 := evalc(Re(subs(z = u+I*v,Z3)));
>       X := [X1,X2,X3];
>       end:
```

```
> Weierpol := proc(F)
>       local Z1,Z2,X1,X2,X3,Z3,X;
>       Z1 := int(F*(1-z^2),z);
>       Z2 := int(I*F*(1+z^2),z);
>       Z3 := int(2*F*z,z);
>       X1 := evalc(Re(subs(z = exp(u)*(cos(v)+I*sin(v)),Z1)));
>       X2 := evalc(Re(subs(z = exp(u)*(cos(v)+I*sin(v)),Z2)));
>       X3 := evalc(Re(subs(z = exp(u)*(cos(v)+I*sin(v)),Z3)));
>       X := [X1,X2,X3];
>       end:
```

These procedures are reasonably self-explanatory. The "int" command takes an integral with respect to the variable (i.e. z here) listed at the end. In Weierpol, the "subs" command is used to plug the polar representation $z = u(\cos v + i \sin v)$ into each of the computed integrals, while "evalc" and "Re" take the real part of the complex expression. For example,

```
> catenoid:=Weier(1/(2*z^2));
```

$$catenoid := \left[-\frac{1}{2}\frac{u(u^2+v^2+1)}{u^2+v^2}, \ -\frac{1}{2}v - \frac{1}{2}\frac{v}{u^2+v^2}, \ \ln(\sqrt{u^2+v^2}) \right]$$

Now, this isn't a particularly pleasing form, so we can substitute for u and v as follows.

```
> simplify(subs({u=exp(r)*cos(s),v=exp(r)*sin(s)},catenoid));
```

$$\left[-\frac{1}{2}\cos(s)e^{(-r)}(e^{(2r)}+1), \ -\frac{1}{2}\sin(s)e^{(-r)}(e^{(2r)}+1), \ r \right]$$

Of course, this *is* a usual catenoid parametrization when the multiplication of exponentials is carried out. This simple illustrates the main difficulty with calculating the Weierstrass representation in MAPLE. Namely, even if the integrals are able to be given in closed form, their exact form may not be one which is immediately recognizable as a standard surface. Of course this is true when the representation is worked out by hand as well. This is but one more instance of the general rule that, while computer algebra systems can do amazing and beautiful things, they are not a panacea.

EXERCISE 6.1. Carry out the following MAPLE commands related to the helicoid.
```
> helicoid:=Weier(I/(2*z^2));
> simplify(subs({u=exp(r)*cos(s),v=exp(r)*sin(s)},helicoid));
> hel:=simplify(Weierpol(I/(2*z^2),trig);
> plot3d(hel,u=-1.2..1.2,v=0..2*Pi,scaling=constrained,grid=[10,30], shading=z);
```

EXERCISE 6.2. Carry out the following, simplify if necessary and plot the resulting surfaces.

> enneper:=Weier(1);
> ennepol:=Weierpol(1);
> trinoid:=simplify(Weierpol(1/(2∗z∗(z^3-1)^2)));

EXERCISE 6.3. For Henneberg's surface, write

> Henne:=Weierpol(1-1/z^4);

Show that the expressions you obtain may be simplified to the usual parametrization for Henneberg's surface. Hint: the identity $\cos^3(v) = \frac{1}{4}\cos(3v) + \frac{3}{4}\cos(v)$ and its sin analogue may be useful here. Now plot Henneberg's surface by

> plot3d(Henne,u=-2..2,v=0..2∗Pi,scaling=constrained, grid=[40,40],shading=XY);

EXERCISE 6.4. Find Weierstrass-Enneper representations for $F(z) = 1/z$ and $F(z) = 1/(4z^3)$. Plot the resulting surfaces.

Now let's see how MAPLE may be used to visualize and calculate in a specific problem. Namely, Exercise 5.2 asks you to show that, for certain boundary curves, Enneper's surface, although minimal, is not the spanning surface of *least* area. The problem puts forth a candidate for smaller area — a generalized cylinder which is not even minimal. Here's what MAPLE can do for this problem.

> with(plots):

> enneper:=[u-u^3/3+u∗v^2,-v+v^3/3-v∗u^2,u^2-v^2];

The following procedure creates a parametrization for the cylinder spanning the boundary curve of Enneper's surface with $R = 1.5$. In order to graph the cylinder, we need two cases — when yv is positive and when yv is negative. The second coordinate yv is negative when v is between 0 and π, so here u must vary from 0 to 2. When v goes from π to 2π, u must vary between -2 and 0.

```
> CylEnn := proc(r)
>       local xv,yv,zv,n,X;
>       xv := r∗cos(v)-1/3∗r^3∗cos(3∗v);
>       yv := -r∗sin(v)-1/3∗r^3∗sin(3∗v);
>       zv := r^2∗cos(2∗v);
>       n:= abs(yv);
>       X := [xv,yv+u∗n,zv]
>       end:
```

Now we must plot the two cases individually. Notice again that when we want a plot structure to be saved for later plotting, we just give the plot3d command a name (such as cyl1 below) *and* we use a colon at the end to suppress any output (which generally would run on for pages and pages).

```
> cyl1:=plot3d(CylEnn(1.5),u=0..2,v=0..Pi,scaling=constrained,
          grid=[5,50]):
> cyl2:=plot3d(CylEnn(1.5),u=-2..0,v=Pi..2*Pi,scaling=constrained,
          grid=[5,50]):
```

We can get Enneper's surface in polar coordinates together with its plot as follows (or from Weierpol).

```
> ennpolar:=simplify(subs({u=r*cos(theta),v=r*sin(theta)}, enneper));
```

```
> enn:=plot3d(ennpolar,r=0..1.5,theta=0..2*Pi,scaling=constrained,
          grid=[5,50]):
```

The following represents the Jordan curve which is the boundary curve for both Enneper's surface with $R = 1.5$ and the cylinder above.

```
> jorcurve:=subs(r=1.5,ennpolar);
```

We can plot the boundary curve using the "spacecurve" command.

```
> bound:=spacecurve(jorcurve,theta=0..2*Pi,color=black,thickness=2):
```

Now we can display the surfaces separately with the same boundary curve.

```
> display({bound,enn},scaling=constrained,style=wireframe);
```

```
> display({bound,cyl1,cyl2},scaling=constrained,style=wireframe);
```

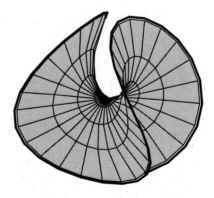

FIGURE 7.1.   Enneper's surface inside the Jordan curve

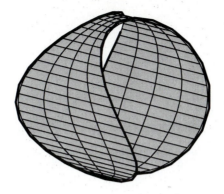

FIGURE 7.2. The cylinder inside the Jordan curve

Now let's compute the surface area of the cylinder. I shall use the notation of Exercise 5.2.

```
> ytheta:=subs(r=1.5,ennpolar[2]);
> x1:=diff(subs(r=1.5,ennpolar[1]),theta);
> z1:=diff(subs(r=1.5,ennpolar[3]),theta);
```

Now we can evaluate the surface area integral numerically. The necessary MAPLE command is "evalf" (for "floating point evaluation") appended to "int."

```
> evalf(int(2*abs(ytheta)*sqrt(x1^2+z1^2),theta=0..Pi));
```

$$31.66323514$$

This is then the area of the cylinder. On the other hand, Enneper's surface has surface area which may be computed explicitly to be $\pi r^2(1 + r^2 + (r^4)/3)$ for radius r. Therefore, for $r = 1.5$,

```
> evalf(subs(r=1.5, Pi*r^2*(1+r^2+(r^4)/3)));
```

$$34.90113089$$

Hence, Enneper's surface is minimal, but not area minimizing within the bounds of the Jordan curve above. Of course, the surface area of the cylinder could've been (and was) computed numerically without MAPLE. Yet, what recommends computer algebra systems to the mathematical and educational community is the convenience, the simplicity of the programming relative to the results obtained and the beauty of instant visualization. As final exercises for this section,

EXERCISE 6.5. Compare the areas of cylinders of varying radii with Enneper's surface.

EXERCISE 6.6. Now do **Exercise 4.10**. That is, show that the domain defined in the problem gives a Möbius strip lying on Henneberg's surface. Hence, Henneberg's surface is nonorientable.

Chapter 8

THE CALCULUS OF VARIATIONS AND GEOMETRY

8.1 THE EULER-LAGRANGE EQUATIONS

In previous chapters, we have seen that various geometric entities have a tendency to minimize some quantity. For instance, geodesics try to be paths of shortest arclength while minimal surfaces (including their physical representatives, soap films) try to be surfaces of least area. Of course we have also seen that there are geodesics which do *not* minimize arclength and minimal surfaces which *are not* least area. In this chapter we will try to put these results into perspective by giving a quick tour of the calculus of variations — the subject whose principles embody this geometric (and physical) tendency toward minimization. Now, whole books have been written about this subject (see [Sag], [Wei] and [Pin] for example) and the subject itself is rife with essential technicalities, so we shall stick with what true variationalists call *the naive theory*. Nevertheless, we shall see that variational principles and methods are intimately connected to geometry.

Here's the general set-up for the calculus of variations: let $x = x(t)$ denote a function of t with fixed endpoints $x(t_0) = x_0$ and $x(t_1) = x_1$.

FIGURE 8.1. Curve with fixed endpoints

Because of this picture, we often refer to $x(t)$ as a curve joining the endpoints. Indeed, the function x may, in fact, be a curve in n-space, so it pays to keep in mind this possibility. In particular, we might like to think of a curve $\mathbf{x}(t) = (x^1(t), \ldots, x^n(t))$ as being the path of a particle with respect to time t. This is the *mechanics* point of view. One potential source of confusion must be mentioned right away. While we have used the standard mechanics notation $x = x(t) = (x^1(t), \ldots, x^n(t))$ above, when we deal with the geometry of the plane, we will use the coordinates which most of us feel comfortable with. Namely, we will write the independent variable as x and the function in question $y = y(x)$. The context of the problem should make everything clear.

Now, there are many choices of curves $x(t)$ joining the given endpoints. It is only when we add some sort of condition to be satisfied that we can pick special $x(t)$ out of this collection. For example, we have the

Fixed Endpoint Problem: 1.1. Find the curve $x = x(t)$ with $x(t_0) = x_0$ and $x(t_1) = x_1$ such that the following integral is minimized

$$J = \int_{t_0}^{t_1} f(t, x(t), \dot{x}(t)) \, dt$$

where $f(t, x, \dot{x})$ is a function of t, x and $\dot{x} = dx/dt$ and the latter two are thought of as independent variables.

Example 1.2. Let $T = 1/2 \, m\dot{x}^2$ denote the kinetic energy of a particle moving along the x-axis. A typical potential energy function depends on the distance of the particle from a specified point x_0, so we can write $V = V(|x - x_0|)$. As we shall see later, Hamilton's principle says that the motion of the particle $x = x(t)$ will be such that the integral

$$J = \int T - V \, dt = \int 1/2 \, m\dot{x}^2 - V(|x - x_0|) \, dt$$

is minimized. (In fact, as we shall see below, this is not quite correct. Hamilton's principle only requires that the integral be *extremized*.) Because this example is so important, it is traditional to keep the dot notation $\dot{x}$ for derivatives in even more general situations.

Example 1.3. We have seen before that the shortest distance between two points in the plane is attained by a straight line. If we use $x \, y$-coordinates, then the problem of determining the curve $y = y(x)$ of minimum arclength is simply to

$$\text{Minimize} \int \sqrt{1 + y'^2} \, dx.$$

In this case, $f(x, y, y') = \sqrt{1 + y'^2}$. Also, note that here we have used the usual prime notation for derivatives.

To approach the fixed endpoint problem, we should first recognize that, just as in ordinary calculus, our methods are best suited to finding *local minima*, not global minima. With this in mind, let's start to analyze the problem.

Suppose $x(t)$ is a curve which minimizes the integral $J = \int_{t_0}^{t_1} f(t, x(t), \dot{x}(t)) \, dt$ and let $x^*(t) = x(t) + \epsilon \eta(t)$ be a variation of x. That is, we think of ϵ as being small and we require that $\eta(t_0) = 0$ and $\eta(t_1) = 0$.

Therefore, the curve $x^*(t)$ still joins x_0 and x_1 as well as being "close" to x. Note too that $\dot{x}^* = \dot{x} + \epsilon\dot{\eta}$. With this notation, we can think of the integral J as a function of the parameter ϵ

$$J(\epsilon) = \int_{t_0}^{t_1} f(t, x^*, \dot{x}^*)\, dt = \int_{t_0}^{t_1} f(t, x + \epsilon\eta, \dot{x} + \epsilon\dot{\eta})\, dt.$$

In order to recognize the minimum $x(t)$, just as we do in ordinary calculus, we take the derivative of J with respect to ϵ and note that, since $x(t)$ is a minimum by hypothesis, this derivative is zero for $\epsilon = 0$. (Note that, speaking informally, we can take the derivative inside the integral because the integral is taken with respect to t and t is independent of ϵ.) Now,

$$\frac{dJ}{d\epsilon} = \int_{t_0}^{t_1} \frac{\partial f}{\partial x^*}\frac{\partial x^*}{\partial \epsilon} + \frac{\partial f}{\partial \dot{x}^*}\frac{\partial \dot{x}^*}{\partial \epsilon}\, dt$$

by the chain rule, so

$$\frac{dJ}{d\epsilon} = \int_{t_0}^{t_1} \frac{\partial f}{\partial x^*}\eta + \frac{\partial f}{\partial \dot{x}^*}\dot{\eta}\, dt$$

with

$$0 = \frac{dJ}{d\epsilon}\bigg|_{\epsilon=0} = \int_{t_0}^{t_1} \frac{\partial f}{\partial x}\eta + \frac{\partial f}{\partial \dot{x}}\dot{\eta}\, dt$$

since, at $\epsilon = 0$, $x^*(t) = x(t)$. Now, the second term inside the integral may be integrated by parts as follows. Let

$$u = \frac{\partial f}{\partial \dot{x}} \qquad dv = \dot{\eta}\, dt$$

$$du = \frac{d}{dt}\left(\frac{\partial f}{\partial \dot{x}}\right) dt \qquad v = \dot{\eta}$$

and compute

$$\int_{t_0}^{t_1} \frac{\partial f}{\partial \dot{x}}\dot{\eta}\, dt = \bigg|_{t_0}^{t_1} \eta\frac{\partial f}{\partial \dot{x}} - \int_{t_0}^{t_1} \eta\frac{d}{dt}\left(\frac{\partial f}{\partial \dot{x}}\right) dt$$

$$= -\int_{t_0}^{t_1} \eta\frac{d}{dt}\left(\frac{\partial f}{\partial \dot{x}}\right) dt$$

since $\eta(t_0) = 0 = \eta(t_1)$ implies the vanishing of the first term. Putting this in the equation above, we have

$$0 = \int_{t_0}^{t_1} \frac{\partial f}{\partial x}\eta - \int_{t_0}^{t_1} \eta\frac{d}{dt}\left(\frac{\partial f}{\partial \dot{x}}\right) dt$$

(*)

$$= \int_{t_0}^{t_1} \eta\left[\frac{\partial f}{\partial x} - \frac{d}{dt}\left(\frac{\partial f}{\partial \dot{x}}\right)\right] dt.$$

This equation must hold for every function η with $\eta(t_0) = 0 = \eta(t_1)$. How can this be? The following exercise provides an answer.

EXERCISE 1.1. Suppose that a continuous function $y = f(x)$ is positive at a point x_0. Show that it is possible to choose a function $\eta(x)$ so that on some interval $[a, b]$

$$\int_a^b \eta(x) \, f(x) \, dx > 0.$$

Hints: (1) draw a graph of $f(x)$ about x_0 and say something about the values of $f(x)$ for all points x near x_0; (2) create a "bump" function $\eta(x)$ which guarantees the positivity of the inetgral above by drawing its graph on the same axes as the graph of $f(x)$ near x_0; (3) remember that positive integrands produce positive integrals.

The exercise tells us that there is only one way equation $(*)$ can hold for all choices of η. Namely, we must have

$$\frac{\partial f}{\partial x} - \frac{d}{dt}\left(\frac{\partial f}{\partial \dot{x}}\right) = 0.$$

This equation is called the *Euler-Lagrange equation* and it gives us a necessary condition for $x(t)$ to be a minimum.

Theorem 1.1. *If $x = x(t)$ is a minimum for the fixed endpoint problem, then x satisfies the Euler-Lagrange equation.*

Notice that we are *not* saying that a solution to the Euler-Lagrange equation is a solution to the fixed endpoint problem. The Euler-Lagrange equation is simply a first step toward solving the fixed endpoint problem, akin to finding critical points in calculus. Nevertheless, because we deal with an unimaginably huge collection of possible solution curves for the fixed endpoint problem, the Euler-Lagrange equation is a powerful tool which is indispensable. Indeed, it is sometimes the case that *only* solutions to the Euler-Lagrange equations may be found with little or no other information to guide us to a solution of the fixed endpoint problem. For this reason solutions to the Euler-Lagrange equation are given the special name *extremals* and the fixed endpoint problem, for example, is often rephrased to say that a curve $x(t)$ is desired which joins the given endpoints and *extremizes* the integral J. In this case, solutions to the Euler-Lagrange equation solve the problem.

EXERCISE 1.2. Suppose that J depends on two functions $x(t)$ and $y(t)$,

$$J = \int_{t_0}^{t_1} f(t, x(t), y(t), \dot{x}(t), \dot{y}(t)) \, dt.$$

Show that variations $x^* = x + \epsilon\eta$ and $y^* = y + \epsilon\tau$ lead to

$$0 = \frac{dJ}{d\epsilon}\bigg|_{\epsilon=0}$$

$$= \int_{t_0}^{t_1} \frac{\partial f}{\partial x}\frac{\partial x}{\partial \epsilon} + \frac{\partial f}{\partial \dot{x}}\frac{\partial \dot{x}}{\partial \epsilon} + \frac{\partial f}{\partial y}\frac{\partial y}{\partial \epsilon} + \frac{\partial f}{\partial \dot{y}}\frac{\partial \dot{y}}{\partial \epsilon} \, dt$$

$$= \int_{t_0}^{t_1} \frac{\partial f}{\partial x}\eta + \frac{\partial f}{\partial \dot{x}}\dot{\eta} + \frac{\partial f}{\partial y}\tau + \frac{\partial f}{\partial \dot{y}}\dot{\tau} \, dt$$

$$= \int_{t_0}^{t_1} \eta\left[\frac{\partial f}{\partial x} - \frac{d}{dt}\left(\frac{\partial f}{\partial \dot{x}}\right)\right] dt + \int_{t_0}^{t_1} \tau\left[\frac{\partial f}{\partial y} - \frac{d}{dt}\left(\frac{\partial f}{\partial \dot{y}}\right)\right] dt$$

and that this equation can hold for all η, τ if and only if the following **Euler-Lagrange equations** hold,

$$\frac{\partial f}{\partial x} - \frac{d}{dt}\left(\frac{\partial f}{\partial \dot{x}}\right) = 0 \qquad \frac{\partial f}{\partial y} - \frac{d}{dt}\left(\frac{\partial f}{\partial \dot{y}}\right) = 0.$$

EXERCISE 1.3. Suppose that J depends on two independent variables t and s,

$$J = \int\int_{\mathcal{R}} f(t, s, x(t, s), x_t(t, s), x_s(t, s)) \, dt \, ds.$$

Show that a variation $x^*(t, s) = x(t, s) + \epsilon\eta(t, s)$ with $\eta|_C = 0$, where C is the boundary of the region of integration $\mathcal{R}$, leads to

$$0 = \frac{dJ}{d\epsilon}\bigg|_{\epsilon=0} = \int\int \frac{\partial f}{\partial x}\eta + \frac{\partial f}{\partial x_t}\eta_t + \frac{\partial f}{\partial x_s}\eta_s \, dt \, ds.$$

Further, recalling Green's theorem $\int -P\,dt + Q\,ds = \int\int \partial Q/\partial t + \partial P/\partial s \, dt \, ds$, letting $Q = \eta(\partial f/\partial x_t)$, $P = \eta(\partial f/\partial x_s)$ and using $\eta|_C = 0$, show that the last two terms of the integral give

$$\int\int \frac{\partial f}{\partial x_t}\eta_t + \frac{\partial f}{\partial x_s}\eta_s \, dt \, ds = -\int\int \eta\left[\frac{\partial^2 f}{\partial x_t \partial t} + \frac{\partial^2 f}{\partial x_s \partial s}\right] dt \, ds$$

and substitution then gives

$$0 = \int\int_{\mathcal{R}} \eta\left[\frac{\partial f}{\partial x} - \frac{\partial^2 f}{\partial x_t \partial t} - \frac{\partial^2 f}{\partial x_s \partial s}\right] dt \, ds.$$

Finally, argue that this implies that the **Euler-Lagrange equation** for two independent variables is

$$\frac{\partial f}{\partial x} - \frac{\partial}{\partial t}\left(\frac{\partial f}{\partial x_t}\right) - \frac{\partial}{\partial s}\left(\frac{\partial f}{\partial x_s}\right) = 0.$$

EXERCISE 1.4. Suppose $f(t, x, \dot{x}) = f(x, \dot{x})$ does not depend on t explicitly. By this, we mean that f only depends on t through the curves x and $\dot{x}$, so that $\partial f / \partial t = 0$. Show that a nonconstant $x(t)$ satisfies the Euler-Lagrange equation if and only if

$$f - \dot{x} \frac{\partial f}{\partial \dot{x}} = c$$

where c is a constant. Hint: compute the derivative with respect to t of the left-handside. Don't forget the chain rule.

EXERCISE 1.5. Suppose $f(t, x, y, \dot{x}, \dot{y}) = f(x, y, \dot{x}, \dot{y})$ does not depend on t explicitly. By this, we mean that f only depends on t through the curves x, y, $\dot{x}$ and $\dot{y}$, so that $\partial f / \partial t = 0$. Show that nonconstant $x(t)$, $y(t)$ satisfy the Euler-Lagrange equations if and only if

$$f - \dot{x} \frac{\partial f}{\partial \dot{x}} - \dot{y} \frac{\partial f}{\partial \dot{y}} = c$$

where c is a constant. Hint: compute the derivative with respect to t of the left-handside. Don't forget the chain rule.

EXERCISE 1.6. Show that, if $\partial f / \partial x = 0$, then $x(t)$ satisfies the Euler-Lagrange equation if and only if $\partial f / \partial \dot{x} = c$, a constant.

EXERCISE 1.7. Show that the integrals

$$\int f(t, x, \dot{x}) \, dt \qquad \text{and} \qquad \int f(t, x, \dot{x}) + \frac{dg(t, x)}{dt} \, dt$$

have the same Euler-Lagrange equation. In particular, note that when $g(t, x) = ct$, with c constant, then $\int f \, dt$ and $\int f + c \, dt$ have the same Euler-Lagrange equation. We will use this fact when we discuss Jacobi's theorem in Section 8.5.

EXERCISE 1.8. Find the extremal $x(t)$ for the fixed endpoint problem

$$J = \int_0^\pi x \sin t + \frac{1}{2} \dot{x}^2 \, dt$$

with $x(0) = 0$, $x(\pi) = \pi$.

EXERCISE 1.9. Find the extremal $x(t)$ for the fixed endpoint problem

$$J = \int_0^1 \dot{x} \, x^2 + \dot{x}^2 \, x \, dt$$

with $x(0) = 1$, $x(1) = 4$.

EXERCISE 1.10. Find the extremal $x(t)$ for the fixed endpoint problem

$$J = \int_0^{\pi/2} \dot{x}^2 - x^2 \, dt$$

with $x(0) = 0$, $x(\pi/2) = 1$.

EXERCISE 1.11. Consider the fixed endpoint problem with $x(t_0) = x_0$, $x(t_1) = x_1$ and integral to be minimized

$$J = \int_{t_0}^{t_1} p^2 \, \dot{x}^2 + q^2 \, x^2 \, dt$$

where $p = p(t)$ and $q = q(t)$ are arbitrary smooth functions.

(1) Show that if $x = x(t)$ is an extremal for J, then
$J = p^2 \, x \, \dot{x}|_{t_0}^{t_1} = p^2 \, x \, \dot{x}(t_1) - p^2 \, x \, \dot{x}(t_0)$.

(2) Show that if $x = x(t)$ is an extremal for J and $\eta = \eta(t)$ is a function with $\eta(t_0) = 0 = \eta(t_1)$, then

$$\widetilde{J} = \int_{t_0}^{t_1} p^2 \, \dot{x} \, \dot{\eta} + q^2 \, x \, \eta \, dt = 0.$$

(3) From (1) and (2) show that an extremal $x(t)$ for J is a solution to the fixed endpoint problem. That is, $x(t)$ minimizes J.

Hints: For (1), differentiate $p^2 \, x \, \dot{x}$ with respect to t. For (2), compute $\int p^2 \, \dot{x} \, \dot{\eta} \, dt$ by parts. For (3), vary x by $x + \eta$ (where the usual ϵ is absorbed into η) and compute

$$\widehat{J} = \int_{t_0}^{t_1} p^2 \, (\dot{x} + \dot{\eta})^2 + q^2 \, (x + \eta)^2 \, dt.$$

Compare $\widehat{J}$ and J.

EXERCISE 1.12. In the fixed endpoint problem, suppose the time t_1 is fixed (as well as t_0 and $x(t_0)$ of course), but the final position $x(t_1)$ is undetermined. From the derivation of the Euler-Lagrange equation, show that not only the Euler-Lagrange equation must be satisfied, but that the extra condition

$$\frac{\partial f}{\partial \dot{x}}(t_1) = 0$$

must be satisfied as well. Hint: everything in the derivation works as before except that, after integrating by parts, the first term no longer vanishes. Argue that the Euler-Lagrange equation is satisfied because the eventual extremal may be considered as an extremal for the fixed endpoint problem with $x(t_1)$ being whatever point the extremal hits at $t = t_1$.

Remark 1.4. The preceding exercise hints that not all variational problems have fixed endpoints. In fact, there are many problems where either the final "time" t_1 or the final "state" $x(t_1)$ are undetermined. As the exercise suggests, in these cases the Euler-Lagrange equation holds *together* with an extra condition which is sometimes called *a transversality condition*. For example, suppose we say that, while t_1 is unspecified, it is required that $x(t_1)$ lie on a given curve $\alpha(t)$. The generalization of the condition above is then that the following condition (see [Pin] or [Sag] for example) must hold for an extremal $x(t)$ at t_1,

$$f(t_1) + (\dot{\alpha}(t_1) - \dot{x}(t_1)) \frac{\partial f}{\partial \dot{x}}(t_1) = 0.$$

EXERCISE 1.13. When either t_1 is specified or $x(t_1)$ is specified, the preceding discussion is made simpler.

(1) Suppose t_1 is specified, but $x(t_1)$ must only lie on $\alpha(t)$. Show that the transversality condition above reduces to

$$\frac{\partial f}{\partial \dot{x}}(t_1) = 0.$$

Hint: factor out $1/\dot{\alpha}(t_1)$ and note that t_1 fixed makes $\alpha(t)$ a vertical line.

(2) Suppose $x(t_1)$ is specified, but t_1 is free. Show that the transversality condition above reduces to

$$f(t_1) - \dot{x}(t_1)\frac{\partial f}{\partial \dot{x}}(t_1) = 0.$$

Hint: $x(t_1)$ fixed makes $\alpha(t)$ a horizontal line.

8.2 THE BASIC EXAMPLES

No discussion of the calculus of variations would be complete without mentioning the problem which launched the subject, the *brachistochrone*. The name is taken from the Greek words *brachist*, which means shortest, and *chronos*, which means time. The problem itself is this:

Example 2.1: The Brachistochrone Problem. Given a point (a, b) in the $x\,y$-plane, find the curve $y(x)$ joining (a, b) and the origin so that a bead, under the influence of gravity, sliding along a frictionless wire in the shape $y(x)$ from (a, b) to $(0, 0)$, minimizes the time of travel.

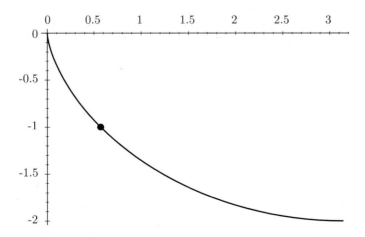

FIGURE 8.2.   Bead on a wire

The key to setting up this problem is the simple formula $D = RT$, distance equals rate times time. The "infinitesimal" distance travelled by the bead is just the arclength of the wire, $D = \sqrt{1 + y'^2}$. The rate of descent may be determined by noting that potential energy is given by the formula mgh where m is the mass of the bead, $g = 9.8\,\mathrm{m/sec^2}$ is the acceleration due to gravity (near the surface of the Earth) and h is the height of the bead above a fixed reference height (which is often taken to be zero on the Earth's surface). If we start the bead at (a, b) with initial velocity zero, then conservation of energy requires that the kinetic energy of the bead be equal to the loss of potential energy due to decreasing height. In other words,

$$mg(b - y) = \frac{1}{2}mv^2$$

where v is the speed of the bead. We then have that $v = \sqrt{2g(b - y)}$. So, $D = RT$ tells us that time T is a function of y and y',

$$T(y, y') = \frac{1}{\sqrt{2g}} \frac{\sqrt{1 + y'^2}}{\sqrt{b - y}}.$$

The total time is then found by integrating $T(y, y')$ with respect to x from a to 0. The problem of the brachistochrone now becomes the fixed endpoint problem

$$\text{Minimize } T = \frac{1}{\sqrt{2g}} \int_a^0 \frac{\sqrt{1 + y'^2}}{\sqrt{b - y}}\, dx$$

with $y(a) = b$ and $y(0) = 0$.

To solve this problem, first substitute $u = b - y$ and ignore the constant factor $1/\sqrt{2g}$ to get a new integral to be extremized,

$$J = \int_a^0 \frac{\sqrt{1 + u'^2}}{\sqrt{u}}\, dx.$$

Because the independent variable does not appear explicitly in J, we may use Exercise 1.4 to get

$$\frac{\sqrt{1 + u'^2}}{\sqrt{u}} - u' \frac{u'}{\sqrt{u(1 + u'^2)}} = c.$$

Now, finding a common denominator, simplifying and replacing $1/c$ by a new constant c gives a separable differential equation $c = u(1 + u'^2)$ which leads to

$$x = \int \frac{\sqrt{u}}{\sqrt{c - u}}\, du$$

$$= -2 \int \sqrt{c^2 - w^2}\, dw$$

where $w = \sqrt{c - u}$ and c is replaced by c^2

$$= -2 \int c \cos \theta \, c \cos \theta \, d\theta$$

where $w = c \sin \theta$

$$= -2c^2 \int \cos^2 \theta \, d\theta$$

$$= -2c^2 \int \frac{1 + \cos \theta}{2} \, d\theta$$

$$= -\frac{c^2}{2} [2\theta + \sin 2\theta] + r$$

and the final answer

$$x(\phi) = k(\phi + \sin \phi) + r$$

where $2\theta = \phi$. Now, $b - y = u = c^2 \cos^2 \theta = c^2 \cos^2 \frac{\phi}{2}$ by the substitutions above, so again using $\cos^2 \frac{\phi}{2} = (1 + \cos \phi)/2$, we obtain

$$y(\phi) = k(1 + \cos \phi) + b$$

where $k = -c^2/2$ as for $x(\phi)$. Note that $y(\pi) = b$, so that it must also be true that $x(\pi) = a$. This implies that $r = a - k\pi$. If we let $\zeta = \phi - \pi$, then we obtain the parametric formulas for x and y,

$$x(\zeta) = k(\zeta - \sin \zeta) + a \quad \text{and} \quad y(\zeta) = k(1 - \cos \zeta) + b.$$

This is the parametrization of a cycloid. Hence, the solution to the problem of the brachistochrone is one of our standard curves from Chapter 1. Also, note that, by Exercise 1.1.6, the brachistochrone is the tautochrone as well. Finally, for the *brachistochrone with friction*, see [HK] and for a very different approach, see [L].

EXERCISE 2.1. Johann Bernoulli solved the brachistochrone problem ingeniously by employing Fermat's principle that light travels to minimize time together with Snell's law of refraction (see [Wei]). Here's a problem in the same vein. Suppose that a light photon moves in the upper half-plane to minimize time. Suppose also that the plane is made of a medium whose properties imply that the speed of the photon is always proportional to its height above the x-axis. That is, $v = ky$ for $k > 0$. What path does the photon take? Hint: set up the time integral and extremize it.

Example 2.2: Shortest distance curves in the plane. Given points (a, b) and (c, d) in the plane, we can ask for the curve $x(t)$ which joins them and which minimizes the arclength integral

$$J = \int_{t_0}^{t_1} \sqrt{\dot{x}^2(t) + \dot{y}^2(t)}\, dt.$$

Here we have chosen a two-variable formulation to show how these types of problems work. Because the integrand does not depend on either x or y, the Euler-Lagrange equations reduce to

$$\frac{\partial f}{\partial \dot{x}} = \frac{\dot{x}}{\sqrt{\dot{x}^2(t) + \dot{y}^2(t)}} = c \quad \text{and} \quad \frac{\partial f}{\partial \dot{y}} = \frac{\dot{y}}{\sqrt{\dot{x}^2(t) + \dot{y}^2(t)}} = d.$$

Solving for the square root in each equation and setting the results equal, we obtain

$$\frac{\dot{y}}{d} = \frac{\dot{x}}{c}$$

which gives, under integration with respect to t,

$$y = \frac{d}{c}x + r.$$

This is, of course, the equation of a straight line in the plane. Therefore, extremals of the arclength integral are straight lines. We will see shortly that these extremals are, in fact, minimizers for arclength.

EXERCISE 2.2. Solve the shortest distance in the plane problem using one variable. That is, arclength is given by $\int \sqrt{1 + y'(x)^2}\, dx$.

Example 2.3: Least area surfaces of revolution. We have already seen that a minimal surface of revolution is a catenoid, so, since least area surfaces are minimal, this example offers another approach to the question. A surface of revolution has surface area given by

$$A = 2\pi \int y\sqrt{1 + y'^2}\, dx.$$

EXERCISE 2.3. Use the fact that the integrand of the surface area integral above is independent of x to reduce the Euler-Lagrange equation to $y' = \sqrt{y^2 - c^2}$. Then solve this separable differential equation to show $y = c\cosh(x/c - d)$, verifying that a catenoid is the only minimal surface of revolution. We have already seen that a catenoid is least area for a given boundary only under certain circumstances, so this example shows again that extremals are not always minimizers.

EXERCISE 2.4. Let $z = f(x, y)$ be a function of two variables. Surface area is given by the integral

$$A = \int \int \sqrt{1 + f_x^2 + f_y^2} \, dx \, dy.$$

Use the two-independent-variable Euler-Lagrange equation of Exercise 1.3 to show that an extremal for this integral is a solution to the *minimal surface equation*.

EXERCISE 2.5. Let $z = \phi(x, y)$ be a function of two variables. The Dirichlet integral is given by

$$A = \int \int \phi_x^2 + \phi_y^2 \, dx \, dy.$$

Use the two-independent-variable Euler-Lagrange equation of Exercise 1.3 to show that an extremal for this integral is a *harmonic function*. Dirichlet's integral arises in many areas of physics and engineering (see Exercise 4.5.2). In particular, the two variable version above could represent the capacity (per unit length) of a cylindrical condenser with annular cross-sections and the analogous three variable version could represent the potential energy associated with an electric field. For the latter, a stable equilibrium is attained when potential energy is minimized, so a minimizer for the Dirichlet integral is identified with the potential for the field. For more information see [Wei].

Example 2.4. *Hamilton's principle* is a principle of mechanics which states that the equations of motion of a physical system may be found by extremizing the so-called *action* integral

$$J = \int T - V \, dt,$$

where T and V represent kinetic and potential energy respectively. Of course T and V may take many forms depending on the problem at hand. To understand why such a principle might have been developed, let's extremize the action for the particular one-dimensional case of a single particle moving along the x-axis under the influence of a conservative force field F. Recall that this means that F is the (negative of the) gradient (which here reduces to a single partial derivative) of some potential function $V(x)$, $F = -\text{grad}\, V = -\partial V/\partial x$. (The negative sign is traditional in physics.) The particle has kinetic energy $T = (1/2)m\dot{x}^2$, so the action integral becomes

$$J = \int \frac{1}{2}m\dot{x}^2 - V(x) \, dt.$$

The Euler-Lagrange equation for this integral is then

$$-\frac{\partial V(x)}{\partial x} - \frac{d}{dt}(m\dot{x}) = 0.$$

Replacing the first term by F and carrying out the differentiation in the second term gives Newton's Law

$$F = m\ddot{x}.$$

Furthermore, notice that the integrand $T - V$ does not depend on t, so we may use the *first integral* $f - \dot{x}(\partial f/\partial \dot{x}) = c$ of Exercise 1.4 to get

$$T - V - \dot{x}\frac{\partial(T - V)}{\partial \dot{x}} = c$$

$$\frac{1}{2}m\dot{x}^2 - V - \dot{x}m\dot{x} = c$$

$$-\frac{1}{2}m\dot{x}^2 - V = c$$

$$\frac{1}{2}m\dot{x}^2 + V = \bar{c}$$

$$T + V = \bar{c}$$

$$E = \bar{c}$$

where $E = T + V$ is the total energy of the particle. This calculation means that the total energy in a conservative force field is actually *conserved* (i.e. is a constant). The fact that both Newton's Law and the conservation of energy may be derived by Hamilton's principle indicates why it is a cornerstone of classical mechanics*. We will see the connection between Hamilton's principle and geometry in 8.4 below.

8.3 THE WEIERSTRASS E-FUNCTION

So how in the world do we ever know if an extremal is truly a minimizing curve for a given integral J? Besides *ad hoc* means (as in Exercise 1.11), there are various sufficiency conditions for minimizers which may be used. None of them, however, seems particularly convenient for large problems, so we will stick with one which is rather straightforward. This condition is due to Weierstrass (whose work on minimal surfaces we've already met in Chapter 7).

Let's say that $x = x(t)$ is a minimizing curve with fixed endpoints for the integral $J = \int f(t, x, \dot{x})\, dt$ and $\widehat{x} = \widehat{x}(t)$ is another curve joining the same endpoints. Denote the values which has along the curves x and $\widehat{x}$ by $J[x]$ and $J[\widehat{x}]$ respectively. Then, since x is minimizing for J, it must be the case that

$$\Delta J \stackrel{\text{def}}{=} J[\widehat{x}] - J[x] \geq 0.$$

This condition, as it stands, is impossible to check for an extremal x against a comparison curve $\widehat{x}$. So, in order to compare $J[\widehat{x}]$ to $J[x]$, we compare

---

*As noted in [Wei], however, it should not be thought that Hamilton's principle is somehow more fundamental than Newton's Law. The very definition of mass is a consequence of $F = ma$ and mass is certainly needed to form the integrand $T - V$ of the action integral.

$\widehat{x}$ (or more precisely, its f-value and its slope) to other extremals in that particular region of space. Of course, the very existence of these extremals is not guaranteed, so we need a definition.

Definition 3.1. For the standard fixed endpoint problem with integral J, a *field of extremals* is a collection of extremals for J (i.e. solutions of the Euler-Lagrange equation) which, in a given region, satisfies the condition that, for each point in the region, there is *precisely one* extremal of the collection passing through it.

Example 3.2. Consider the fixed endpoint problem with $x(0) = 0$, $x(2) = 2$ and integral

$$J = \int_0^2 \frac{1}{2}\dot{x}^2 + x\dot{x} + x + \dot{x}\, dt.$$

The Euler-Lagrange equation gives a general extremal

$$x(t) = \frac{t^2}{2} + ct + d$$

which, upon applying the initial conditions, becomes $x(t) = t^2/2$. A field of extremals may be defined by using one of the constants of integration in the formula for a general extremal. In this way we are assured of obtaining an extremal for the original J. For example, take the field to be

$$x_d(t) = \frac{t^2}{2} + d.$$

Now, if two of these extremals, x_{d_1} and x_{d_2}, pass through a given (t, x), then $t^2/2 + d_1 = t^2/2 + d_2$, resulting in $d_1 = d_2$ and the equality of the extremals. Therefore, we really do have a field of extremals.

EXERCISE 3.1. For the fixed endpoint problem of Exercise 1.8,

$$J = \int_0^\pi x\,\sin t + \frac{1}{2}\dot{x}^2\, dt$$

with $x(0) = 0$, $x(\pi) = \pi$, find a field of extremals.

EXERCISE 3.2. For the fixed endpoint problem of Exercise 1.9,

$$J = \int_0^1 \dot{x}\,x^2 + \dot{x}^2\,x\, dt$$

with $x(0) = 1$, $x(1) = 4$, find a field of extremals.

EXERCISE 3.3. For the fixed endpoint problem of Exercise 1.10,

$$\int_0^{\pi/2} \dot{x}^2 - x^2\, dt$$

with $x(0) = 0$, $x(\pi/2) = 1$, find a field of extremals.

Once we have a field of extremals for a problem, we can compare the field to any curve joining the endpoints. More precisely, given an integral J, endpoint conditions $x(t_0) = x_0$, $x(t_1) = x_1$ and a curve $\widehat{x}$ joining the endpoints, at an arbitrary point (t, x) define $p = p(t, x)$ to be the derivative of *the unique extremal* in the field which passes through (t, x). Since the function p is defined in terms of derivatives of extremals, it called the *slope function of the field*. In order to compare values of J along curves, we make the following

Definition 3.3. Given a field of extremals with slope function $p = p(t, x)$ associated to an integral $J = \int f(t, x, \dot{x}) \, dt$, the Weierstrass $E(xcess)$-function is

$$E(t, x, \dot{x}, p) = f(t, x, \dot{x}) - f(t, x, p) - (\dot{x} - p)\frac{\partial f}{\partial p}(t, x, p).$$

Why is this E-function defined the way it is? Here's one rather informal explanation. Suppose we fix (t, x) and consider f as a function of the variable $\dot{x}$. Taylor's theorem for f *at the point* (t, x, p) gives

$$f(t, x, \dot{x}) = f(t, x, p) + (\dot{x} - p)\frac{\partial f}{\partial p}(t, x, p) + (\dot{x} - p)^2 \frac{\partial^2 f}{\partial p^2}(\mathfrak{t}, \mathfrak{x}, \dot{\mathfrak{x}}),$$

where the second partial is evaluated at $(\mathfrak{t}, \mathfrak{x}, \dot{\mathfrak{x}})$ between $(t, x, \dot{x})$ and (t, x, p). If we subtract the first two terms of the righthand side from both sides of the equation, we obtain

$$E(t, x, \dot{x}, p) = (\dot{x} - p)^2 \frac{\partial^2 f}{\partial p^2}(\mathfrak{t}, \mathfrak{x}, \dot{\mathfrak{x}}).$$

If we knew, for example, that the second partial of f with respect to p was always positive, then we could say that the E-function was always positive as well. Furthermore, if we integrate the E-function (evaluated on an arbitrary curve $\widehat{x}$), we see that there is a natural splitting

$$\int E(t, \widehat{x}, \dot{\widehat{x}}, p) \, dt = \int f(t, \widehat{x}, \dot{\widehat{x}}) - f(t, \widehat{x}, p) - (\dot{\widehat{x}} - p)\frac{\partial f}{\partial p}(t, \widehat{x}, p) \, dt$$

$$= \int f(t, \widehat{x}, \dot{\widehat{x}}) \, dt - \int f(t, \widehat{x}, p) + (\dot{\widehat{x}} - p)\frac{\partial f}{\partial p}(t, \widehat{x}, p) \, dt$$

$$= J[\widehat{x}] - K[\widehat{x}]$$

where $K[\widehat{x}]$ is the integral

$$K[\widehat{x}] = \int f(t, \widehat{x}, p) + (\dot{\widehat{x}} - p)\frac{\partial f}{\partial p}(t, \widehat{x}, p) \, dt.$$

Now,

(1) if E were always nonnegative (so that $\int E \, dt \geq 0$ as well)
(2) *and if $K[\widehat{x}] = J[x]$ for the extremal of the fixed endpoint problem x,*
    then we would have

$$\Delta J = J[\widehat{x}] - J[x]$$

$$= J[\widehat{x}] - K[\widehat{x}]$$

$$= \int E \, dt$$

$$\geq 0,$$

and $x(t)$ would be a minimizer for J. We will prove that (2) always holds. In fact, we will now show that K is path independent in the sense that any curve $\widetilde{x}(t)$ produces the same value for $K[\widetilde{x}]$. For this reason — and because David Hilbert discovered this integral K — K is called *Hilbert's Invariant Integral*.

Lemma 3.1. *The integral*

$$K[\widehat{x}] = \int f(t, \widehat{x}, p) + (\dot{\widehat{x}} - p)\frac{\partial f}{\partial p}(t, \widehat{x}, p) \, dt$$

is independent of the path $\widehat{x}(t)$ joining the endpoints x_0 and x_1.

Remark 3.4. Notice right away that $K[x] = J[x]$ for an extremal x since, in this case, $p = \dot{x}$.

Instead of proving the lemma directly, we are going to consider the two-variable case because it has its uses in geometry and is rarely presented explicitly. Consider the fixed endpoint problem with endpoint conditions $x(t_0) = x_0$, $y(t_0) = y_0$, $x(t_1) = x_1$, $y(t_1) = y_1$ and integral to be minimized

$$J[x, y] = \int_{t_0}^{t_1} f(t, x, y, \dot{x}, dy) \, dt.$$

Suppose that $(x, y) = (x(t), y(t))$ is an extremal for the problem. Further, assume that a field of extremals exists with two-variable slope function

$$p = (p_1, p_2) \stackrel{\text{def}}{=} (\dot{x}, \dot{y}).$$

The associated Weierstrass E-function is

$$E(t, \widehat{x}, \widehat{y}, \dot{\widehat{x}}, \dot{\widehat{y}}, p_1, p_2) = f(t, \widehat{x}, \widehat{y}, \dot{\widehat{x}}, \dot{\widehat{y}}) - f(t, \widehat{x}, \widehat{y}, p_1, p_2)$$

$$- (\dot{\widehat{x}} - p_1)\frac{\partial f}{\partial p_1}(t, \widehat{x}, \widehat{y}, p_1, p_2) - (\dot{\widehat{y}} - p_2)\frac{\partial f}{\partial p_2}(t, \widehat{x}, \widehat{y}, p_1, p_2)$$

and Hilbert's invariant integral is

$$K[\widehat{x}, \widehat{y}] = \int f(t, \widehat{x}, \widehat{y}, p_1, p_2) + (\dot{\widehat{x}} - p_1)\frac{\partial f}{\partial p_1}(t, \widehat{x}, \widehat{y}, p_1, p_2)$$

$$+ (\dot{\widehat{y}} - p_2)\frac{\partial f}{\partial p_2}(t, \widehat{x}, \widehat{y}, p_1, p_2) \, dt$$

$$= \int f - p_1\frac{\partial f}{\partial p_1} - p_2\frac{\partial f}{\partial p_2} \, dt + \frac{\partial f}{\partial p_1} \, d\widehat{x} + \frac{\partial f}{\partial p_2} \, d\widehat{y},$$

where we have used $\dot{\widehat{x}}\,dt = d\widehat{x}$ and $\dot{\widehat{y}}\,dt = d\widehat{y}$. The last equality turns K into a line integral and we can now use the standard fact about line integrals that they are independent of path (in a simply connected region) if and only if the associated vector field has a potential function. Recall that a vector field on an open simply connected region $\mathcal{O} \subseteq \mathbb{R}^3$ is a mapping $F : \mathcal{O} \to \mathbb{R}^3$, $F(t, x, y) = (A(t, x, y), B(t, x, y), C(t, x, y))$ and a line integral of F along a path $\alpha(t) = (t, x(t), y(t))$ with differential $d\alpha = (dt, dx, dy)$ is given by $\int_\alpha F \cdot d\alpha = \int A\,dt + B\,dx + C\,dy$. The vector field F has a potential function if

$$F = (A, B, C) = \operatorname{grad} \phi = (\phi_t, \phi_x, \phi_y),$$

where grad denotes the gradient and ϕ_x etc. denotes partial derivative with respected to the subscript variable. Clearly then, the condition for having a potential function is

$$\frac{\partial A}{\partial \widehat{x}} = \frac{\partial B}{\partial t} \qquad \frac{\partial A}{\partial \widehat{y}} = \frac{\partial C}{\partial t} \qquad \frac{\partial B}{\partial \widehat{y}} = \frac{\partial C}{\partial \widehat{x}}.$$

In particular, for $K[\widehat{x}, \widehat{y}]$ to be path independent, we need

$$\frac{\partial \left(f - p_1 \frac{\partial f}{\partial p_1} - p_2 \frac{\partial f}{\partial p_2} \right)}{\partial \widehat{x}} = \frac{\partial}{\partial t} \frac{\partial f}{\partial p_1}$$

$$= \frac{\partial^2 f}{\partial p_1 \partial t} + \frac{\partial^2 f}{\partial p_1^2} \frac{\partial p_1}{\partial t} + \frac{\partial^2 f}{\partial p_1 \partial p_2} \frac{\partial p_2}{\partial t}$$

$$\frac{\partial \left(f - p_1 \frac{\partial f}{\partial p_1} - p_2 \frac{\partial f}{\partial p_2} \right)}{\partial \widehat{y}} = \frac{\partial}{\partial t} \frac{\partial f}{\partial p_2}$$

$$= \frac{\partial^2 f}{\partial p_2 \partial t} + \frac{\partial^2 f}{\partial p_2^2} \frac{\partial p_2}{\partial t} + \frac{\partial^2 f}{\partial p_1 \partial p_2} \frac{\partial p_1}{\partial t}$$

$$\frac{\partial^2 f}{\partial p_1 \partial \widehat{y}} = \frac{\partial^2 f}{\partial p_2 \partial \widehat{x}}.$$

The third condition is a condition on the field of extremals and we will make the tacit assumption it always holds. Such a field of extremals is called a *Mayer field*. See [Sag] for an in-depth discussion. Note that, in the one-variable situation, the third condition holds vacuously. In the following, we shall prove that the first condition holds. The second condition follows similarly. Also, in the proof below we will use the "hatless" notation x instead of $\widehat{x}$ for the sake of simplicity. This does not mean however that x is an extremal, but rather that we are interested only in formal partial derivatives without regard to particular curves.

Theorem 3.2. $K[\widehat{x}, \widehat{y}]$ *is independent of path.*

Proof. The proof is a tedious, but instructive exercise in the chain rule. Let's carry out the differentiations listed above for (1), set the quantities which we want to be equal equal to each other and search for a true identity. Case (2) will follow similarly.

$$\frac{\partial\left(f - p_1\frac{\partial f}{\partial p_1} - p_2\frac{\partial f}{\partial p_2}\right)}{\partial x} = \frac{\partial f}{\partial x} - p_1\left[\frac{\partial^2 f}{\partial x \partial p_1} + \frac{\partial^2 f}{\partial p_1^2}\frac{\partial p_1}{\partial x} + \frac{\partial^2 f}{\partial p_1 \partial p_2} + \frac{\partial p_2}{\partial x}\right]$$
$$- p_2\left[\frac{\partial^2 f}{\partial y \partial p_1} + \frac{\partial^2 f}{\partial p_1^2}\frac{\partial p_1}{\partial y} + \frac{\partial^2 f}{\partial p_1 \partial p_2} + \frac{\partial p_2}{\partial y}\right]$$

which we set equal to

$$= \frac{\partial^2 f}{\partial p_1 \partial t} + \frac{\partial^2 f}{\partial p_1^2}\frac{\partial p_1}{\partial t} + \frac{\partial^2 f}{\partial p_1 \partial p_2}\frac{\partial p_2}{\partial t}$$

which, by simplifying and isolating $\partial f/\partial x$ gives

$$\frac{\partial f}{\partial x} = \frac{\partial^2 f}{\partial t \partial p_1} + \frac{\partial^2 f}{\partial p_1^2}\frac{\partial p_1}{\partial t} + \frac{\partial^2 f}{\partial p_1 \partial p_2}\frac{\partial p_2}{\partial t} + p_1\frac{\partial^2 f}{\partial x \partial p_1}$$
$$+ p_2\frac{\partial^2 f}{\partial y \partial p_1} + p_1\frac{\partial^2 f}{\partial p_1^2}\frac{\partial p_1}{\partial x} + p_1\frac{\partial^2 f}{\partial p_1 \partial p_2}\frac{\partial p_2}{\partial x}$$
$$+ p_2\frac{\partial^2 f}{\partial p_1^2}\frac{\partial p_1}{\partial y} + p_2\frac{\partial^2 f}{\partial p_1 \partial p_2}\frac{\partial p_2}{\partial y}$$
$$= \frac{\partial^2 f}{\partial t \partial p_1} + p_1\frac{\partial^2 f}{\partial x \partial p_1} + p_2\frac{\partial^2 f}{\partial y \partial p_1} + \frac{\partial^2 f}{\partial p_1^2}\left(\frac{\partial p_1}{\partial t} + p_1\frac{\partial p_1}{\partial x} + p_2\frac{\partial p_1}{\partial y}\right)$$
$$+ \frac{\partial^2 f}{\partial p_1 \partial p_2}\left(\frac{\partial p_2}{\partial t} + p_1\frac{\partial p_2}{\partial x} + p_2\frac{\partial p_2}{\partial y}\right).$$

Now, along an extremal, $p_1 = \dot{x}$ and $p_2 = \dot{y}$, so differentiation produces the following general relationships among p_1 and p_2.

$$\dot{p}_1 = \frac{\partial p_1}{\partial t} + p_1\frac{\partial p_1}{\partial x} + p_2\frac{\partial p_1}{\partial y}, \qquad \dot{p}_2 = \frac{\partial p_2}{\partial t} + p_1\frac{\partial p_2}{\partial x} + p_2\frac{\partial p_2}{\partial y}.$$

Substituting in, we obtain

$$\frac{\partial f}{\partial x} = \frac{\partial^2 f}{\partial t \partial p_1} + p_1\frac{\partial^2 f}{\partial x \partial p_1} + p_2\frac{\partial^2 f}{\partial y \partial p_1} + \frac{\partial^2 f}{\partial p_1^2}\dot{p}_1 + \frac{\partial^2 f}{\partial p_1 \partial p_2}\dot{p}_2.$$

Note that this is an expression in formal partial derivatives of f *and* in components of the slope function associated to a field of *extremals*. We can find another relationship between p_1 and p_2 by using the fact that they are associated to extremals. Namely, along an extremal the Euler-Lagrange equations hold, so we have (upon replacing the usual $\dot{x}$ notation by p_1) and carrying out the chain rule,

$$\frac{\partial}{\partial t}\left(\frac{\partial f}{\partial p_1}\right) = \frac{\partial^2 f}{\partial t \partial p_1} + p_1\frac{\partial^2 f}{\partial x \partial p_1} + p_2\frac{\partial^2 f}{\partial y \partial p_1} + \frac{\partial^2 f}{\partial p_1^2}\dot{p}_1 + \frac{\partial^2 f}{\partial p_1 \partial p_2}\dot{p}_2 = \frac{\partial f}{\partial x}$$

which is true by the Euler-Lagrange equation. Comparing this equation with the one above, we see that they are identical, so we have found the required identity. Therefore, working backwards up the chain, we see that all the equations are true. Hence, we have proven condition (1) and (since (2) follows the same way) K is independent of path. □

Corollary 3.3: The Weierstrass Condition. *If a trajectory* $(x, y) = (x(t), y(t))$ *is a member of a field of extremals for a fixed endpoint problem with integral* J *and*

$$E(t, x, y, \dot{x}, \dot{y}, p_1, p_2) \geq 0,$$

then (x, y) *is a minimizer for* J.

Proof. Although we have explained what the proof should be earlier, we repeat it here for convenience and completeness. Let $\hat{x}$ denote any other curve joining the endpoints. Again, recall that $K[(x, y)] = J[(x, y)]$ for an extremal (x, y) since the terms involving $(\dot{x} - p_1)$ and $(\dot{y} - p_2)$ vanish.

$$\begin{aligned}\Delta J &= J[(\hat{x}, \hat{y})] - J[(x, y)] \\ &= J[(\hat{x}, \hat{y})] - K[(x, y)] \\ &= J[(\hat{x}, \hat{y})] - K[(\hat{x}, \hat{y})]\end{aligned}$$

since K is invariant, so

$$\begin{aligned}\Delta J &= \int E(t, x, y, \dot{x}, dy, p_1, p_2)\, dt \\ &\geq 0\end{aligned}$$

since $E \geq 0$. Hence, $J[(\hat{x}, \hat{y})] \geq J[(x, y)]$. □

EXERCISE 3.4. In the case of one variable x, show that $K[x]$ is independent of path. Then show that if x is a member of a field of extremals and $E(t, x, \dot{x}, p) \geq 0$, then x is a minimizer.

Example 3.5. Consider the fixed endpoint problem with $x(0) = 0$, $x(2) = 2$ and integral

$$J = \int_0^2 \frac{1}{2}\dot{x}^2 + x\dot{x} + x + \dot{x}\, dt.$$

An extremal is $x(t) = t^2/2$ and a field of extremals may be defined by $x_d(t) = \frac{t^2}{2} + d$. Let's compute the Weierstrass E-function.

$$
\begin{aligned}
E(t, x, \dot{x}, p) &= \frac{1}{2}\dot{x}^2 + x\dot{x} + x + \dot{x} - \frac{1}{2}p^2 - xp - x - p - (\dot{x} - p)(p + x + 1) \\
&= \frac{1}{2}\dot{x}^2 - \frac{1}{2}p^2 - \dot{x}\,p + p^2 \\
&= \frac{1}{2}\dot{x}^2 + \frac{1}{2}p^2 - \dot{x}\,p \\
&= \frac{1}{2}[\dot{x}^2 - 2\dot{x}\,p + p^2] \\
&= \frac{1}{2}(\dot{x} - p)^2 \\
&\geq 0.
\end{aligned}
$$

Hence, $x(t) = t^2/2$ is a minimizer for J.

Example 3.6: The Brachistochrone. Assume that the brachistochrone problem has a field of extremals. Then, the E-function has the form

$$
\begin{aligned}
E &= \frac{\sqrt{1 + u'^2}}{\sqrt{u}} - \frac{\sqrt{1 + p^2}}{\sqrt{u}} - (u' - p)\frac{p}{\sqrt{u(1 + p^2)}} \\
&= \frac{\sqrt{1 + u'^2}\sqrt{1 + p^2} - (1 + p^2) - pu' + p^2}{\sqrt{u(1 + p^2)}} \\
&= \frac{\sqrt{1 + u'^2}\sqrt{1 + p^2} - 1 - pu'}{\sqrt{u(1 + p^2)}} \\
&= \frac{|(1, u')|\,|(1, p)| - (1, u') \cdot (1, p)}{\sqrt{u(1 + p^2)}}
\end{aligned}
$$

where $|\ |$ denotes length and $\cdot$ denotes dot product. We then have

$$
E = \frac{|(1, u')|\,|(1, p)| - (1, u') \cdot (1, p)}{\sqrt{u(1 + p^2)}} \geq 0
$$

by the Schwarz inequality. Hence, the cycloid is a true minimizer for the brachistochrone problem. Compare [HK] and, especially, [L].

EXERCISE 3.5. We have seen that straight lines are extremals for the arclength integral

$$
J = \int \sqrt{\dot{x}^2(t) + \dot{y}^2(t)}\, dt.
$$

Clearly, by simply translating the line up and down we obtain a field of extremals. Compute the two-variable Weierstrass E-function and show that straight lines really do give the shortest distance between two points in the plane. Hint: follow the brachistochrone example.

EXERCISE 3.6. Show that the Weierstrass E-function is always nonnegative for the extremal of the surface of revolution of least area (see Example 2.3 and Exercise 2.3). The reason that the catenoid does not always minimize area is that, beyond a certain point, it no longer sits in a field of extremals.

EXERCISE 3.7. For the fixed endpoint problem of Exercise 1.8,

$$J = \int_0^\pi x \sin t + \frac{1}{2} \dot{x}^2 \, dt$$

with $x(0) = 0$, $x(\pi) = \pi$, compute the Weierstrass E-function to show that the extremal is a minimizer.

EXERCISE 3.8. For the fixed endpoint problem of Exercise 1.9,

$$J = \int_0^1 \dot{x} \, x^2 + \dot{x}^2 \, x \, dt$$

with $x(0) = 1$, $x(1) = 4$, compute the Weierstrass E-function to show that the extremal is a minimizer.

EXERCISE 3.9. For the fixed endpoint problem of Exercise 1.10,

$$\int_0^{\pi/2} \dot{x}^2 - x^2 \, dt$$

with $x(0) = 0$, $x(\pi/2) = 1$, compute the Weierstrass E-function to show that the extremal is a minimizer.

EXERCISE 3.10. For the fixed endpoint problem,

$$\int_0^1 t^2 - \dot{x}^2 \, dt$$

with $x(0) = 0$, $x(1) = 1$, compute the Weierstrass E-function to show that the extremal is a maximizer.

The Weierstrass E-function provides a powerful tool for identifying minimizing curves. Nevertheless, it does not provide a panacea. For one thing, we are usually able to show $E \geq 0$ only when some sort of special algebraic structure is present. In particular, we usually prove $E \geq 0$ for an arbitrary *variable* p as opposed to actually using specific properties of a particular slope function p. This is so simply because information about p rarely helps. Therefore, while it may not be hard to show E nonnegative for relatively simple problems, for more complicated problems, the algebraic complexity goes up exponentially. Secondly, the existence of a field of extremals is by no means something which is guaranteed. Indeed, there is a *Jacobi differential equation* akin to that of Chapter 5 and a notion of *conjugate point* which detects whether an extremal sits in a field. Specialized to geodesics, this explains why going past a conjugate point on a geodesic implies that the geodesic is not minimizing. This general Jacobi condition is too complicated to be presented fully in our naive context, but the interested reader should see [Pin] for an elementary (and brief) discussion, as well as [Bli], [Ew] and [Sag]. We will only present a sketch of the situation for geodesics in the following example. For details, see [Hsi].

Example 3.7: The Jacobi Equation Revisited.
Suppose that the v-parameter curve $\alpha(v) = \mathbf{x}(0, v)$ is a shortest distance geodesic joining two fixed endpoints and the u-parameter curves $\mathbf{x}(u, v_0)$ are geodesics orthogonal to α. In fact, it can be shown that such a patch always exists in a region near α and, in this case, the parameters u and v are called *Fermi coordinates*. By taking v to be the α arclength parameter, we obtain the metric coefficients $E = 1$, $F = 0$ and $G > 0$ with

$$G(0, v) = 1 \qquad \text{and} \qquad G_u(0, v) = 0.$$

The first equality follows from the fact that the geodesic $\alpha = \mathbf{x}(0, v)$ is parametrized by arclength, so $G(0, v) = \mathbf{x}_v \cdot \mathbf{x}_v = \alpha' \cdot \alpha' = 1$. The second equality follows from the formula developed for geodesic curvature in Theorem 5.4.1. We have $\kappa_g = 0$ since α is a geodesic and the angle from α' to $\mathbf{x}_u$ as well as the metric coefficient $E = 1$ are constants. Hence, $G_u(0, v) = 0$. The equalities above are saying that, along the curve α, the metric looks like a Euclidean metric up to first order derivatives. This allows us to carry out a "second variation" below as if we were in the plane.

EXERCISE 3.11. Use the equalities $G(0, v) = 1$ and $G_u(0, v) = 0$ to show that, along the curve $\alpha = \mathbf{x}(0, v)$, the Gauss curvature of the surface is given by $K = -G_{uu}(0, v)/2$. Hint: use Theorem 3.4.1.

Now let's choose a variation of α of the form $\mathbf{x}(\epsilon\eta(v), v)$ where ϵ is small and $\eta(a) = 0$, $\eta(b) = 0$ (and $\alpha(a)$, $\alpha(b)$ are fixed endpoints on α). The arclength integral is then

$$J(\epsilon) = \int_a^b \sqrt{\epsilon^2\eta'^2 + G(\epsilon\eta(v), v)}\, dv$$

with ϵ-derivative equal to

$$0 = \left.\frac{dJ}{d\epsilon}\right|_{\epsilon=0} = \left.\int_a^b \frac{2\epsilon\eta'^2 + G_u\eta}{2\sqrt{\epsilon^2\eta'^2 + G(\epsilon\eta(v), v)}}\, dv\right|_{\epsilon=0}$$

since we assume α to be shortest distance. Also, since α gives a minimum for arclength, the second derivative of the arclength function (at $\epsilon = 0$) must be nonnegative. Therefore, we obtain $d^2 J / d\epsilon^2 =$

$$\int_a^b \frac{[2\eta'^2 + G_{uu}\eta^2]\sqrt{\epsilon^2\eta'^2 + G} - [2\epsilon\eta'^2 + G_u\eta][2\epsilon\eta'^2 + G_u\eta]/2\sqrt{\epsilon^2\eta'^2 + G}}{\epsilon^2\eta'^2 + G} \, dv$$

$$= \int_a^b \frac{2[2\eta'^2 + G_{uu}\eta^2][\epsilon^2\eta'^2 + G] - [2\epsilon\eta'^2 + G_u\eta]^2}{2(\epsilon^2\eta'^2 + G)^{3/2}} \, dv.$$

Upon evaluation at $\epsilon = 0$, we get

$$\frac{d^2 J}{d\epsilon^2}\bigg|_{\epsilon=0} = \int_a^b \frac{2[2\eta'^2 + G_{uu}(0, v)\eta^2] \, G(0, v) - G_u^2(0, v)\eta^2}{2G^{3/2}(0, v)} \, dv$$

$$= \int_a^b 2\eta'^2 + G_{uu}(0, v)\eta^2 \, dv$$

since $G(0, v) = 1$ and $G_u(0, v) = 0$. If α truly is a shortest length geodesic joining the endpoints, then, as a minimum of the function $J(\epsilon)$, the integral above (i.e. the second derivative of J) must be nonnegative *for all choices of η with $\eta(a) = 0 = \eta(b)$*. But the integrand of J's second derivative is a function $f(v, \eta, \eta')$ and so an η giving the minimum of the integral (which we want to be zero) *must satisfy the Euler-Lagrange equation for this integral*. Thus, we arrive at Jacobi's *secondary variational problem*. Namely, we must calculate the Euler-Lagrange equation of J's second derivative.

$$2\,G_{uu}(0, v)\,\eta - \frac{d}{dv}\,(4\eta') = 0$$

$$\eta'' - \frac{G_{uu}(0, v)}{2}\eta = 0$$

$$\eta'' + K(0, v)\,\eta = 0$$

since $G_{uu}(0, v) = 2K$ by Exercise 3.11 above. Of course this is the *Jacobi equation* of Chapter 6. Now, here's the connection between solutions of the Jacobi equation and geodesics of shortest length. Recall that a point $c \in [a, b]$ is *conjugate* to a if there is a (not identically zero) solution of the Jacobi equation η such that $\eta(a) = 0$ and $\eta(c) = 0$. (We assume c is the first such point in the interval $[a, b]$.) We shall show that the existence of a conjugate point implies that there is an $\bar{\eta}$ for which the second derivative is negative

$$\frac{d^2 J}{d\epsilon^2}\bigg|_{\epsilon=0} < 0.$$

By the discussion above, this means that α could not possibly be a *minimum* for the arclength function. In other words, *a geodesic is shortest length only up to the first conjugate point*. (Technically speaking, we can actually only say that, for all curves sufficiently close to the geodesic, the geodesic is shortest length only up to the first conjugate point.)

To prove that a solution η to Jacobi's equation with $\eta(c) = 0$ gives a negative second derivative, we must use a technical tool from the calculus of variations called the *Weierstrass-Erdmann corner condition* (see [Pin] or [Sag] for example). This condition says that, along a piecewise smooth extremal $x(t)$ (i.e. an extremal which is smooth except at a finite number of corners) for an integral $J = \int f(t, x, \dot{x}) \, dt$, the partial derivative $\partial f / \partial \dot{x}$ must be continuous at a corner T. That is, if we take the limit of partials on both sides of the corner as we approach the corner T, we must get the same answer. With this in mind, let η be a nonzero solution to the Jacobi equation $\eta'' + K\eta = 0$ (along α) with $\eta(c) = 0$ and $\eta(p) \neq 0$ for all $p \in (a, c)$. Define

$$\xi(p) = \begin{cases} \eta(p) & \text{for } p \in [a, c] \\ 0 & \text{for } p \in [c, b]. \end{cases}$$

For this variation ξ (and using $K = -G_{uu}/2$), the second derivative is

$$\left. \frac{d^2 J}{d\epsilon^2} \right|_{\epsilon=0} = 2 \int_a^b \xi'^2 - K\xi^2 \, dv$$

$$= 2 \int_a^b \xi'^2 \, dv - 2 \int_a^b K\xi^2 \, dv$$

$$= 2 \int_a^c \eta'^2 \, dv - 2 \int_a^c K\eta^2 \, dv$$

since $\xi = \eta$ on $[a, c]$ and $\xi = 0$ otherwise. Integrating the first term by parts gives

$$\left. \frac{d^2 J}{d\epsilon^2} \right|_{\epsilon=0} = 2\eta\eta'|_a^c - 2 \int_a^c \eta\eta'' \, dv - 2 \int_a^c K\eta^2 \, dv$$

and since η is zero at a and c,

$$= -2 \int_a^c \eta[\eta'' + K\eta] \, dv$$

$$= 0$$

since η satisfies the Jacobi equation. So, ξ gives a value of zero for the second derivative. Could ξ in fact be a minimizing curve for the second derivative? If so, then the Weierstrass-Erdmann corner condition would have to be satisfied. On one side of the corner c, the η' derivative of J's second derivative's integrand $\eta'^2 - K\eta^2$ is $2\eta'$ while on the other side (where $\xi = 0$), the η' derivative is zero. Hence, we obtain $\eta'(c) = 0$. But we already know that η is a solution to a second order linear differential equation (the Jacobi equation) and $\eta(c) = 0$. By the uniqueness of solutions of such differential equations with given initial data, it must be the case that $\eta = 0$ identically. This contradicts our assumption about η. Therefore, the Weierstrass-Erdmann condition cannot hold for η and, consequently, *it cannot be a minimum for the second derivative.* Since η gives a value of zero for the second derivative, there must exist a piecewise smooth function $\bar{\eta}$ which makes the second derivative negative,

$$\left. \frac{d^2 J}{d\epsilon^2} \right|_{\epsilon=0} [\bar{\eta}] < 0.$$

In fact, any corners of such a function may be "rounded off" to produce a smooth such $\bar{\eta}$. The existence of such an $\bar{\eta}$ then implies that the original geodesic α is *not* a minimum of arclength for any point past c. This example illustrates the importance of the Jacobi equation in determining whether or not extremals actually minimize. Also, from the definition of the patch $\mathbf{x}(u, v)$ itself, we see that the Jacobi equation is intimately connected with the existence of fields of extremals.

Besides these questions about fields of extremals, the E-function and conjugate points, there is a question which we have avoided up to now. While the Euler-Lagrange equations provide a necessary condition for an extremum, there may in fact be *no* solution to a given variational problem. For instance, if we consider the plane with the origin removed and ask for the curve of shortest length joining $(1, 0)$ and $(-1, 0)$, then there is no solution because the straight line extremal joining the endpoints passes through $(0, 0)$ and, therefore, is not allowed. In the calculus of variations, each problem must be analyzed separately to determine if a solution exists. Of course, the question of what types of functions may be allowed as solutions arises as well. Usually, as in Example 3.7 above, an *admissible* class os functions for variational problems is the class of piecewise smooth functions. These are continuous functions whose derivative(s) exist and are continuous everywhere except at a finite set of points where corners occur. The following exercise indicates some of the subtleties of the subject.

EXERCISE 3.12. For the fixed endpoint problem with $x(0) = 1$, $x(1) = 0$ and

$$J[x] = \int_0^1 x^4 + x\,\dot{x} + \frac{1}{2}\; dt,$$

a.) Show that the only extremal is $x(t) = 0$, so that there is *no* minimizer.
b.) Show that $J[x] \geq 0$. Hint: show that $J[x] = \int_0^1 x^4\, dt$. Use $\dot{x}\, dt = dx$.
c.) Show $J[x]$ gets arbitrarily close to 0. Hint: consider the piecewise linear function defined by

$$x(t) = \begin{cases} 1 - \epsilon t & 0 \leq t \leq \frac{1}{\epsilon} \\ 0 & \frac{1}{\epsilon} \leq t \leq 1 \end{cases}$$

and compute $J[x]$ for these functions.

There are various methods for analyzing whether extremals are minimizers, many of which can be found in texts such as [Bli], [Sag] and [Ew] for example. In particular there is a general approach which depends on the so-called *second variation* and, while we do not have space to discuss this in generality, we point out that Example 3.7 above and the method of Schwarz considered in Chapter 7.5 exemplify this method. See [DoC1] for a good discussion of the second variation as applied to geodesics.

8.4 PROBLEMS WITH CONSTRAINTS

Often, variational problems come with extra side conditions which must also be satisfied. For example, we saw in Chapter 4 that we can minimize the surface area of a compact surface enclosing a region of space subject to the constraint that the enclosed volume remain constant. Recall that the minimizer surface turned out to be a sphere. (For a variation of this constrained problem designed to describe the shape of a mylar balloon, see [Pau].) In fact, the most ancient of all these types of problems is the following two-dimensional version: given a closed curve of fixed arclength, what is the shape the curve should assume to *maximize* the enclosed area? Because of this, constrained variational problems are sometimes called *isoperimetric* (i.e. same perimeter) problems. In this section we shall see what modifications are necessary in our previous work to handle variational problems with constraints. The standard problem shall be,

$$\text{Minimize } J = \int_{t_0}^{t_1} f(t, x, \dot{x}) \, dt$$

subject to the endpoint conditions $x(t_0) = x_0$, $x(t_1) = x_1$ and the requirement that

$$I = \int_{t_0}^{t_1} g(t, x, \dot{x}) \, dt = c$$

where c is a constant.

We will now derive a form of the Euler-Lagrange equation which is a necessary condition for the solution of the constrained problem. Just as before, assume $x = x(t)$ is a minimizer for the problem and take a variation $\hat{x} = x + \epsilon(a\eta + b\xi)$ where $\eta(t_0) = 0 = \eta(t_1)$ and $\xi(t_0) = 0 = \xi(t_1)$. We take two "perturbations" η and ξ because taking only one would not allow us to vary J while holding I constant. By taking η *and* ξ, we can vary J while offsetting the effects of one perturbation with the other in I. The usual Euler-Lagrange argument gives

$$0 = \left. \frac{dJ}{d\epsilon} \right|_{\epsilon=0} = \int_{t_0}^{t_1} (a\eta + b\xi) \left(\frac{\partial f}{\partial x} - \frac{d}{dt} \left(\frac{\partial f}{\partial \dot{x}} \right) \right) \, dt.$$

Now, however, the usual argument must be modified because $a\eta + b\xi$ is *not* arbitrary. The requirement $I = c$ puts restrictions on a and b. If we carry through the argument above for I, we get

$$0 = \left.\frac{dI}{d\epsilon}\right|_{\epsilon=0} = \int_{t_0}^{t_1} (a\eta + b\xi)\left(\frac{\partial g}{\partial x} - \frac{d}{dt}\left(\frac{\partial g}{\partial \dot{x}}\right)\right)\,dt$$

where the derivative is zero since $I = c$ is a constant. But x is *not* an extremal for I, so the Euler-Lagrange equation with g does not hold. Instead, we have

$$\int_{t_0}^{t_1}(a\eta+b\xi)\left(\frac{\partial f}{\partial x} - \frac{d}{dt}\left(\frac{\partial f}{\partial \dot{x}}\right)\right)\,dt = 0 = \int_{t_0}^{t_1}(a\eta+b\xi)\left(\frac{\partial g}{\partial x} - \frac{d}{dt}\left(\frac{\partial g}{\partial \dot{x}}\right)\right)\,dt$$

which produces

$$-\frac{\int_{t_0}^{t_1}\xi\left(\frac{\partial g}{\partial x} - \frac{d}{dt}\left(\frac{\partial g}{\partial \dot{x}}\right)\right)\,dt}{\int_{t_0}^{t_1}\eta\left(\frac{\partial g}{\partial x} - \frac{d}{dt}\left(\frac{\partial g}{\partial \dot{x}}\right)\right)\,dt} = \frac{a}{b} = -\frac{\int_{t_0}^{t_1}\xi\left(\frac{\partial f}{\partial x} - \frac{d}{dt}\left(\frac{\partial f}{\partial \dot{x}}\right)\right)\,dt}{\int_{t_0}^{t_1}\eta\left(\frac{\partial f}{\partial x} - \frac{d}{dt}\left(\frac{\partial f}{\partial \dot{x}}\right)\right)\,dt}.$$

Upon rearranging we obtain

$$\frac{\int_{t_0}^{t_1}\xi\left(\frac{\partial f}{\partial x} - \frac{d}{dt}\left(\frac{\partial f}{\partial \dot{x}}\right)\right)\,dt}{\int_{t_0}^{t_1}\xi\left(\frac{\partial g}{\partial x} - \frac{d}{dt}\left(\frac{\partial g}{\partial \dot{x}}\right)\right)\,dt} = \frac{\int_{t_0}^{t_1}\eta\left(\frac{\partial f}{\partial x} - \frac{d}{dt}\left(\frac{\partial f}{\partial \dot{x}}\right)\right)\,dt}{\int_{t_0}^{t_1}\eta\left(\frac{\partial g}{\partial x} - \frac{d}{dt}\left(\frac{\partial g}{\partial \dot{x}}\right)\right)\,dt}.$$

The lefthand side is a function of ξ while the right is a function of η, so the only way for these expressions to be identical is for both expressions to be equal to the same constant λ (*the Lagrange multiplier*). Simplifying the expression

$$\frac{\int_{t_0}^{t_1}\eta\left(\frac{\partial f}{\partial x} - \frac{d}{dt}\left(\frac{\partial f}{\partial \dot{x}}\right)\right)\,dt}{\int_{t_0}^{t_1}\eta\left(\frac{\partial g}{\partial x} - \frac{d}{dt}\left(\frac{\partial g}{\partial \dot{x}}\right)\right)\,dt} = \lambda$$

gives the equation

$$\int_{t_0}^{t_1}\eta\left[\frac{\partial(f - \lambda g)}{\partial x} - \frac{d}{dt}\left(\frac{\partial(f - \lambda g)}{\partial \dot{x}}\right)\right]\,dt = 0.$$

Now the usual argument may be continued. Because η is arbitrary, the previous equation can hold only if the term in brackets vanishes. We, thus, obtain the following *Euler-Lagrange necessary condition* for the constrained problem.

Theorem 4.1. *If $x = x(t)$ is a solution to the standard constrained problem, then*

$$\frac{\partial(f - \lambda g)}{\partial x} - \frac{d}{dt}\left(\frac{\partial(f - \lambda g)}{\partial \dot{x}}\right) = 0.$$

Example 4.1. The *bending energy* of a plane curve is defined to be the integral of squared curvature over the length of the curve, $\int \kappa^2(s)\,ds$. Recall from Chapter 1 that the curvature of a plane curve also is given by $\dot{\theta} = d\theta/ds$, where $\theta(s)$ is the angle the unit tangent $T(s)$ makes with the x-axis. As we shall see when we discuss Hamilton's principle, it is not unreasonable to assume that a wire will try to assume a shape which *minimizes* bending energy. (For a biological application of this idea see [Can] as well as [DH1] and [DH2].) Therefore, we can ask the following question. What shape will a wire take if the total turning of its tangent is fixed and the turning is zero at the endpoints? As a constrained variational problem, we write

$$\text{Minimize} \quad J = \int_0^1 \frac{1}{2}\dot{\theta}^2\,ds \qquad \text{subject to} \qquad I = \int_0^1 \theta\,ds = \frac{1}{6}$$

$$\text{with } \theta(0) = 0,\ \theta(1) = 0.$$

The factor $1/2$ in the first integral is only put in to make calculations cleaner. Minimizing bending energy is clearly equivalent to minimizing one-half of bending energy. Similarly, the $1/6$ value of the second integral is chosen only to avoid fractions later on. This said, the Euler-Lagrange equation for the constrained problem is $-\lambda - \frac{d}{ds}\dot{\theta} = 0$ with solution

$$\theta(s) = \frac{\lambda}{2}(-s^2 + s).$$

An application of the constraint gives

$$\frac{1}{6} = I = \frac{\lambda}{2}\Big|_0^1 \left(-\frac{s^3}{3} + \frac{s^2}{2}\right) = \frac{\lambda}{12}.$$

Hence, $\lambda = 2$ and $\theta(s) = -s^2 + s$. Further, a field of extremals is given by $\theta_d = -s^2 + s + d$ and the Weierstrass E-function is $E = 1/2(\dot{\theta} - p)^2 \geq 0$. Hence, θ is a true minimizer for bending energy subject to the constraint. We have met the curve corresponding to θ previously in Exercise 1.5.11. Recall that a plane curve $\beta(s)$ with curvature $\kappa = d\theta/ds$ may be reconstructed by the formula

$$\beta(s) = \left(\int_0^s \cos(\theta(u))\,du, \int_0^s \sin(\theta(u))\,du\right).$$

In the case at hand, $\theta(s) = -s^2 + s$ and we have a form of Euler's spiral.

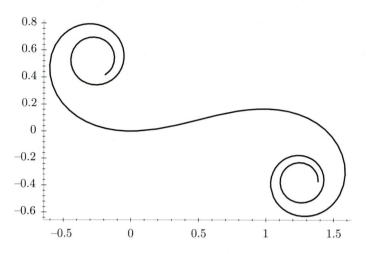

FIGURE 8.3.   Euler's spiral (the spiral of Cornu)

Remark 4.2. The reader should compare and contrast the preceding example with the classical problem of the *elastic rod* (see [Sag]) of fixed length L. Suppose a rod is clamped at points (a, y_a) and (b, y_b) at fixed angles θ_a and θ_b respectively. The shape the rod assumes at equilibrium is determined by minimizing the potential energy of the rod and this quantity is proportional to $\int \kappa^2(s)\, ds$. The constraints of the problem are determined by the fixed angles. Namely, projections onto the x and y axes give, $\int_0^L \cos\theta\, ds = b - a$ and $\int_0^L \sin\theta\, ds = y_b - y_a$. Therefore, using two Lagrange multipliers, we extremize the integral

$$J = \int_0^L \left(\frac{d\theta}{ds}\right)^2 - \lambda_1 \cos\theta - \lambda_2 \sin\theta\ ds$$

to obtain the Euler-Lagrange equation

$$\lambda_1 \sin\theta - \lambda_2 \cos\theta - \frac{d}{ds}\left(2\frac{d\theta}{ds}\right) = 0.$$

EXERCISE 4.1. Integrate the Euler-Lagrange equation with respect to s to obtain

$$\frac{d\theta}{ds} = -\frac{\lambda_1}{2}y + \frac{\lambda_2}{2}x + C.$$

Explain why all inflection points of the elastic rod lie on a straight line.

If the line determined in Exercise 4.1, $-\frac{\lambda_1}{2}y + \frac{\lambda_2}{2}x + C = 0$, is rotated to become the new y axis, then the shape of the elastic rod has the expression

$$y = \int \frac{x^2/2 + a}{\sqrt{1 - (x^2/2 + a)^2}}\ dx + b.$$

EXERCISE 4.2. Solve the constrained problem

$$\text{Minimize} \quad J = \int_0^1 \frac{1}{2}\dot{x}^2 + x\dot{x} \, dt \qquad \text{subject to} \qquad I = \int_0^1 x \, dt = \frac{7}{12}$$

with endpoint conditions $x(0) = 0$ and $x(1) = 0$.

EXERCISE 4.3. Solve the constrained problem

$$\text{Minimize} \quad J = \int_0^1 xt^2 - \frac{1}{2}\dot{x}^2 \, dt \qquad \text{subject to} \qquad I = \int_0^1 x \, dt = 1$$

with endpoint conditions $x(0) = 0$ and $x(1) = 1$.

EXERCISE 4.4. In the notation of this section, show that (under the assumption that the Lagrange multiplier λ is nonzero) a minimizer $x(t)$ for J subject to fixed I is a minimizer for I subject to fixed J.

EXERCISE 4.5. From Exercise 1.1.10 we know that a freely hanging chain of fixed length hangs in the shape of a catenary $y = c \cos(x/c + d)$. Show that this follows not only by the force diagram of Exercise 1.1.10, but by energy considerations as well. Namely, the potential energy of the hanging chain is proportional to $\int_{x_0}^{x_1} y\sqrt{1 + y'^2} \, dx$ and equilibrium position of the chain is assumed when potential energy is minimized. The problem is constrained, however, by the fixed length of the chain $L = \int_{x_0}^{x_1} \sqrt{1 + y'^2} \, dx$. Therefore, the equilibrium shape y will be attained when the following integral is extremized:

$$J = \int_{x_0}^{x_1} y\sqrt{1 + y'^2} - \lambda\sqrt{1 + y'^2} \, dx.$$

Extremize this integral to obtain a catenary. Hint: the integrand is independent of x.

Example 4.3: The Isoperimetric Problem. Suppose we have a closed curve of fixed length $L = \int \sqrt{\dot{x}^2 + \dot{y}^2} \, dt$ and we wish to enclose the *maximum* amount of area. Green's theorem allows us to express area as a line integral $\int (1/2)(-y\dot{x} + x\dot{y}) \, dt$ so that the variational problem becomes:

$$\text{Maximize} \quad J = \int (1/2)(-y\dot{x} + x\dot{y}) - \lambda\sqrt{\dot{x}^2 + \dot{y}^2} \, dt.$$

The two variable Euler-Lagrange equations are

$$\frac{\dot{y}}{2} - \frac{d}{dt}\left(-\frac{y}{2} - \frac{\lambda\dot{x}}{\sqrt{\dot{x}^2 + \dot{y}^2}}\right) = 0 \qquad -\frac{\dot{x}}{2} - \frac{d}{dt}\left(\frac{y}{2} - \frac{\lambda\dot{y}}{\sqrt{\dot{x}^2 + \dot{y}^2}}\right) = 0$$

which may be rewritten as

$$\frac{d}{dt}\left(y + \frac{\lambda\dot{x}}{\sqrt{\dot{x}^2 + \dot{y}^2}}\right) = 0 \qquad \frac{d}{dt}\left(-x + \frac{\lambda\dot{y}}{\sqrt{\dot{x}^2 + \dot{y}^2}}\right) = 0.$$

We then have

$$y + \frac{\lambda \dot{x}}{\sqrt{\dot{x}^2 + \dot{y}^2}} = C \quad \text{and} \quad -x + \frac{\lambda \dot{y}}{\sqrt{\dot{x}^2 + \dot{y}^2}} = -D$$

and a little algebra produces

$$(x - D)^2 + (y - C)^2 = \lambda^2,$$

a circle of radius $r = \lambda$ centered at (D, C). Because, for a circle, $L = 2\pi r$, we have $r = \lambda = L/2\pi$.

EXERCISE 4.6. Show that the two variable Weierstrass E-function is always non-*positive* for this problem. Thus, assuming it may be embedded in a field of extremals, the circle provides a *maximum* for the integral J and is, therefore, the curve of fixed length which encloses the largest area.

EXERCISE 4.7. Show by variational methods that the circle also solves the problem of finding the curve of shortest length which encloses a fixed area. Then show the same thing by the following argument. Let α be a closed curve of fixed area A and arclength L. Since the circle solves the problem of finding a curve of fixed length having maximum area, we must have $A \leq L^2/4\pi$. Take a circle of area A and show by simple algebra that the arclength of this circle is smaller than L. Hence, for given area A, the circle has smallest arclength.

Integral constraints such as those above are not the only types of constraints which arise in variational problems. We shall consider one more class of constraints which pertain to geometry and mechanics, the so-called *holonomic constraints*. Consider the following situation: a marble is placed inside a frictionless bowl in the shape of a hemisphere and the marble is given a horizontal push. What is then the trajectory of the marble inside the bowl? There are two forces to be dealt with here, gravity and the force which constrains the marble to remain in the bowl. It is a principle due to D'Alembert that the force of constraint is a normal force to the bowl and so does no work on the marble. See [Arn] for a discussion of holonomic constraints and D'Alembert's principle. What this all means is this. In order to discover the equations of motion of the marble, it suffices to consider the gravitational potential $V(x, y, z) = mgz$ and kinetic energy $T = (1/2)m(\dot{x}^2 + \dot{y}^2 + \dot{z}^2)$ *restricted to the surface itself*. In other words, because a parametrization of the sphere of radius R is given by

$$\mathbf{x}(u, v) = (R \cos u \cos v, \ R \sin u \cos v, \ R \sin v),$$

we obtain the restricted forms of V and T,

$$V = mgR \sin v, \qquad T = \frac{1}{2} m R^2 (\cos^2 v \, \dot{u}^2 + \dot{v}^2).$$

To show the latter, we can either compute $\dot{x}$, $\dot{y}$ and $\dot{z}$ directly (by the chain rule) in terms of $\dot{u}$ and $\dot{v}$ or we can reason as follows. The path of the marble is given in three space by a curve $\alpha(t) = (x(t), y(t), z(t))$ with velocity vector $\alpha'(t) = (\dot{x}, \dot{y}, \dot{z})$. The kinetic energy may therefore be written $T = (1/2)|\alpha'|^2$. As we have seen many times before, the chain rule gives $\alpha' = \mathbf{x}_u\, \dot{u} + \mathbf{x}_v\, \dot{v}$ with

$$|\alpha'|^2 = \mathbf{x}_u \cdot \mathbf{x}_u\, \dot{u}^2 + \mathbf{x}_v \cdot \mathbf{x}_v\, \dot{v}^2$$

since the metric coefficient F is zero. Hence,

$$|\alpha'|^2 = E\, \dot{u}^2 + G\, \dot{v}^2$$

and the kinetic energy becomes

$$T = \frac{1}{2} m (E\, \dot{u}^2 + G\, \dot{v}^2).$$

Of course, for the R-sphere, $E = \cos^2 v$ and $G = 1$, so we obtain the expression for T above. What we have just done embodies the essence of the term "holonomic constraints." In fact, we may say that constraints are *holonomic* when they arise by simply inserting the parametrization of the constraint surface into the usual mechanical quantities such as kinetic and potential energy. By the discussion above, the action integral which is relevant to the marble in the hemispherical bowl is

$$J = \int \frac{1}{2} m R^2 (\cos^2 v\, \dot{u}^2 + \dot{v}^2) - mgR \sin v \; dt.$$

The Euler-Lagrange equations for this two variable problem are

$$mR^2 \cos^2 v\, \dot{u} = c$$
$$-mgR \cos v - mR^2 \sin v \cos v\, \dot{u}^2 - mR^2 \ddot{v} = 0,$$

or, more simply,

(†) $\dot{u} = \dfrac{c}{mR^2 \cos^2 v}$ and $\ddot{v} + \dfrac{g}{R} \cos v + \sin v \cos v\, \dot{u}^2 = 0.$

These are then the equations of motion of the marble. The trajectory corresponding to these equations (and plotted by MAPLE with $g = m = R = 1$) is shown below.

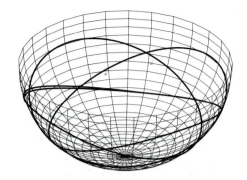

FIGURE 8.4.   Marble in a hemispherical bowl

EXERCISE 4.8. Find the equations of motion of a marble in a paraboloid shaped bowl. Take the paraboloid's parametrization to be $\mathbf{x}(u, v) = (u \cos v, \ u \sin v, \ u^2)$.

We can see the influence of geometric ideas on physics if we consider the case of a particle constrained to a surface $M \colon \mathbf{x}(u, v)$ *with zero potential,* $V = 0$. In physics it would be said that the particle moves *inertially* on the surface — that is, only under the influence of the constraint forces. According to D'Alembert's principle, these forces are normal to the surface, so the particle's acceleration is likewise in the normal direction. In geometry we recognize the path of such a particle as a geodesic. Indeed, we have (taking, for simplicity, an orthogonal patch)

Theorem 4.2. *If the potential is zero, then, according to Hamilton's principle, constrained particles move along geodesics.*

Proof. The kinetic energy is given by

$$T = \frac{1}{2} m |\alpha'|^2 = \frac{1}{2} m (E \, \dot{u}^2 + G \, \dot{v}^2)$$

with action $J = \int T \, dt$. The Euler-Lagrange equations (with m divided out) for this action are

$$\frac{E_u}{2} \dot{u}^2 + \frac{G_u}{2} \dot{v}^2 - \frac{d}{dt}(E \dot{u}) = 0 \quad \text{and} \quad \frac{E_v}{2} \dot{u}^2 + \frac{G_v}{2} \dot{v}^2 - \frac{d}{dt}(G \dot{v}) = 0.$$

The t-derivatives are

$$\frac{d}{dt}(E \dot{u}) = (E_u \dot{u} + E_v \dot{v})\dot{u} + E \ddot{u}$$

$$\frac{d}{dt}(G \dot{v}) = (G_u \dot{u} + G_v \dot{v})\dot{v} + G \ddot{v}$$

so, rearranging and simplifying, we obtain

$$\ddot{u} + \frac{E_u}{2E} \dot{u}^2 + \frac{E_v}{E} \dot{u}\dot{v} - \frac{G_u}{2E} \dot{v}^2 = 0$$

and

$$\ddot{v} - \frac{E_v}{2G}\dot{u}^2 + \frac{G_u}{G}\dot{u}\dot{v} + \frac{G_v}{2G}\dot{v}^2 = 0.$$

By Hamilton's principle, these are the equations of motion of the particle. By our work in Chapter 5, they are also the equations describing a geodesic. □

EXERCISE 4.9. Show that the two variable Weierstrass E-function for the action $J = \int T\, dt$ is always nonnegative. Therefore, whether or not a geodesic minimizes the action is simply a question about the existence of a field of extremals.

EXERCISE 4.10. If $V = V(u, v)$ is a nonzero potential for the problem of a particle constrained to a surface, show that the equations of motion for the particle are

$$\ddot{u} + \frac{E_u}{2E}\dot{u}^2 + \frac{E_v}{E}\dot{u}\dot{v} - \frac{G_u}{2E}\dot{v}^2 = -V_u$$

and

$$\ddot{v} - \frac{E_v}{2G}\dot{u}^2 + \frac{G_u}{G}\dot{u}\dot{v} + \frac{G_v}{2G}\dot{v}^2 = -V_v.$$

Recalling how the geodesic equations arose, we may write these equations as

$$\alpha_{\text{tan}}'' = \nabla_{\alpha'}\alpha' = -\operatorname{grad} V,$$

where the gradient of V is taken with respect to u and v. Thus, the gradient of the potential is the obstruction to the particle travelling along a geodesic.

8.5 FURTHER APPLICATIONS TO GEOMETRY AND MECHANICS

In this section we will look at geodesics as potential minimizers of arclength. Furthermore, we will generalize Theorem 4.2 to say that the motions of particles are *always* geodesics — with respect to a metric on the surface *conformal* to the original one. We will also see how the surfaces of Delaunay of Chapter 3.6 arise from a variational principle.

We have said all along that, philosophically, geodesics should be curves of shortest arclength. We took the definition of geodesic to be a curve whose tangential component of acceleration vanishes, however, because this characterized lines in 3-space and seemed a more directly verifiable condition. We learned in Theorem 6.7.3 (and Example 3.7) that geodesics do not always *minimize* arclength, but we may still ask if they *extremize* arclength. As usual, we assume for convenience that the metric has $F = 0$.

Theorem 5.1. *Let $M : \mathbf{x}(u,v)$ be a surface with metric E and G. Then, extremals for the arclength integral*

$$J = \int \sqrt{E\,\dot{u}^2 + G\,\dot{v}^2}\, dt$$

are geodesics on the surface M.

Proof. We calculate the two variable Euler-Lagrange equations to be

$$\frac{E_u\,\dot{u}^2 + G_u\,\dot{v}^2}{2\sqrt{E\,\dot{u}^2 + G\,\dot{v}^2}} - \frac{d}{dt}\left(\frac{E\,\dot{u}}{\sqrt{E\,\dot{u}^2 + G\,\dot{v}^2}}\right) = 0$$

$$\frac{E_v\,\dot{u}^2 + G_v\,\dot{v}^2}{2\sqrt{E\,\dot{u}^2 + G\,\dot{v}^2}} - \frac{d}{dt}\left(\frac{G\,\dot{v}}{\sqrt{E\,\dot{u}^2 + G\,\dot{v}^2}}\right) = 0.$$

Now we introduce the arclength parameter $s(t)$ since we know geodesics must be parametrized to have constant speed. This means that we have the relations

$$\frac{d}{dt} = \frac{d}{ds} \cdot \sqrt{E\,\dot{u}^2 + G\,\dot{v}^2}$$

$$u' = \frac{du}{ds} = \frac{du}{dt}\frac{dt}{ds} = \frac{\dot{u}}{\sqrt{E\,\dot{u}^2 + G\,\dot{v}^2}} \qquad v' = \frac{dv}{ds} = \frac{dv}{dt}\frac{dt}{ds} = \frac{\dot{v}}{\sqrt{E\,\dot{u}^2 + G\,\dot{v}^2}}.$$

Putting these into the equations above (and dividing out by the factor $\sqrt{E\,\dot{u}^2 + G\,\dot{v}^2}$) gives

$$\frac{E_u}{2}u'^2 + \frac{G_u}{2}v'^2 - \frac{d}{ds}(Eu') = 0$$

$$\frac{E_v}{2}u'^2 + \frac{G_v}{2}v'^2 - \frac{d}{ds}(Gv') = 0.$$

Carrying out the differentiation and simplifying produces

$$u'' + \frac{E_u}{2E}u'^2 + \frac{E_v}{E}u'v' - \frac{G_u}{2E}v'^2 = 0$$

and

$$v'' - \frac{E_v}{2G}u'^2 + \frac{G_u}{G}u'v' + \frac{G_v}{2G}v'^2 = 0.$$

These are the geodesic equations. $\square$

Remark 5.1. The Weierstrass E-function for the arclength integral is computed to be

$$E(t,u,v,\dot{u},\dot{v},p_1,p_2) = \frac{\sqrt{E\,\dot{u}^2 + G\,\dot{v}^2}\sqrt{E\,p_1^2 + G\,p_2^2} - (E\dot{u}p_1 + G\dot{v}p_2)}{\sqrt{E\,\dot{u}^2 + G\,\dot{v}^2}}.$$

The fact that $E \geq 0$ follows because the metric E, G gives an inner product to each tangent plane defined by

$$\mathbf{v} \circ \mathbf{w} = E\, v_1 w_1 + G\, v_2 w_2$$

and the Schwarz inequality $|\mathbf{v} \circ \mathbf{w}| \leq |\mathbf{v}||\mathbf{w}|$ holds for all inner products. Therefore, once again we see that the shortest distance property of existing geodesics depends entirely on the existence of a field of extremals. This in turn, recall, depends on the nonexistence of zeros for solutions to the Jacobi equation.

EXERCISE 5.1. Consider the (arclength parametrized) Euler-Lagrange equations for the arclength integral above,

$$\frac{d}{ds}(Eu') = \frac{E_u}{2}u'^2 + \frac{G_u}{2}v'^2$$
$$\frac{d}{ds}(Gv') = \frac{E_v}{2}u'^2 + \frac{G_v}{2}v'^2$$

and let $\alpha(s)$ be an extremal (i.e. a solution curve for these equations). Show directly that the tangential component of α'' is zero. Hints: write $\alpha' = \mathbf{x}_u\, u' + \mathbf{x}_v\, v'$ so that $Eu' = \alpha' \cdot \mathbf{x}_u$ and $Gv' = \alpha' \cdot \mathbf{x}_v$. Differentiate the righthand sides of these equations and compare with the Euler-Lagrange equations.

Now that we know that geodesics are extremals of the arclength integral, we can give a beautiful interpretation due to Jacobi of the equations of motion of a constrained conservative system. Recall that kinetic and potential energies have the form

$$T = \frac{1}{2}\left|\frac{d\alpha}{dt}\right|^2 = \frac{1}{2}(E\dot{u}^2 + G\dot{v}^2) \quad \text{and} \quad V = V(u, v),$$

where α is a particle trajectory. The action integral $\int T - V\, dt$ is independent of t, so the two variable first integral of Exercise 1.5 implies that energy $H = T + V$ is a constant along paths of motion. Specifically,

$$T - V - \dot{u}\frac{\partial(T - V)}{\partial\dot{u}} - \dot{v}\frac{\partial(T - V)}{\partial\dot{v}} = c$$
$$\frac{1}{2}(E\dot{u}^2 + G\dot{v}^2) - V - \dot{u}(E\dot{u}) - \dot{v}(G\dot{v}) = c$$
$$\frac{1}{2}(E\dot{u}^2 + G\dot{v}^2) - V - E\dot{u}^2 - G\dot{v}^2 = c$$
$$-\frac{1}{2}(E\dot{u}^2 + G\dot{v}^2) - V = c$$
$$T + V = -c.$$

Here and in the following we use H for energy to avoid confusion with the E of the metric.

Now let's take a curve $\alpha\colon I \to M$ representing the motion of a particle determined by Hamilton's principle on a constraint surface M. We take the metric of the surface to be E, $F = 0$ and G. Because we know that the path of the particle must conserve energy and because this condition is incompatible with the unit speed condition, we must explicitly reparametrize the curve to have constant energy. To do this, let H_0 denote the constant energy along the path of the particle and define

$$\tau = \int \frac{1}{\sqrt{2(H_0 - V)}} \sqrt{E\dot{u}^2 + G\dot{v}^2} \; dt.$$

Lemma 5.2. *The reparametrized curve $\alpha(\tau)$ has constant energy.*

Proof. First, note that the fundamental theorem of calculus implies

$$\frac{d\tau}{dt} = \frac{1}{\sqrt{2(H_0 - V)}} \sqrt{E\dot{u}^2 + G\dot{v}^2}$$

so that the chain rule then gives

$$\frac{du}{d\tau} = \frac{du}{dt}\frac{dt}{d\tau} = \frac{du}{dt}\frac{\sqrt{2(H_0 - V)}}{\sqrt{E\dot{u}^2 + G\dot{v}^2}}$$

$$\frac{dv}{d\tau} = \frac{dv}{dt}\frac{dt}{d\tau} = \frac{dv}{dt}\frac{\sqrt{2(H_0 - V)}}{\sqrt{E\dot{u}^2 + G\dot{v}^2}}.$$

Using this, the following calculation then shows that this reparametrization conserves energy at the value H_0 along the curve.

$$T(\tau) + V(\tau) = \frac{1}{2}\left(E(\tau)\left(\frac{du}{d\tau}\right)^2 + G(\tau)\left(\frac{dv}{d\tau}\right)^2\right) + V(\tau)$$

$$= \frac{1}{2}\left(E\left(\frac{du}{dt}\right)^2 \frac{2(H_0 - V)}{E\dot{u}^2 + G\dot{v}^2} + G\left(\frac{dv}{dt}\right)^2 \frac{2(H_0 - V)}{E\dot{u}^2 + G\dot{v}^2}\right) + V$$

$$= 2(H_0 - V)\frac{1}{2}\left(\frac{E\dot{u}^2 + G\dot{v}^2}{E\dot{u}^2 + G\dot{v}^2}\right) + V$$

$$= H_0 - V + V$$

$$= H_0.$$

$\square$

Furthermore, note that this calculation shows that $T(\tau) = H_0 - V$. Also, the general energy equation $T + V = H$ gives $2T - H = 2T - (T + V) = T - V$. We shall use both of these relations in the following to determine the equations of motion of a particle as extremals of $\int T(\tau) - V(\tau)\, d\tau$. Note that, because of conservation of energy, these extremals must be found within the class of all curves with constant energy H_0. The first step is to observe that, from what we have said above,

$$\int T(\tau) - V(\tau)\, d\tau = \int 2T(\tau) - H_0\, d\tau.$$

Furthermore, since the derivative of a constant vanishes, adding a constant to an integrand has no effect on the Euler-Lagrange equations. Therefore, finding extremals for either side of this equation is equivalent to finding extremals for

$$
\begin{aligned}
\int 2T(\tau)\, d\tau &= \int 2(H_0 - V)\, d\tau \\
&= \int 2(H_0 - V)\frac{d\tau}{dt}\, dt \\
&= \int 2(H_0 - V)\frac{1}{\sqrt{2(H_0 - V)}}\sqrt{E\dot{u}^2 + G\dot{v}^2}\, dt \\
&= \int \sqrt{2(H_0 - V)E\dot{u}^2 + 2(H_0 - V)G\dot{v}^2}\, dt \\
&= \int \sqrt{\bar{E}\dot{u}^2 + \bar{G}\dot{v}^2}\, dt
\end{aligned}
$$

where $\bar{E} = 2(H_0 - V)E$ and $\bar{G} = 2(H_0 - V)G$ *defines a new metric on M conformal with respect to the the original E and G.* (Here, we assume that $H_0 > V$ in this region of the parameter domain.) It is now clear that finding an extremal for $\int T - V\, d\tau$ corresponds to finding an extremal for $\int \sqrt{\bar{E}\dot{u}^2 + \bar{G}\dot{v}^2}\, dt$. But, by Theorem 5.1, $\int \sqrt{\bar{E}\dot{u}^2 + \bar{G}\dot{v}^2}\, dt$ has extremals which are *geodesics on M with respect to the metric* $\bar{E} = 2(H_0 - V)E$ *and* $\bar{G} = 2(H_0 - V)G$. Therefore, we have the following special $F = 0$ case of a result of Jacobi.

Theorem 5.3. *Let α denote the path of a particle with constant energy H_0 under the influence of a potential V constrained to lie on a surface M with metric E, $F = 0$ and G. Then, for $H_0 > V$, α is a geodesic on M with respect to a conformal metric*

$$\bar{E} = 2(H_0 - V)\, E \qquad \bar{F} = 0 \qquad \bar{G} = 2(H_0 - V)\, G.$$

This theorem, while hardly the means to an explicit description of equations of motion, nevertheless allows theoretical results about geodesics to be applied to mechanical systems. In particular, one of the most important questions about a mechanical system is whether or not there is a periodic orbit — that is, does some particle, say, return to its initial position after a certain amount of time. By the theorem above, this question is equivalent to asking whether a surface M (with any metric) has a closed geodesic — a geodesic $\alpha\colon [a, b] \to M$

with $\alpha(a) = \alpha(b)$. In 1951, Lusternik and Fet [LF] (also see [Kl]) proved a general theorem which, in particular, says that any compact (i.e. closed and bounded) surface does indeed have a closed geodesic. Applying this result to a surface in conjunction with Jacobi's theorem gives

Corollary 5.4. *If M is a compact surface and $V(u, v)$ is a potential function with $H_0 > V(u, v)$ for all (u, v) in the parameter domain of M, then there exists a periodic solution to the equations of motion of a particle constrained to move on the surface under the influence of V.*

Example 5.2: The two body problem. The following comes from [P]. Suppose that two bodies are mutually attracted with potential V depending only on the distance between the bodies. Mechanics tells us that the problem may be reduced to the case of single body planar motion in a central force field. The Jacobi metric for the motion along a constant energy curve has conformal scaling factor (see Chapter 5.4) $f = 1/\sqrt{2(H_0 - V)}$ so that Exercise 5.4.1 gives the Gauss curvature,

$$K = \frac{1}{4}\left[\frac{V_{uu} + V_{vv}}{(H_0 - V)^2} + \frac{V_u^2 + V_v^2}{(H_0 - V)^3}\right].$$

EXERCISE 5.2. Use Exercise 5.4.1 to verify the formula for Gauss curvature above. This curvature is called the *mechanical curvature*. Further, show that if V has a local minimum at (u_0, v_0), then $K(u, v) > 0$ for all (u, v) sufficiently close to (u_0, v_0). Similarly, $K(u, v) < 0$ near a local maximum.

EXERCISE 5.3. Let $V = -1/\sqrt{u^2 + v^2}$ be the Newtonian potential in a plane. Show that

$$V_u^2 + V_v^2 = \frac{1}{(u^2 + v^2)^2} \quad\text{and}\quad V_{uu} + V_{vv} = -\frac{1}{(u^2 + v^2)^{\frac{3}{2}}}.$$

By Exercise 5.3, the Newtonian potential $V = -1/\sqrt{u^2 + v^2}$ gives

$$K = -\frac{H_0}{4(H_0\sqrt{u^2 + v^2} + 1))^3} = -\frac{H_0}{4(H_0 r + 1))^3}$$

where $r = \sqrt{u^2 + v^2}$ is the radial distance from the origin. Now, the constant energy equation $(1/2)\nu^2 - 1/r = H_0$ gives $\nu^2 = 2(H_0 r + 1)/r > 0$. Hence, the denominator of K is positive. Therefore,

$$H_0 > 0 \Leftrightarrow K < 0$$
$$H_0 = 0 \Leftrightarrow K = 0$$
$$H_0 < 0 \Leftrightarrow K > 0.$$

For the two body problem, orbits are characterized in terms of energy (see [PC]). Putting this together with the above gives

$$K < 0 \quad \Leftrightarrow \quad H_0 > 0 \quad \Leftrightarrow \quad \text{the orbit is hyperbolic}$$
$$K = 0 \quad \Leftrightarrow \quad H_0 = 0 \quad \Leftrightarrow \quad \text{the orbit is parabolic}$$
$$K > 0 \quad \Leftrightarrow \quad H_0 < 0 \quad \Leftrightarrow \quad \text{the orbit is elliptical.}$$

Therefore, orbits are characterized in terms of their Gauss curvatures. In particular, a point where $K > 0$ guarantees a periodic orbit.

Before we end our discussion of Jacobi's theorem, we should point out explicitly that everything we have said holds true in higher dimensions as well. Also, it should be noted that one of the main features of Einstein's geometrization of gravity is his assertion that particles, planets and light photons, say, travel along paths which are geodesics in a region of spacetime whose metric is determined by the mass-energy in that region. Considerable philosophical hindsight, therefore, allows us to view Einstein's approach as a natural descendent of Jacobi's theorem.

To end this section, we want to exhibit yet another of our earlier examples as a constrained variational problem. In Chapter 3.6 we considered the surfaces of revolution with constant mean curvature — the so-called surfaces of Delaunay. Recall that these surfaces were characterized by a certain differential equation,

$$h^2 \pm \frac{2ah}{\sqrt{1 + h'^2}} = \pm b^2$$

where a and b are constants. Here, $h = h(u)$ denotes the profile curve (or meridian in the plane) of the surface of revolution. Let's consider the following problem. For surfaces of revolution, fix a volume $V = \pi \int h(u)^2 \, du$ and minimize the surface area $S = 2\pi \int h(u)\sqrt{1 + h'(u)^2} \, du$. What are the resulting surfaces? The usual variational setup for such a constrained problem (neglecting π in the formula) gives

$$\text{Minimize } J = \int 2h(u)\sqrt{1 + h'(u)^2} - \lambda h(u)^2 \, du.$$

Theorem 5.5. *The extremals for the variational problem above are the surfaces of Delaunay.*

Proof. Since the integrand does not depend on the independent variable u, we may use the first integral $f - \dot{x}(\partial f / \partial \dot{x}) = c$ in place of the Euler-Lagrange equation.

$$2h\sqrt{1+h'^2} - \lambda h^2 - h'\left(\frac{2hh'}{\sqrt{1+h'^2}}\right) = c$$

$$2h(1+h'^2) - \lambda h^2\sqrt{1+h'^2} - 2hh' = c\sqrt{1+h'^2}$$

$$2h = (c+\lambda h^2)\sqrt{1+h'^2}$$

$$\frac{2h}{\sqrt{1+h'^2}} = c+\lambda h^2$$

$$-\lambda h^2 + \frac{2h}{\sqrt{1+h'^2}} = c$$

$$h^2 \pm \frac{2ah}{\sqrt{1+h'^2}} = \pm b^2$$

where $a = -1/\lambda$ and $b = -c/\lambda$. Of course, this is precisely the equation above characterizing surfaces of revolution of constant mean curvature. □

As we mentioned in Chapter 3, many one-celled creatures exhibit rotational symmetry and shapes which seem close to surfaces of Delaunay. D'Arcy Wentworth Thompson asked whether the minimization of surface area under the constraint of fixed volume together with a certain biological tendency toward symmetry might be responsible for the observed shapes. Of course we have seen in Chapter 4 that compact surfaces of constant mean curvature are spheres, so Thompson needed other biological considerations to allow for the physical existence of rounded-off "compactified" versions of surfaces of Delaunay. More information on this approach to cellular morphology may be found (of course) in [DWT] as well as in [HT].

8.6 THE PONTRYAGIN MAXIMUM PRINCIPLE

The subject of optimal control theory provides a sightly different perspective on variational questions from the calculus of variations. We will only give a brief outline of one of the main results in optimal control — the Pontryagin maximum principle — and how this result may be applied in geometry. The true power of the Pontryagin maximum principle is seen in its application to discontinuous problems (i.e. the so-called bang-bang controls) such as those found in rocket motor control or trajectory control, but we shall only consider simple geometric applications here. For an elementary introduction to the principle's nongeometric applications, see [Pin]. Also, we shall restrict ourselves to two dimensional problems as we have done so far.

The main problem in optimal control is to find a path (or trajectory) $\mathbf{x} = (x_1(t), x_2(t))$ which satisfies the system of differential equations

$$\dot{x}_1 = f_1(\mathbf{x}, \mathbf{u}) \qquad\qquad \dot{x}_2 = f_2(\mathbf{x}, \mathbf{u})$$

(where $\mathbf{u} = (u_1(t), u_2(t))$ is the *control*) and which minimizes *a given cost function*

$$J = \int_{t_0}^{t_1} f_0(\mathbf{x}, \mathbf{u}) \, dt.$$

Note that we have reverted to the control theorist's subscripts in place of the geometer's superscripts for coordinate functions. This should ease the transition to further study in optimal control. Also, a typical optimal control problem may well fix the initial and final *states* of the system, $\mathbf{x}(t_0) = \mathbf{x}_0$ and $\mathbf{x}(t_1) = \mathbf{x}_1$, as well as the initial time t_0. The final time t_1 is typically undetermined. For example, if a spaceship is to make a landing on the moon in the shortest time, it must satisfy a system of differential equations derived from Newton's Law $F = ma$ together with the thrust control of its rockets. Because time is to be minimized, and starting from $t_0 = 0$, the cost is $J = \int_{t_0}^{t_1} 1 \, dt = t_1$, so the final time t_1 is not fixed initially, but is itself determined by solving the problem. See [Kir p. 247] for a discussion of this problem.

Now, how would a variationalist approach this problem? We have already seen that various types of constraints may be placed on variational problems, so it seems reasonable to simply incorporate the system of differential equations into the cost integral as a constraint. Because $\dot{x}_i(t) = f_i(\mathbf{x}, \mathbf{u})$, it is certainly true that

$$\int_{t_0}^{t_1} \psi_i(t)(\dot{x}_i(t) - f_i(\mathbf{x}, \mathbf{u})) \, dt = 0$$

for time dependent Lagrange multipliers ψ_i. Therefore, in order to solve the optimal control problem above, instead solve the constrained variational problem

$$\text{Minimize } \tilde{J} = \int_{t_0}^{t_1} f_0 + \psi_1(\dot{x}_1(t) - f_1(\mathbf{x}, \mathbf{u})) + \psi_2(\dot{x}_2(t) - f_2(\mathbf{x}, \mathbf{u})) \, dt.$$

To solve *this* problem, we will write down the usual first integral for t-independent integrands, the Euler-Lagrange equations (for four functions $x_1(t)$, $x_2(t)$, $u_1(t)$, $u_2(t)$) and then take into account the endpoint conditions, t_1 undetermined and $\mathbf{x}(t_1)$ fixed.

Because the integrand of $\tilde{J}$ does not involve t explicitly, we may write

$$f_0 + \psi_1(\dot{x}_1(t) - f_1(\mathbf{x}, \mathbf{u})) + \psi_2(\dot{x}_2(t) - f_2(\mathbf{x}, \mathbf{u})) - \dot{x}_1\psi_1 - \dot{x}_2\psi_2 = c$$

or, rather

(1) $$f_0 - \psi_1 f_1 - \psi_2 f_2 = c.$$

The Euler-Lagrange equations with respect to x_1, x_2, u_1 and u_2 are

(2)
$$\dot{\psi}_1 = \frac{\partial f_0}{\partial x_1} - \psi_1 \frac{\partial f_1}{\partial x_1} - \psi_2 \frac{\partial f_2}{\partial x_1}$$

(3)
$$\dot{\psi}_2 = \frac{\partial f_0}{\partial x_2} - \psi_1 \frac{\partial f_1}{\partial x_2} - \psi_2 \frac{\partial f_2}{\partial x_2}$$

(4)
$$0 = \frac{\partial f_0}{\partial u_1} - \psi_1 \frac{\partial f_1}{\partial u_1} - \psi_2 \frac{\partial f_2}{\partial u_1}$$

(5)
$$0 = \frac{\partial f_0}{\partial u_2} - \psi_1 \frac{\partial f_1}{\partial u_2} - \psi_2 \frac{\partial f_2}{\partial u_2}.$$

EXERCISE 6.1. Verify the Euler-Lagrange equations above.

In Remark 1.4 we mentioned that the general condition for extremizing $J = \int f \, dt$ subject to the terminal endpoint at t_1 lying on a given curve $\alpha(t)$ is

$$f(t_1) + (\dot{\alpha}(t_1) - \dot{x}(t_1)) \frac{\partial f}{\partial \dot{x}}(t_1) = 0.$$

In our case $\alpha(t) = \mathbf{x}(t_1)$, so $\dot{\alpha} = 0$ and we have

$$f(t_1) - \dot{x}(t_1) \frac{\partial f}{\partial \dot{x}}(t_1) = 0.$$

We must use a multivariable version of this condition of course. Noting that the integrand of $\tilde{J}$ does not depend on $\dot{u}_1$ or $\dot{u}_2$ and using $F = f_0 + \psi_1(\dot{x}_1(t) - f_1(\mathbf{x}, \mathbf{u})) + \psi_2(\dot{x}_2(t) - f_2(\mathbf{x}, \mathbf{u}))$, we obtain an endpoint condition

$$F(t_1) - \dot{x}_1(t_1) \frac{\partial F}{\partial \dot{x}_1}(t_1) - \dot{x}_2(t_1) \frac{\partial F}{\partial \dot{x}_2}(t_1) = 0$$

which is then written out as

(6)
$$f_0(t_1) - \psi_1(t_1) f_1(t_1) - \psi_2(t_1) f_2(t_1) = 0$$

Now, for convenience, define the *Hamiltonian* associated to the problem to be

$$H = -f_0 + \psi_1 f_1 + \psi_2 f_2$$

and note that equations $(1) - (6)$ imply the following relationships.

(i) $H(t, \mathbf{x}, \mathbf{u}) = 0$, by (1) and (6).

(ii) $\dot{\psi}_1 = -\dfrac{\partial H}{\partial x_1}$ $\dot{\psi}_2 = -\dfrac{\partial H}{\partial x_2}.$

(iii) $\partial H / \partial u_1 = 0$ and $\partial H / \partial u_2 = 0$ along an optimal trajectory.

This heuristic discussion may be strengthened considerably to give the following necessary conditions for optimality.

Theorem 6.1: The Pontryagin Maximum Principle. *Suppose* $\mathbf{u}(t)$ *is a control which transfers the system with*

State Equations $\qquad \dot{x}_1 = f_1(\mathbf{x}, \mathbf{u}) \qquad\qquad \dot{x}_2 = f_2(\mathbf{x}, \mathbf{u})$

from fixed state $\mathbf{x}_0$ *to fixed state* $\mathbf{x}_1$ *along a path* $\mathbf{x}(t)$. *If* $\mathbf{u}(t)$ *and* $\mathbf{x}(t)$ *minimize the cost*

$$J = \int_{t_0}^{t_1} f_0(\mathbf{x}, \mathbf{u}) \, dt,$$

then there exist ψ_1 *and* ψ_2 *so that the Hamiltonian* $H = -f_0 + \psi_1 f_1 + \psi_2 f_2$ *obeys*

(i) $\qquad H(\psi_1, \psi_2, x_1, x_2, u_1, u_2) = 0$ *along the optimal trajectory.*

(ii) Co-state Equations $\qquad \dot{\psi}_1 = -\dfrac{\partial H}{\partial x_1} \qquad \dot{\psi}_2 = -\dfrac{\partial H}{\partial x_2}.$

(iii) *For each* t, H *attains its maximum with respect to* u *at* $\mathbf{u}(t)$. *In particular,*

$$\frac{\partial H}{\partial u_1} = 0 \qquad\qquad \frac{\partial H}{\partial u_2} = 0.$$

Example 6.1: Geodesics. Suppose we want to find geodesics on a surface $M \colon \mathbf{x}(u, v)$ with orthogonal metric E and G. We can formulate the problem by saying that, from a starting point $\mathbf{x}_0$, we wish to control our journey to minimize arclength. Because we can write a curve on M in the form $\alpha(t) = \mathbf{x}(u(t), v(t))$, we see that we are really interested in controlling the functions $u(t)$ and $v(t)$. Therefore, the optimal control problem becomes, find a curve $(u(t), v(t))$ so that

$$\dot{u} = u_1 \qquad\qquad \dot{v} = u_2$$

and the integral

$$J = \int \sqrt{E\dot{u}^2 + G\dot{v}^2} \, dt = \int \sqrt{Eu_1^2 + Gu_2^2} \, dt$$

is minimized. The Hamiltonian associated to the problem is

$$H = -\sqrt{Eu_1^2 + Gu_2^2} + \psi_1 u_1 + \psi_2 u_2$$

with co-state equations

$$\dot{\psi}_1 = -\frac{\partial H}{\partial u} = \frac{E_u u_1^2 + G_u u_2^2}{2\sqrt{Eu_1^2 + Gu_2^2}} \qquad \dot{\psi}_2 = -\frac{\partial H}{\partial v} = \frac{E_v u_1^2 + G_v u_2^2}{2\sqrt{Eu_1^2 + Gu_2^2}}$$

and u-critical point equations for H,

$$\frac{\partial H}{\partial u_1} = -\frac{Eu_1}{\sqrt{Eu_1^2 + Gu_2^2}} + \psi_1 = 0 = -\frac{Gu_2}{\sqrt{Eu_1^2 + Gu_2^2}} + \psi_2 = \frac{\partial H}{\partial u_2}.$$

These last equations say that $\psi_1 = Eu_1/\sqrt{Eu_1^2 + Gu_2^2}$ and $\psi_2 = Gu_2/\sqrt{Eu_1^2 + Gu_2^2}$. Replacing ψ_1 and ψ_2 by these quantities and u_1 and u_2 by $\dot{u}$ and $\dot{v}$ in the co-state equations gives

$$\frac{d}{dt}\left(\frac{E\,\dot{u}}{\sqrt{E\,\dot{u}^2 + G\,\dot{v}^2}}\right) = \frac{E_u\,\dot{u}^2 + G_u\,\dot{v}^2}{2\sqrt{E\,\dot{u}^2 + G\,\dot{v}^2}}$$

$$\frac{d}{dt}\left(\frac{G\,\dot{v}}{\sqrt{E\,\dot{u}^2 + G\,\dot{v}^2}}\right) = \frac{E_v\,\dot{u}^2 + G_v\,\dot{v}^2}{2\sqrt{E\,\dot{u}^2 + G\,\dot{v}^2}}.$$

These are, of course, the "geodesic equations" arising in the proof of Theorem 5.1. Hence, the optimal trajectories for this problem are geodesics.

Remark 6.2. With respect to the metric on M, the amount of effort expended to move from one point to another along an optimal path (i.e. a geodesic) is precisely $\int |\mathbf{u}|_{E,G}^2\,dt = \int \sqrt{Eu_1^2 + Gu_2^2}\,dt = J$. Therefore, geodesics may be considered to be paths along which a minimum amount of control is needed. Compare this with fuel optimal control problems in [Pin] and [Kir]. Also, note that we have directly obtained geodesics as *locally determined* minimizers of arclength here. The controls u_1 and u_2 (and thus the cost $J = \int \sqrt{Eu_1^2 + Gu_2^2}\,dt$) are *local* in the sense that we recreate the correct $u(t)$ and $v(t)$ from the state equations and cost, both of which rely on local control information. In a way, this is very different, philosophically, from trying to find a minimizer among all curves joining given endpoints — the underlying concept of the calculus of variations.

EXERCISE 6.2. Show that an arclength minimizer between $(1, 2)$ and $(2, 3)$ in the plane is the straight line $y = x + 1$ by forming the optimal control problem

$$\dot{x} = u_1 \qquad\qquad \dot{y} = u_2$$

$$\text{Minimize } J = \int \sqrt{u_1^2 + u_2^2}\,dt$$

and solving by the Pontryagin maximum principle.

EXERCISE 6.3. Show that a least area surface of revolution is a catenoid by forming the one dimensional optimal control problem

$$y' = u, \qquad \text{Minimize } J = \int y \sqrt{1 + u^2}\, dx$$

and solving by the Pontryagin maximum principle. Hint: derive the differential equation $y\, y'' - y'^2 - 1 = 0$ and let $y' = w$ so that $y'' = w\,(dw/dy)$.

EXERCISE 6.4: Newton's Aerodynamical Problem. Newton asked what the optimal shape was for a surface of revolution passing through, what he called, a *rare* medium. Here, the word optimal is taken to mean "offers the least resistance to movement through the medium". Also, it should be noted that usual media such as water or the lower atmosphere do not qualify as rare while the stratosphere does. Newton's problem may be cast in the language of optimal control (see [Tik], [ATF, p. 15] for a derivation). Solve

$$y' = u,\ u \ge 0; \qquad y(0) = 0, \qquad y(a) = b,$$

to minimize cost

$$J = \int_0^a \frac{x}{1 + u^2}\, dx,$$

where $y = y(x)$ represents the profile curve of the surface of revolution. Solve Newton's problem by determining y using the Pontryagin maximum principle. Hint: after writing down the conditions of the principle, consider the *minimum* of the function

$$\phi = \frac{w}{1 + u^2} + u.$$

What is the minimum for various w's? MAPLE may be useful. Note $u \ge 0$. Show $u = 0$ is a local minimum at least. How do minima come about? After x is determined parametrically in terms of u, find y, using

$$\frac{dy}{du} = \frac{dy}{dx}\frac{dx}{du} = y'\frac{dx}{du}.$$

Remember to find the constant of integration. The answer to the problem may be found in Exercise 1.1.4.

Since optimal control theory arose in the 1950's in response to the problems of rocket and satellite trajectory control, it has revitalized the calculus of variations. We have presented only the barest hint of the power and beauty of this subject, but surely, within it hide many more applications to the geometry of surfaces.

8.7 THE CALCULUS OF VARIATIONS AND MAPLE

In this section, I want to present some MAPLE procedures which are useful in finding and solving the Euler-Lagrange equations. Further, I'll apply these to the problem of a marble rolling inside a (frictionless) hemispherical bowl under the influence of gravity and the result will be the plot of the motion of the marble. This final stage will be a modification of the "plotgeo" procedure used to plot geodesics on surfaces. First, let's take a look at the Euler-Lagrange equation in one variable.

```
> EL1:=proc(f)
>       local part1,part2,p2,dfdx,dfddx;
>       part1:=subs({x=x(t),dx=diff(x(t),t)},diff(f,x));
>       dfdx:=diff(f,dx);
>       dfddx:=subs({x=x(t),dx=dx(t)},dfdx);
>       p2:=diff(dfddx,t);
>       part2:=subs(dx(t)=diff(x(t),t),p2);
>       RETURN(simplify(part1-part2=0));
>       end:
```

Notice that we have to do a bit of substituting to ensure that the answer comes out in a form MAPLE recognizes as a differential equation. You might ask, why didn't we input the function f in differential form? The answer is just that we become used to writing $f(t, x, \dot{x}) = x^2 - \dot{x}^2 - 2x\sin(t)$ for the integrand of a variational integral. Rather that write $\dot{x}$ as $d(x(t))/dt$ (where we also have to tell MAPLE that x is a function of t), it seems easier to write

```
> f:=x^2-dx^2-2*x*sin(t);
```

$$f := x^2 - dx^2 - 2x\sin(t)$$

and understand "dx" to mean $\dot{x}$. Further, the procedure above, once we take the time to write it, takes care of making x a function of t and writing the final version in differential equations form. Here's an example.

```
> EL1(f);
```

$$2x(t) - 2\sin(t) + 2\left(\frac{d^2}{dt^2}x(t)\right) = 0$$

```
> dsolve({EL1(f)=0,x(0)=1,x(Pi/2)=2},x(t));
```

$$x(t) = -\arctan\left(\frac{\sin(t)}{1 + \cos(t)}\right)\cos(t) + 2\sin(t) + \cos(t)$$

This is another instance where MAPLE produces output which is not in the best form. If you recall that

$$\tan(t/2) = \frac{\sin(t)}{1 + \cos(t)}$$

then you will see that the true solution of the Euler-Lagrange equation is

$$x(t) = -\frac{t}{2}\cos(t) + 2\sin(t) + \cos(t).$$

EXERCISE 7.1 Carry out the following MAPLE commands to extremize $h(t, x, \dot{x})$ $= \dot{x}^2/t^3$ subject to the conditions $x(1) = 2$ and $x(2) = 17$.

```
> h:=dx^2/t^3;
> EL1(h);
> dsolve({"=0,x(1)=2,x(2)=17},x(t));
```

EXERCISE 7.2 Extremize $h(t, x, \dot{x}) = \dot{x}^2 + 2x\sin(t)$ subject to the conditions $x(0) = 0$ and $x(\pi) = 0$.

The following procedure handles the case where f does not depend on t.

```
> EL2:=proc(f)
>       local part1,part2,dfddx ;
>       part1:=subs({x=x(t),dx=diff(x(t),t)},f);
>       dfddx:=diff(f,dx);
>       part2:=diff(x(t),t)*subs({x=x(t), dx=diff(x(t),t)},dfddx);
>       RETURN(simplify(part1-part2=c));
>       end:
```

For example,

```
> g:=sqrt(1+dx^2)/sqrt(x);
```

$$g := \frac{\sqrt{1 + dx^2}}{\sqrt{x}}$$

```
> EL2(g);
```

$$\frac{1}{\sqrt{1 + \left(\frac{d}{dt}x(t)\right)^2}\sqrt{x(t)}} = c$$

A small problem now arises because we have set the equation equal to c. MAPLE has no idea what c is, so if we now try to solve the differential equation, MAPLE doesn't know what to do. There are two ways to get around this difficulty. The first is to use the fact that c is a constant to form a new differential equation obtained by differentiating with respect to t and

thus annihilating c. To do this, we must peel off the left side of the output of
EL2(g). This can be accomplished by noting that the output is made up of
two *operands*, the left side (operand 1) and the right side (operand 2). The
command "op(1,EL2(g))" takes the left side. Hence,

> dsolve(diff(op(1,EL2(g)),t)=0,x(t))[1];

$$t = -\sqrt{-1 + \frac{C1}{x(t)}}\, x(t) - C1 \arctan\left(\sqrt{-1 + \frac{C1}{x(t)}}\right) - C2$$

I should point out that MAPLE provides an easier way to take left or right
sides of equations. Namely, the commands "lhs" and "rhs" may be invoked. I
have chosen to use the "op" command above because it comes in handy when
many operands are involved and one of them is to be isolated. Therefore, I
think it should be seen at least once. Also, notice that there is a [1] attached to
the end of the command. This is because MAPLE provides a set of solutions
for the differential equation and I only wanted the first. Try the command
without the [1] to see what I mean. The second way to solve the equation is
to get rid of the c by taking the left side and then put it back in explicitly as
follows. Try this.

> dsolve(op(1,EL2(g))=C,x(t))[1];

I don't happen to like this way because if you want to do anything else with
this result, MAPLE still has trouble knowing what C is.

EXERCISE 7.3 Extremize $k(t, x, \dot{x}) = \dot{x}^2/x^3$ subject to the conditions $x(0) = 1$ and
$x(2) = 4$. Do the problem by hand and then carry out the following MAPLE commands. This will give you an idea of some pitfalls involved in the use of computers
to symbolically solve diffential equations. (By the way, when you see a double quote
"" in a command, this means that MAPLE will take the previous output as input.)

> k:=dx^2/x^3;
> EL2(k);
> dsolve(op(1,EL2(k))=C,x(t));
> dsolve({diff(op(1,EL2(k)),t)=0,x(0)=1,x(2)=4},x(t))[1];
> dsolve({diff(op(1,EL2(k)),t)=0,x(0)=1,x(2)=4},x(t))[2];
> dsolve(op(1,EL2(k))=C,x(t))[1];
> dsolve(op(1,EL2(k))=C,x(t));
> subs(x(t)=x,");
> ss:=solve(",x);
> eq1:=subs(t=0,ss)=1;
> eq2:=subs(t=2,ss)=4;
> sols:=solve({eq1,eq2});
> simplify(subs(sols[1],ss));
> simplify(subs(sols[2],ss));

The following example shows how isoperimetric problems may be handled. The problem is to minimize $J = \int \dot{x}^2 \, dt$ with $x(0) = 2$, $x(1) = 4$ and subject to the constraint $\int x \, dt = 1$.

```
> k1:=dx^2-lambda*x;
```

$$k1 := dx^2 - \lambda x$$

```
> EL1(k1);
```

$$-\lambda - 2 \left(\frac{d^2}{dt^2} x(t) \right) = 0$$

```
> dsolve({EL1(k1),x(0)=2,x(1)=4},x(t));
```

$$x(t) = -\frac{1}{4}\lambda t^2 + 2 + \left(\frac{1}{4}\lambda + 2 \right) t$$

```
> op(2,dsolve({EL1(k1),x(0)=2,x(1)=4},x(t)));
```

$$-\frac{1}{4}\lambda t^2 + 2 + \left(\frac{1}{4}\lambda + 2 \right) t$$

```
> int(",t=0..1);
```

$$\frac{1}{24}\lambda + 3$$

```
> solve("=1,lambda);
```

$$-48$$

```
> subs(lambda=",dsolve({EL1(k1),x(0)=2,x(1)=4},x(t)));
```

$$x(t) = 12t^2 + 2 - 10t$$

Now let's look at the two variable case. In fact, I should say *cases* since we may want to choose different forms of the Euler-Lagrange equations depending on whether the integrand is x or y independent. The following procedures deal with lagrangians having 2 t-dependent functions x and y and their derivatives $\dot{x}$ and $\dot{y}$. The first case is when both Euler-Lagrange equations should be computed.

```
> ELsys1:=proc(f)
>       local el1,el2,part1,part2,part3,part4,p2,p4,dfddx,dfddy;
>       part1:=subs({x=x(t),y=y(t),dx=diff(x(t),t),dy=diff(y(t),t)},
>                 diff(f,x));
>       dfddx:=subs({x=x(t),dx=dx(t),y=y(t),dy=dy(t)},diff(f,dx));
>       p2:=diff(dfddx,t);
```

```
>       part2:=subs({dx(t)=diff(x(t),t),dy(t)=diff(y(t),t)},p2);
>       el1:=simplify(part1-part2=0);
>
>       part3:=subs({x=x(t),y=y(t),dx=diff(x(t),t),dy=diff(y(t),t)},
                      diff(f,y));
>       dfddy:=subs({x=x(t),dx=dx(t),y=y(t),dy=dy(t)},diff(f,dy));
>       p4:=diff(dfddy,t);
>       part4:=subs({dx(t)=diff(x(t),t),dy(t)=diff(y(t),t)},p4);
>       el2:=simplify(part3-part4=0);
>       RETURN(el1,el2);
>       end:
```

The third case is when the second Euler-Lagrange equation should be computed, but the lagrangian is independent of x.

```
> ELsys3:=proc(f)
>       local el1,el2,part3,part4,p2,p4,dfddy;
>       p2:=simplify(subs({x=x(t),dx=dx(t),y=y(t),dy=dy(t)},
                      diff(f,dx))=c);
>       el1:=subs({dx(t)=diff(x(t),t),dy(t)=diff(y(t),t)},p2);
>
>       part3:=subs({x=x(t),y=y(t),dx=diff(x(t),t),dy=diff(y(t),t)},
                      diff(f,y));
>       dfddy:=subs({x=x(t),dx=dx(t),y=y(t),dy=dy(t)},diff(f,dy));
>       p4:=diff(dfddy,t);
>       part4:=subs({dx(t)=diff(x(t),t),dy(t)=diff(y(t),t)},p4);
>       el2:=simplify(part3-part4=0);
>       RETURN(el1,el2);
>       end:
```

EXERCISE 7.4. Write MAPLE procedures which handle the second and fourth cases of the Euler-Lagrange equations for two variables. Namely, handle the cases when the integrand f is independent of y and dependent on x and when it is independent of both variables.

We might also wish to return a portion of an equation unevaluated. For example, we may want d/dt unevaluated. In this case, we can replace "diff(-,t)" with "Diff(-,t)" in the expressions for "dfddx" and "dfddy" above. Now let's apply the Euler-Lagrange procedure to the situation of a marble rolling in a hemispherical bowl. Recall that the action integral is

$$J = \int \frac{1}{2}mR^2(\cos^2 v\, \dot{u}^2 + \dot{v}^2) - mgR\sin v \; dt.$$

where the bowl has radius R, the marble has mass m and g is the gravitational constant of the Earth. The Euler-Lagrange equations for this two variable problem are

$$\dot{u} = \frac{c}{mR^2 \cos^2 v} \qquad \text{and} \qquad \ddot{v} + \frac{g}{R}\cos v + \sin v \cos v\, \dot{u}^2 = 0.$$

These are then the equations of motion of the marble. In Figure 8.4 the trajectory corresponding to these equations with $g = m = R = 1$ is shown. Now let's see how this trajectory was obtained by using MAPLE.

```
> with(plots):
```

The following is a parametrization for the sphere ($R = 1$).

```
> sph:=[cos(u)*cos(v),sin(u)*cos(v),sin(v)];
```

The following series of procedures calculates the Euler-Lagrange equations for the action integral constrained to the surface of the sphere. Of course, since we need the metric, we must also include the usual procedures necessary for its calculation.

```
> dp := proc(X,Y) X[1]*Y[1]+X[2]*Y[2]+X[3]*Y[3] end:
>
> nrm := proc(X) sqrt(dp(X,X)) end:
>
> xp := proc(X,Y)
>       local a,b,c;
>       a := X[2]*Y[3]-X[3]*Y[2];
>       b := X[3]*Y[1]-X[1]*Y[3];
>       c := X[1]*Y[2]-X[2]*Y[1];
>       [a,b,c];
>       end:

> Jacf := proc(X)
>       local Xu,Xv;
>       Xu := [diff(X[1],u),diff(X[2],u),diff(X[3],u)];
>       Xv := [diff(X[1],v),diff(X[2],v),diff(X[3],v)];
>       simplify([Xu,Xv]);
>       end:
> EFG := proc(X)
>       local E,F,G,Y;
>       Y := Jacf(X);
>       E := dp(Y[1],Y[1]);
>       F := dp(Y[1],Y[2]);
>       G := dp(Y[2],Y[2]);
>       simplify([E,F,G]);
>       end:
```

The following procedure gives the Euler-Lagrange equations for the action integral. By Hamilton's principle, these equations must be the equations of motion of the associated physical system. Notice that we use ELsys3 for the Euler-Lagrange equations since the constrained integrand $T - V$ has no dependence on u. In other situations, other cases of the ELsys procedures may be necessary.

```
> Equat:=proc(X)
>       local Metric,TT,VV,LL,Eul,eq1,eq2;
>       Metric:=subs({u=x,v=y},EFG(X));
>       TT:=1/2*(Metric[1]*dx^2+ Metric[3]*dy^2);
>       VV:=subs({u=x,v=y},X[3]);
>       LL:=TT-VV;
>       Eul:=ELsys3(LL);
>       eq1:=subs({x(t)=u(t),y(t)=v(t)},Eul[1]);
>       eq2:=subs({x(t)=u(t),y(t)=v(t)},Eul[2]);
>       RETURN(eq1,eq2);
>       end:
```

For example, the equations of motion for the sphere-marble problem are obtained by

```
> Equat(sph);
```

$$\cos(v(t))^2 \left(\frac{d}{dt} u(t) \right) = c$$

$$-\cos(v(t)) \left(\frac{d}{dt} u(t) \right)^2 \sin(v(t)) - \cos(v(t)) - \left(\frac{d^2}{dt^2} v(t) \right) = 0$$

Here c is a constant dependent on the initial speed at which we propel the marble. Different c's should be expected to produce different motions. The next procedure allows us to plug the numerical solution of the equations of motion into the definition of a surface X and plot both the surface and the motion. This procedure is almost identical to "plotgeo" which allowed us to plot geodesics on surfaces. The only real differences are that we must be able to plug in a value for c (and this necessitates some finagling with the definition of "sys" and the way we provide a *set* of equations for "desys") and that the u-equation is *first order*. Hence, for "plotmotion," there is no "Du0" input. Again, different problems may require modification here.

> plotmotion:=proc(X,const,ustart,uend,vstart,vend,u0,v0,Dv0,n,gr,
>   theta,phi)
>   local sys,desys,dequ,deqv,listp,j,motion,plotX;
>   sys:=[subs(c=const,Equat(X)[1]),Equat(X)[2]];
>   desys:=dsolve({sys[1],sys[2],u(0)=u0,v(0)=v0,D(v)(0)=Dv0},
>   {u(t),v(t)},type=numeric,output=listprocedure);
>   dequ:=subs(desys,u(t)); deqv:=subs(desys,v(t));
>   listp:=[seq(subs({u=dequ(j/n[3]),v=deqv(j/n[3])},X), j=n[1]..n[2])];
>   motion:=spacecurve({listp}, color=black,thickness=2):
>   plotX:=plot3d(X,u=ustart..uend,v=vstart..vend,grid=[gr[1],gr[2]]):
>   display({motion,plotX},style=wireframe,scaling=constrained,
>   orientation=[theta,phi]);
>   end:

Again, just as in "plotgeo," note that the inputs "n" and "gr" are vectors. Figure 8.4 comes from the command

> plotmotion(sph,1,0,2*Pi,-Pi/2,0,0,-Pi/4,0,[0,200,20],[30,20],0,60);

EXERCISE 7.5. Carry out the following MAPLE command and explain the result in terms of the chosen c and the equations of motion (for this c!).

> plotmotion(sph,1/(sqrt(2)*2^(1/4)),0,2*Pi,-Pi/2,0,0,-Pi/4,0, [0,150,20],
>   [30,20],0,60);

EXERCISE 7.6. Now do **Exercise 4.8**. Namely, find the equations of motion of a marble in a paraboloid shaped bowl and plot the motion for varying c. Try

> par:=[v*cos(u),v*sin(u),v^2];
> plotmotion(par,1,0,2*Pi,0,2.5,0,2,0,[0,500,20],[30,12],0,60);

Note that the parametrization is chosen so that "plotmotion" can be used without modification.

Finally, let's see how MAPLE may be used in optimal control. Let's take one standard problem and see how MAPLE procedures may allow for calculation and visualization of solutions. Here's the problem. Suppose we are on a block of ice on an ice covered (so frictionless) pond and we want to get to the center of the pond as quickly as possible. It so happens that our ice block has a rocket motor attached to it which can be fired forward or backwards. The question then is, what is the optimal way to fire the rocket so that our ice block ends up at the center of the pond in the least time?

Newton's Law $F = ma$ tells us a fundamental relation between our control — the firing of the rocket — and the acceleration of the ice block. Instead of dealing with this second order equation, we can use a standard trick to transform the problem into a system of first order equations. Namely, let $x1$ denote the distance coordinate (from our position to the pond's center) and $x2 = \dot{x}1$ denote $x1$'s velocity. Then, with $|F| = u$ the control, Newton's Law becomes

$$\dot{x}1 = x2 \qquad \text{and} \qquad \dot{x}2 = u.$$

Since we are trying to minimize *time*, the integral to be minimized is $J = \int 1 \, dt = T$. Also, and very importantly, our rocket has limitations on the thrust it can give. Let's say that

$$|F| = |u| \leq 1.$$

Now, the Hamiltonian is easy to write down, but let's write a MAPLE procedure to do this and (if possible) solve the co-state equations. Indeed, we can solve the whole problem in a conceptual and visual manner by making use of MAPLE's graphics capabilities. First, we need the following packages.

```
> with(DEtools): with(plots): with(linalg):
> Hamilton:=proc(f1,f2)
>       local Ham,part1,part2,psi1,psi2:
>       Ham:=collect(-1+psi1(t)*f1 + psi2(t)*f2,u);
>       part1:=-diff(psi1(t),t)=diff(Ham,x1);
>       part2:=-diff(psi2(t),t)=diff(Ham,x2);
>       RETURN(H=Ham,dsolve({part1,part2},{psi1(t),psi2(t)}));
>       end:

> StateEq:=proc(f1,f2,U)
>       local part1,part2:
>       part1:=diff(x1(t),t)=subs({x1=x1(t),x2=x2(t),u=U},f1);
>       part2:=diff(x2(t),t)=subs({x1=x1(t),x2=x2(t),u=U},f2);
>       RETURN(factor(dsolve({part1,part2},{x1(t),x2(t)},laplace)));
>       end:
```

By the way, note that I solved the system of differential equations above using "dsolve(...,laplace)," which means that the explicit solution is obtained by laplace transform methods. You should look up the help on "dsolve" to see all the options available for this command. Remember, to access MAPLE help, type "?dsolve." Now, let's write the right sides of the state equations and find the Hamiltonian.

```
> f1:=x2; f2:=u;
```

$$f1 := x2$$
$$f2 := u$$

> Hamilton(f1,f2);

$$H = -1 + \psi 1(t)x2 + \psi 2(t)u, \ \{\psi 1(t) = -C1, \psi 2(t) = C1 + C2\,t\}$$

The Pontryagin Maximum Principle tells us that the Hamiltonian along an optimal trajectory must always be a maximum with respect to u. But H is *linear* in u, so the maximum is attained at the endpoints; that is, when $u = 1$ and $u = -1$. Before we use this information to plot solution trajectories, we can actually solve the state equations explicitly. (This is very rare for optimal control problems and only occurs here because this is a rather simple system.)

> StateEq(f1,f2,1);

$$\left\{ x2(t) = x2(0) + t, x1(t) = x1(0) + x2(0)t + \frac{1}{2}t^2 \right\}$$

> StateEq(f1,f2,-1);

$$\left\{ x2(t) = x2(0) - t, x1(t) = x1(0) + x2(0)t - \frac{1}{2}t^2 \right\}$$

Now we can plot the trajectories for $u = 1$ and $u = -1$ separately and then try to discern a combination of trajectories which get us to the origin quickest. Trajectory plots may be accomplished through MAPLE's "phaseportrait" command. The general format includes inputs which are the $f1$ and $f2$ of the state equations ([x2,1] below), the variables x1 and x2, the trajectory parameter (-4..4 below) and a list of points from which the trajectories begin. A point is given by three coordinates. The first coordinate gives a starting value for the trajectory parameter while the second and third coordinates are the usual cartesian coordinates of the point in the plane. For example, [0,2,0] starts the trajectory parameter at $t = 0$ from the plane point $(2, 0)$. Note that we give the plot structure a name and save it to be displayed later. (Remember the *colon.*)

> ph1:=phaseportrait([x2,1],[x1,x2],-4..4,{[0,0,0],[0,2,0], [0,4,0],[0,6,0],
        [0,-2,0],[0,-4,0],[0,-5,0],[0,-1,0],[0,1,0],[0,0,0]}, thickness=2):

Also, it is possible to graph the direction vector field corresponding to the system of differential equations.

> field1:=dfieldplot([x2,1],[x1,x2],-3..3,x1=-5..9,x2=-5..5):

Both of these pictures may now be displayed in a given viewing region.

> display({ph1,field1},view=[-5..9,-5..5],scaling=constrained);

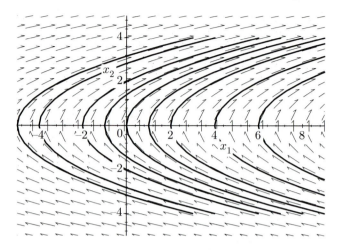

FIGURE 8.5.   Trajectories for $u = 1$

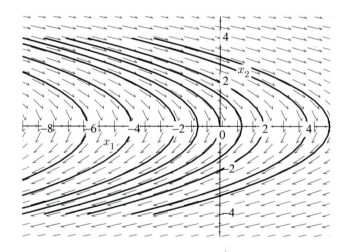

FIGURE 8.6.   Trajectories for $u = -1$

The following commands give the $u = -1$ trajectories.

> ph2:=phaseportrait([x2,-1],[x1,x2],-4..4,{[0,0,0],[0,2,0],[0,4,0], [0,5,0],
        [0,-2,0],[0,-4,0],[0,-1,0], [0,-6,0],[0,1,0],[0,0,0]}, thickness=2):

> field2:=dfieldplot([x2,-1],[x1,x2],-3..3,x1=-9..5,x2=-5..5):

> display({ph2,field2},view=[-9..5,-5..5],scaling=constrained);

> display({ph1,ph2,field1,field2},view=[-9..9,-5..5], scaling=constrained);
But now we can see the trajectories which get us to the origin and we can plot
them alone as follows. We also include a trajectory starting from a point in

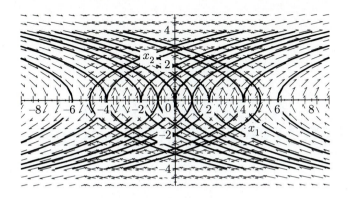

FIGURE 8.7.   Trajectories for $u = 1$ and $u = -1$

the plane to show what the solution to the original problem is. Namely, from any point in the plane, take the trajectory (whether 1 or -1) which travels toward the trajectory leading straight to the origin. Specifically, the picture below shows a starting point $(2, 2)$, so that we take a $u = -1$ trajectory until it intersects the $u = 1$ trajectory going to the origin. We then switch to the latter and go right to $(0,0)$. In fact, for this problem, the switch time (and total time) may be calculated precisely. In general, however, it is the picture which provides the qualitative solution. For this, MAPLE is well suited.

```
> ph01:=phaseportrait([x2,1],[x1,x2],-3..0,{[0,0,0]},thickness=2):
> ph02:=phaseportrait([x2,-1],[x1,x2],-3..0,{[0,0,0]},thickness=2):
> ph11:=phaseportrait([x2,-1],[x1,x2],0..4,{[0,2,2]},thickness=2):
> display({ph01,ph02,ph11},scaling=constrained,view=[-4..4,-3..3]);
```

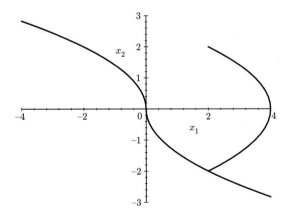

FIGURE 8.8.   The final solution path

Chapter 9

A GLIMPSE AT
HIGHER DIMENSIONS

9.1 INTRODUCTION

Up to this point, we have only considered surfaces $\mathbf{x}(u, v)$ dependent on *two* parameters. In mathematics and the sciences, however, it is often the case that geometric structures depend on many parameters. Indeed, the number of degrees of freedom of a physical system tells us precisely the number of parameters necessary to describe the so-called *configuration space* of the system. Just as we could use differential geometry to understand particles moving on constraint surfaces, so we would like to do the same for systems with many parameters. This means that we must invent a notion of higher dimensional surface which mimics the properties of the geometry we are comfortable with, namely the geometry of two-dimensional surfaces. In this chapter, we will consider these higher dimensional surfaces from a naive point of view with the goal of introducing the relevant notation and making the analogy with the two-dimensional situation. In this sense, this chapter is simply a dictionary for readers making the transition from the geometry of two dimensions to that of n dimensions. In particular, we will not introduce manifolds and their covariant derivatives in their most abstract generality, but instead stick (mostly) to the case of submanifolds of Euclidean space with the induced metric and covariant derivative. In this way, we hope the reader can be introduced to the geometry of higher dimensions while still maintaining some touch with reality. Basic references for this chapter are [DoC2], [Spi], [GHL], [Hi] and (for connections with physics) [CM].

9.2 MANIFOLDS

Just as for two dimensions, we may define a *k-patch*, or *coordinate chart*, $\mathbf{x}: D \to \mathbb{R}^{n+1}$ for a connected open set $D \subseteq \mathbb{R}^k$. Here, we assume that

$$\mathbf{x}(u^1, \dots, u^k) = \left(x^1(u^1, \dots, u^k), \dots, x^{n+1}(u^1, \dots, u^k)\right)$$

is smooth (i.e. all partials of all orders exist and are continuous), injective (with continuous inverse on its image) and the tangent vectors of parameter curves

$$\partial_i \mathbf{x} \stackrel{\text{def}}{=} \left(\frac{\partial x^1}{\partial u^i}, \ldots, \frac{\partial x^{n+1}}{\partial u^i} \right)$$

are linearly independent ($i = 1, \ldots, k$). We shall usually denote these tangent vectors, when the patch is understood, by ∂_i alone. A subset $M^k \subset \mathbb{R}^{n+1}$ is a k-*manifold* if it is covered by k-patches with the property that, for any two patches $\mathbf{x} \colon D_{\mathbf{x}} \to \mathbb{R}^{n+1}$ and $\mathbf{y} \colon D_{\mathbf{y}} \to \mathbb{R}^{n+1}$ with nonempty intersection, the composition *transition map*

$$\mathbf{x}^{-1} \circ \mathbf{y} \colon \mathbf{y}^{-1} \left(\mathbf{x}(D_{\mathbf{x}}) \cap \mathbf{y}(D_{\mathbf{y}}) \right) \to D_{\mathbf{x}}$$

is smooth. Of course, when we say "covered by k-patches," we simply mean that every point of M lies in the image of some patch. Also, the k-dimensional *tangent space* to M at $p \in M$ is simply the subspace of $\mathbb{R}^{n+1}$ spanned by the linearly independent vectors $\partial_1(p), \ldots, \partial_k(p)$. The tangent space is denoted by $T_p M$ just as for surfaces.

Example 2.1: The sphere S^n. The n-sphere in $\mathbb{R}^{n+1}$ is defined as

$$S^n \stackrel{\text{def}}{=} \{ (x^1, \ldots, x^{n+1}) | (x^1)^2 + \ldots + (x^{n+1})^2 = 1 \}.$$

This is the direct analogue of S^2, the 2-sphere in $\mathbb{R}^3$. Note that we call the sphere in $\mathbb{R}^{n+1}$ the n-sphere instead of the $n + 1$-sphere. The reason for this terminology is that the sphere in $\mathbb{R}^{n+1}$ is an n-*manifold*, not an $n+1$-manifold. So, the appellation "n-sphere" denotes the dimension of the manifold itself, not the dimension of the ambient, or surrounding, space. Let us now show that S^n is a manifold of dimension n. We must cover S^n by patches which satisfy the "transition property" above.

Now, stereographic projection $\text{St} \colon S^n \to \mathbb{R}^n$ works in $\mathbb{R}^{n+1}$ just as in $\mathbb{R}^3$. Namely, denoting the North and South poles by $\mathcal{N} = (0, \ldots, 0, 1)$ and $\mathcal{S} = (0, \ldots, 0, -1)$ respectively, we have the North and South projections

$$\text{St}_{\mathcal{N}}(x^1, \ldots, x^{n+1}) = \left(\frac{x^1}{1 - x^{n+1}}, \ldots, \frac{x^n}{1 - x^{n+1}} \right)$$

$$\text{St}_{\mathcal{S}}(x^1, \ldots, x^{n+1}) = \left(\frac{x^1}{1 + x^{n+1}}, \ldots, \frac{x^n}{1 + x^{n+1}} \right).$$

Let $\mathbf{x}(u^1, \ldots, u^n) = \text{St}^{-1}(u^1, \ldots, u^n) \colon \mathbb{R}^n \to S^n - \{\mathcal{N}\}$ and $\mathbf{y}(w^1, \ldots, w^n) = \text{St}^{-1}(w^1, \ldots, w^n) \colon \mathbb{R}^n \to S^n - \{\mathcal{S}\}$ denote the respective inverses. Both $\mathbf{x}$ and $\mathbf{y}$ are smooth maps which are injective with smooth inverses (i.e. St) on their images. Also, it is easy to see that the parameter tangent vectors are linearly independent, so $\mathbf{x}$ and $\mathbf{y}$ are patches. Clearly, $\mathbf{x}$ and $\mathbf{y}$ cover S^n, so we must only show the transition property to see that S^n is an n-manifold. For this, note that $\mathbf{x}$ and $\mathbf{y}$ have the explicit representations

$$\mathbf{x}(u^1, \ldots, u^n) = \left(\frac{2u^1}{\sum u^{i^2} + 1}, \ldots, \frac{2u^n}{\sum u^{i^2} + 1}, \frac{\sum u^{i^2} - 1}{\sum u^{i^2} + 1} \right)$$

$$\mathbf{y}(w^1, \ldots, w^n) = \left(\frac{2w^1}{1 + \sum w^{i^2}}, \ldots, \frac{2w^n}{1 + \sum w^{i^2}}, \frac{1 - \sum w^{i^2}}{1 + \sum w^{i^2}} \right)$$

The transition function is then given by

$$\mathbb{R}^n - \{(0, \ldots, 0)\} \xrightarrow{\text{St}_{\mathcal{N}}^{-1}} S^n - \{\mathcal{N}, \mathcal{S}\} \xrightarrow{\text{St}_{\mathcal{S}}} \mathbb{R}^n$$

and calculated to be

$$\mathbf{y}^{-1} \circ \mathbf{x}(u^1, \ldots, u^n) = \left(\frac{u^1}{\sum u^{i^2}}, \ldots, \frac{u^n}{\sum u^{i^2}} \right) = (w^1, \ldots, w^n).$$

Therefore, each component has the formula

$$u^i \mapsto \frac{u^i}{\sum u^{i^2}} = w^i$$

and, since the origin of $\mathbb{R}^n$ is precluded from the domain of the transition map, each component function is smooth. Thus, S^n is an n-manifold.

EXERCISE 2.1. Verify the formulas above for stereographic projection, $\mathbf{x}$, $\mathbf{y}$ and the transition $\mathbf{y}^{-1} \circ \mathbf{x}$ and explain why the origin of $\mathbb{R}^n$ is not in the domain of the transition.

Now suppose that $\mathbf{x} \colon D_\mathbf{x} \to \mathbb{R}^{n+1}$ and $\mathbf{y} \colon D_\mathbf{y} \to \mathbb{R}^{n+1}$ are two patches with nonempty intersection $\mathbf{x}(D_\mathbf{x}) \cap \mathbf{y}(D_\mathbf{y}) \neq \emptyset$. Then, on the intersection,

$$\mathbf{x}(u^1, \ldots, u^k) = \mathbf{y}(w^1, \ldots, w^k)$$

and $\mathbf{y}^{-1} \circ \mathbf{x}(u^1, \ldots, u^k) = (w^1, \ldots, w^k)$. Since each component function y^s may be considered as a function of the u^i's, we have by the chain rule,

$$\frac{\partial y^s}{\partial u^i} = \frac{\partial x^s}{\partial u^i}$$

$$= \sum_{j=1}^{k} \frac{\partial w^j}{\partial u^i} \frac{\partial y^s}{\partial w^j}.$$

This type of transformation rule is familiar to physicists who are interested in relating the physical quantities of one reference frame to those of another. Indeed, for this reason and because of the coordinate change formula above, physicists in the old days often defined the mathematical objects they were interested in (i.e. "tensors") in terms of such transformation rules. If we write out the coordinate form of the formula above, we obtain

$$\left(\frac{\partial x^1}{\partial u^i}, \ldots, \frac{\partial x^{n+1}}{\partial u^i}\right) = \frac{\partial w^1}{\partial u^i}\left(\frac{\partial y^1}{\partial w^1}, \ldots, \frac{\partial y^{n+1}}{\partial w^1}\right)$$
$$+ \ldots + \frac{\partial w^k}{\partial u^i}\left(\frac{\partial y^1}{\partial w^k}, \ldots, \frac{\partial y^{n+1}}{\partial w^k}\right).$$

Written more compactly, we have

$$\boldsymbol{\partial}_i \mathbf{x} = \sum_{j=1}^{k} \frac{\partial w^j}{\partial u^i} \boldsymbol{\partial}_j \mathbf{y}$$

and this formula shows that the coordinate change rule allows us to change the basis of the tangent space $T_p M$ via the matrix

$$\begin{pmatrix} \frac{\partial w^1}{\partial u^1} & \cdots & \frac{\partial w^1}{\partial u^i} & \cdots & \frac{\partial w^1}{\partial u^k} \\ \cdots & \cdots & \cdots & \cdots & \cdots \\ \cdots & \cdots & \frac{\partial w^j}{\partial u^i} & \cdots & \cdots \\ \cdots & \cdots & \cdots & \cdots & \cdots \\ \frac{\partial w^k}{\partial u^1} & \cdots & \frac{\partial w^k}{\partial u^i} & \cdots & \frac{\partial w^k}{\partial u^k} \end{pmatrix}$$

which we denote by $J(u, w)$ since it is just the *Jacobian matrix* of multivariable calculus associated to a coordinate change. This formulation allows us to make the notion of orientability (see Chapter 2) precise. Say that a manifold M is *orientable* if there is a collection of patches $\{\mathbf{x}_\alpha\}_{\alpha \in \mathcal{A}}$ which cover M such that the Jacobian matrices for all possible transitions $\mathbf{x}_\alpha^{-1} \circ \mathbf{x}_\beta$ have *positive determinant* (evaluated at all points in the overlap). Intuitively, we can think of this requirement as the analogue of saying that a rotation keeps an object oriented the same way while a reflection produces an oppositely oriented mirror image. One thing to notice is the following

Proposition 2.1. *Suppose that the intersection of two patches is path connected. In particular, suppose* $\mathbf{x}(D_\mathbf{x}) \cap \mathbf{y}(D_\mathbf{y}) \neq \emptyset$. *Then the determinant of the associated Jacobian does not change sign when evaluated at any point in the intersection.*

Proof. We will give a rather sophisticated (as opposed to computational) proof here because the same ideas occur often in the geometry of higher dimensions. Choose two arbitrary points p and q in the intersection $\mathbf{x}(D_\mathbf{x}) \cap \mathbf{y}(D_\mathbf{y})$ and take a path $\alpha \colon I \to \mathbf{x}(D_\mathbf{x}) \cap \mathbf{y}(D_\mathbf{y}) \subset M$ with $\alpha(0) = p$ and $\alpha(1) = q$. Now, just as in Lemma 2.1.1, we may write

$$\mathbf{x}(u^1(t), \ldots, u^k(t)) = \alpha(t) = \mathbf{y}(w^1(t), \ldots, w^k(t))$$

with $\mathbf{y}^{-1} \circ \mathbf{x}(u^1(t), \ldots, u^k(t)) = (w^1(t), \ldots, w^k(t))$. In this way the Jacobian

$$J(u, w) = \left(\frac{\partial w^j}{\partial u^i}(t) \right)_{\alpha(t)}$$

is dependent on t as well and the determinant becomes a *continuous* function $\det J(u, w) \colon I \to \mathbb{R}$ (since the partials are continuous). Now, $J(u, w) \in GL(k, \mathbb{R})$, the group of invertible $k \times k$-matrices over $\mathbb{R}$, so $\det J(u, w)(t) \neq 0$ for all t. Therefore, by continuity (and the intermediate value theorem), the continuous function $\det J(u, w)$ cannot take both positive and negative values.
               □

Corollary 2.2. *If a manifold M is covered by two patches whose intersection is path connected, then M is orientable.*

Proof. We must only show that one single $J(u, w)$ has positive determinant. Suppose it does not. Then, by either changing the sign of one coordinate function or interchanging two coordinate functions in one of the patches, we change the sign of the determinant of $J(u, w)$. This follows from the usual properties of determinants; namely, if a row (or column) is multiplied by -1 or if two rows (or columns) are interchanged, then the sign of the determinant is changed. Note that the connectedness of the intersection and the proposition above ensure that we must only check the positivity of $\det J(u, w)$ at a single point.   □

Example 2.2: The sphere S^n. For the patches $\mathbf{x}$ and $\mathbf{y}$ of Example 2.1, the transition function was calculated to be

$$\mathbf{y}^{-1} \circ \mathbf{x}(u^1, \ldots, u^n) = \left(\frac{u^1}{\sum u^{i^2}}, \ldots, \frac{u^n}{\sum u^{i^2}} \right) = (w^1, \ldots, w^n).$$

The Jacobian matrix may be calculated by first observing that

$$\frac{\partial w^j}{\partial u^i} = \frac{(\sum_s u^{s2}) \cdot \delta_i^j - 2u^j u^i}{(\sum_s u^{s2})^2}$$

where $\delta_i^j = 0$ if $i \neq j$ and $\delta_i^j = 1$ if $i = j$. (That is, δ_i^j is the *Kronecker delta*.) By Proposition 2.1, we need test the Jacobian at a single point, so choose $(1, 0, \ldots, 0)$. Then, we have

$$J(u, v)_{(1,0,\ldots,0)} = \begin{pmatrix} -1 & 0 & \cdots & 0 \\ 0 & 1 & \cdots & 0 \\ \cdots & \cdots & \cdots & \cdots \\ 0 & 0 & \cdots & 1 \end{pmatrix}.$$

This matrix has determinant -1, so the given patches *do not orient* S^n. However, by Corollary 2.2, since the patch intersection is the path connected set $S^n - \{\mathcal{N}, \mathcal{S}\}$, S^n must be orientable. Indeed, by the proof of Corollary 2.2, we can redefine $\mathbf{y}$ to be a new patch

$$\bar{\mathbf{y}}(w^1, \ldots, w^k) = \left(\frac{2w^2}{1 + \sum w^{i^2}}, \frac{2w^1}{1 + \sum w^{i^2}}, \frac{2w^3}{1 + \sum w^{i^2}}, \right.$$
$$\left. \ldots, \frac{2w^n}{1 + \sum w^{i^2}}, \frac{1 - \sum w^{i^2}}{1 + \sum w^{i^2}} \right)$$

obtained by switching the first two coordinate functions. The Jacobian matrix at $(1, 0, \ldots, 0)$ now has its first two rows switched,

$$J(u, v)_{(1,0,\ldots,0)} = \begin{pmatrix} 0 & 1 & \cdots & 0 \\ -1 & 0 & \cdots & 0 \\ \cdots & \cdots & \cdots & \cdots \\ 0 & 0 & \cdots & 1 \end{pmatrix}$$

and $\det J(u, v)_{(1,0,\ldots,0)} = 1$. The patches $\mathbf{x}$ and $\bar{\mathbf{y}}$ therefore orient S^n.

9.3 THE COVARIANT DERIVATIVE

From now on we will confine ourselves to the image of a single patch $\mathbf{x} \colon D \subset \mathbb{R}^k \to \mathbb{R}^{n+1}$ on a k-manifold M^k. Hence, from now on, we are doing *local* differential geometry. From Proposition 2.1, we see that $\mathbf{x}$ is automatically orientable since D, and hence its image $\mathbf{x}(D)$ are connected. For the patch $\mathbf{x}$ (which we may also refer to as M), the parameter tangent vectors $\boldsymbol{\partial}_1, \ldots, \boldsymbol{\partial}_k$ form a basis at each point for the tangent space $T_p M$. We may extend this basis for $T_p M$ to a basis for $\mathbb{R}^{n+1}$ by choosing vectors, $U_1, \ldots, U_{n+1-k}$ in $\mathbb{R}^{n+1}$ such that

$$U_s \cdot \boldsymbol{\partial}_i = 0 \qquad \text{for all } s \text{ and } i$$

and

$$U_s \cdot U_t = \delta_s^t \qquad \text{where } \delta_s^t \text{ is the Kronecker delta.}$$

The U_i's are then *normal vectors* for M. If $k = n$, then only one normal vector U exists and, in this case, M is said to be a *hypersurface*. All surfaces in $\mathbb{R}^3$ are thus hypersurfaces. Recall that a *smooth vector field* V on M is simply the assignment of an $\mathbb{R}^{n+1}$ vector to each point of M so that $V \colon M \to \mathbb{R}^{n+1}$ is smooth. If the assigned vectors are always tangent to M, then V is said to be a *tangent* vector field. The vector fields $\boldsymbol{\partial}_1, \ldots, \boldsymbol{\partial}_k$ are tangent vector fields on M^k while the vector fields $U_1, \ldots, U_{n+1-k}$ are *normal* vector fields.

For a vector field $Z = (Z^1, \ldots, Z^{n+1})$ in $\mathbb{R}^{n+1}$, the $\mathbb{R}^{n+1}$-*covariant deriv-ative* is defined just as in the two dimensional situation. Namely,

$$\nabla_{\mathbf{v}}^{\mathbb{R}^{n+1}} Z = \sum_{i=1}^{n+1} \mathbf{v}[Z^i] e_i$$

where e_i is the standard i^{th} basis vector for $\mathbb{R}^{n+1}$ and $\mathbf{v}[\cdot]$ denotes the usual $\mathbb{R}^{n+1}$ directional derivative $\mathbf{v}[f] = \nabla f \cdot \mathbf{v}$. The *covariant derivative* for M is then defined to be the orthogonal projection of $\nabla^{\mathbb{R}^{n+1}}$ onto $T_p M$,

$$\nabla_{\mathbf{v}} Z = \text{proj}_{T_p M} \nabla_{\mathbf{v}}^{\mathbb{R}^{n+1}} \bar{Z}$$

where $\mathbf{v} \in T_p M$, Z is defined on M and $\bar{Z}$ is a local extension of Z to an open set in $\mathbb{R}^{n+1}$ containing M. Similarly, for a tangent vector field V, we may define $\nabla_V Z = \text{proj}_{T_p M} \nabla_{\bar{V}}^{\mathbb{R}^{n+1}} \bar{Z}$ by

$$\nabla_V Z(p) = \text{proj}_{T_p M} \nabla_{\bar{V}(p)}^{\mathbb{R}^{n+1}} \bar{Z},$$

where $\bar{V}$ is a local extension of V as well. The following exercise shows that $\nabla_V Z$ is well defined.

EXERCISE 3.1. Show that the definition of $\nabla_V Z$ does not depend on the extensions chosen for V and Z. Hints: (1) start by showing that $\nabla_{V_1}^{\mathbb{R}^{n+1}} \bar{Z} = \nabla_{V_2}^{\mathbb{R}^{n+1}} \bar{Z}$ on M for V_1 and V_2 extensions of V. Use the fact that $V_1 = V_2$ on M. (2) Now show that $\nabla_V^{\mathbb{R}^{n+1}} Z_1 = \nabla_V^{\mathbb{R}^{n+1}} Z_2$ on M as well.

From now on, for notational convenience, we shall dispense with distin-guishing vector fields on M from their local extensions on open sets in $\mathbb{R}^{n+1}$. It is important to realize that the covariant derivative on M is simply the $\mathbb{R}^{n+1}$ covariant derivative *minus its U_i-components*, $i = 1, \ldots, n+1-k$.

Proposition 3.1. *Let V, Z and W be tangent vector fields on M and $f \colon M \to \mathbb{R}$ be a function on M. The following are properties of the covariant derivative of M:*

(i) $\nabla_V(Z + W) = \nabla_V Z + \nabla_V W$

(ii) $\nabla_{fV} Z = f \nabla_V Z$

(iii) $\nabla_V f Z = V[f] Z + f \nabla_V Z$

(iv) $V[\langle Z, W \rangle] = \langle \nabla_V Z, W \rangle + \langle Z, \nabla_V W \rangle$, *where $\langle \cdot, \cdot \rangle$ denotes the usual dot product in $\mathbb{R}^{n+1}$ restricted to vectors tangent to M.*

(v) $\nabla_V Z - \nabla_Z V = [V, Z]$, *where $[\cdot, \cdot]$ is defined below.*

Proof. To prove (i), it suffices to note that directional derivative and projection both preserve sums. For (ii), note that, by definition,

$$\nabla^{\mathbb{R}^{n+1}}_{fV} Z = \sum_i fV[Z^i]\, e_i = f \sum_i V[Z^i]\, e_i = f\, \nabla^{\mathbb{R}^{n+1}}_{V} Z.$$

Then we take projections to obtain

$$\nabla_{fV} Z = \mathrm{proj}\,{}_{T_p M}\nabla^{\mathbb{R}^{n+1}}_{fV} Z$$
$$= \mathrm{proj}\,{}_{T_p M}(f\nabla^{\mathbb{R}^{n+1}}_{V} Z)$$
$$= f\, \mathrm{proj}\,{}_{T_p M}\nabla^{\mathbb{R}^{n+1}}_{V} Z$$

since $f(p)$ simply multiplies vectors by a scalar and, so, doesn't affect projection. Then, by definition,

$$\nabla_{fV} Z = f\, \nabla_V Z.$$

For (iv), let us compute using the definitions of $\langle \cdot, \cdot \rangle$ (i.e. dot product) and $\nabla^{\mathbb{R}^{n+1}}$.

$$V[\langle Z, W \rangle] = V\left[\sum_i Z^i W^i\right] \qquad \text{by the definition of } \langle \cdot, \cdot \rangle$$

$$= \sum_i V[Z^i]W^i + Z^i V[W^i] \qquad \text{by the Leibniz rule}$$

$$= \langle \nabla^{\mathbb{R}^{n+1}}_{V} Z, W \rangle + \langle Z, \nabla^{\mathbb{R}^{n+1}}_{V} W \rangle \qquad \text{by the definition of } \nabla^{\mathbb{R}^{n+1}}$$

$$= \langle \nabla_V Z, W \rangle + \langle Z, \nabla_V W \rangle$$

because, since Z and W are *tangent* vector fields to M, all dot products with normal vectors $\langle U_j, W \rangle$ and $\langle Z, U_j \rangle$ are zero. Since $\nabla^{\mathbb{R}^{n+1}}$ and ∇ differ only by their normal components, the equality of the last two lines of the calculation above follows. Part (iii) is left to the reader in the exercise below and part (v) will be proved below after we discuss the bracket $[\cdot, \cdot]$. $\qquad\square$

EXERCISE 3.2. Prove the equality $\nabla_V fZ = V[f]\,Z + f\,\nabla_V Z$ in (iii) above. Hint: use (ii) as a guide.

Now let us discuss the *Lie bracket* of vector fields. If V and W are tangent vector fields on M, then we define

$$[V, W][f] = V[W[f]] - W[V[f]].$$

Here we are saying that $[V, W]$ acts on a function f as a vector field should. We will show that this actually defines a tangent vector field on M by writing $[V, W]$ as a linear combination of the ∂_i. Also, to avoid a surfeit of square brackets, we will abuse notation a bit and write $[V, W] = VW[f] - WV[f]$. First, write $V = \sum_i v^i \partial_i$ and $W = \sum_j w^j \partial_j$ and then compute

$$
WV[f] = W \sum_i v^i \partial_i[f]
$$

$$
= \sum_i W[v^i]\partial_i[f] + v^i W[\partial_i[f]]
$$

$$
= \sum_i \sum_j (w^j \partial_j[v^i]\partial_i[f] + v^i w^j \partial_j \partial_i[f])
$$

$$
= \sum_i \sum_j \left(w^j \frac{\partial v^i}{\partial u^j} \frac{\partial f}{\partial u^i} + v^i w^j \frac{\partial^2 f}{\partial u^i \partial u^j} \right)
$$

and similarly,

$$
VW[f] = \sum_i \sum_j \left(v^j \frac{\partial w^i}{\partial u^j} \frac{\partial f}{\partial u^i} + v^j w^i \frac{\partial^2 f}{\partial u^i \partial u^j} \right)
$$

so that

$$
(VW - WV)[f] = \sum_i \left(\sum_j \left(v^j \frac{\partial w^i}{\partial u^j} - w^j \frac{\partial v^i}{\partial u^j} \right) \right) \frac{\partial f}{\partial u^i}.
$$

In other words, we have

$$
[V, W] = \sum_i a^i \partial_i \qquad \text{where} \qquad a^i = \sum_j \left(v^j \frac{\partial w^i}{\partial u^j} - w^j \frac{\partial v^i}{\partial u^j} \right)
$$

and the bracket $[V, W]$ is then a tangent vector field on M.

EXERCISE 3.3. Apply the definition of bracket to prove the following properties.

(1) $[V, W] = -[W, V]$

(2) $[aV + bW, Z] = a[V, Z] + b[W, Z]$ where $a, b \in \mathbb{R}$.

(3) $[[V, W], Z] + [[W, Z], V] + [[Z, V], W] = 0$ (Jacobi Identity)

(4) $[fV, gW] = fg[V, W] + fV[g]W - gW[f]V$ for smooth functions f and g.

Now let us concentrate on proving (v) of Proposition 3.1. First, we will relate the covariant derivative and the bracket by a general formula which will eventually reduce to (v). To begin, note that $[\partial_i, \partial_j] = 0$ for all i, j. This follows because $\partial_i[f] = \partial f / \partial u^i$, so that

$$[\partial_i, \partial_j][f] = \partial_i \partial_j[f] - \partial_j \partial_i[f] = \frac{\partial^2 f}{\partial u^i \partial u^j} - \frac{\partial^2 f}{\partial u^j \partial u^i} = 0$$

since mixed partials are equal. Now, for $V = \sum_i v^i \partial_i$ and $W = \sum_j w^j \partial_j$, we have

$$\nabla_V W = \sum_i v^i \nabla_{\partial_i} \left(\sum_j w^j \partial_j \right)$$

$$= \sum_i \sum_j (v^j \partial_j w^i \partial_i + v^j w^i \nabla_{\partial_j} \partial_i)$$

after switching i and j. Similarly,

$$\nabla_W V = \sum_i \sum_j (w^j \partial_j v^i \partial_i + w^j v^i \nabla_{\partial_j} \partial_i).$$

Then we have

$$\nabla_V W - \nabla_W V = \sum_i \left(\sum_j \left(v^j \frac{\partial w^i}{\partial u^j} - w^j \frac{\partial v^i}{\partial u^j} \right) \right) \partial_i$$

$$+ \sum_i \sum_j \left(v^j w^i - w^j v^i \right) \nabla_{\partial_j} \partial_i$$

$$= [V, W] + \sum_i \sum_j \left(v^j w^i - w^j v^i \right) \nabla_{\partial_j} \partial_i.$$

To prove (v), we will show that the second term in this formula vanishes.

Lemma 3.2. $\nabla_{\partial_i} \partial_j - \nabla_{\partial_j} \partial_i = 0$.

Proof. From the general formula above applied to the $\mathbb{R}^{n+1}$ covariant derivative $\nabla^{\mathbb{R}^{n+1}}$ with $\partial_i = \sum_k a^k e_k$ and $\partial_j = \sum_l b^l e_l$, we obtain

$$\nabla^{\mathbb{R}^{n+1}}_{\partial_i} \partial_j - \nabla^{\mathbb{R}^{n+1}}_{\partial_j} \partial_i = [\partial_i, \partial_j] + \sum_{k,l} (a^l b^k - b^l a^k) \nabla^{\mathbb{R}^{n+1}}_{e_j} e_i = 0 + 0 = 0$$

since $[\partial_i, \partial_j] = 0$ by the discussion above and $\nabla^{\mathbb{R}^{n+1}}_{e_j} e_i = 0$ by definition of $\nabla^{\mathbb{R}^{n+1}}$. Then, since $\nabla^{\mathbb{R}^{n+1}}_{\partial_i} \partial_j = \nabla^{\mathbb{R}^{n+1}}_{\partial_j} \partial_i$, the projections $\nabla_{\partial_i} \partial_j$ and $\nabla_{\partial_j} \partial_i$ must be equal as well. $\square$

Proof of Proposition 3.1 (v). Consider the general formula

$$\nabla_V W - \nabla_W V = [V, W] + \sum_i \sum_j \left(v^j w^i - w^j v^i \right) \nabla_{\partial_j} \partial_i$$

and focus on the last term $\sum_i \sum_j \left(v^j w^i - w^j v^i \right) \nabla_{\partial_j} \partial_i$. In this term, the factor in parentheses is antisymmetric in i and j while, by Lemma 3.2, the factor $\nabla_{\partial_j} \partial_i$ is symmetric in i and j. Hence, summing over all i and j produces zero since a term and its negative always appear. Therefore,

$$\nabla_V W - \nabla_W V = [V, W].$$

$\square$

Suppose $M^n \subset \mathbb{R}^{n+1}$ is a hypersurface, so that there is only one (outward) unit normal vector field U on M. For tangent vector fields V and Z, we may write

$$\nabla_V^{\mathbb{R}^{n+1}} Z = \nabla_V Z + \langle \nabla_V^{\mathbb{R}^{n+1}} Z, U \rangle U$$

since $\nabla_V Z$ is the tangential projection of $\nabla_V^{\mathbb{R}^{n+1}} Z$. Again note that V and Z must be extended to an open set in $\mathbb{R}^{n+1}$ to make sense of $\nabla_V^{\mathbb{R}^{n+1}} Z$.

EXERCISE 3.4. Show that the coefficient of the normal component above is $\langle \nabla_V^{\mathbb{R}^{n+1}} Z, U \rangle$.

We may write the equation above as

$$\nabla_V Z = \nabla_V^{\mathbb{R}^{n+1}} Z - \langle \nabla_V^{\mathbb{R}^{n+1}} Z, U \rangle U$$
$$= \nabla_V^{\mathbb{R}^{n+1}} Z + \langle Z, \nabla_V^{\mathbb{R}^{n+1}} U \rangle U$$

since, by (iv), $V \langle Z, U \rangle = \langle \nabla_V^{\mathbb{R}^{n+1}} Z, U \rangle + \langle Z, \nabla_V^{\mathbb{R}^{n+1}} U \rangle$ and $\langle Z, U \rangle = 0$. Then by symmetry of $\langle \cdot, \cdot \rangle$,

$$\nabla_V Z = \nabla_V^{\mathbb{R}^{n+1}} Z - \langle S(V), Z \rangle U$$

where $S(V) = -\nabla_V^{\mathbb{R}^{n+1}} U$ is the *shape operator* of M in $\mathbb{R}^{n+1}$. The properties of Proposition 3.1 (especially (iv) and (v)) may be used to prove the following generalization of Theorem 2.3.1.

Theorem 3.3. *The shape operator is a symmetric linear transformation on the tangent space.*

Proof. S is clearly linear by the properties of $\nabla^{\mathbb{R}^{n+1}}$. Because $\langle U, U \rangle = 1$ is constant,

$$0 = V\langle U, U \rangle = \langle \nabla_V^{\mathbb{R}^{n+1}} U, U \rangle + \langle U, \nabla_V^{\mathbb{R}^{n+1}} U \rangle = 2\langle \nabla_V^{\mathbb{R}^{n+1}} U, U \rangle$$

and hence, $\langle \nabla_V^{\mathbb{R}^{n+1}} U, U \rangle = 0$. Thus, $\langle S(V), U \rangle = \langle -\nabla_V^{\mathbb{R}^{n+1}} U, U \rangle = 0$, so $S(V)(p) \in T_p M$ for all $p \in M$. Now we must show that the linear transformation S is symmetric. Throughout the following, we use the fact that $\langle U, Z \rangle = 0$ for any tangent vector field Z as well as Proposition 3.1 (iv) and (v).

$$
\begin{aligned}
\langle S(V), W \rangle &= \langle -\nabla_V^{\mathbb{R}^{n+1}} U, W \rangle \\
&= \langle U, \nabla_V^{\mathbb{R}^{n+1}} W \rangle && \text{by (iv) and } \langle U, W \rangle = 0 \\
&= \langle U, [V, W] + \nabla_W^{\mathbb{R}^{n+1}} V \rangle && \text{by (v)} \\
&= \langle U, [V, W] \rangle + \langle U, \nabla_W^{\mathbb{R}^{n+1}} V \rangle \\
&= 0 - \langle \nabla_W^{\mathbb{R}^{n+1}} U, V \rangle && \text{since } [V, W] \text{ is tangent and by (iv)} \\
&= \langle V, -\nabla_W^{\mathbb{R}^{n+1}} U \rangle \\
&= \langle V, S(W) \rangle.
\end{aligned}
$$

$\square$

Example 3.1: The sphere S^n. In coordinates $x^1, \dots, x^{n+1}$, the unit normal vector field for the n-sphere in $\mathbb{R}^{n+1}$ is

$$U = \frac{1}{R} \left(x^1, \dots, x^{n+1} \right).$$

The Euclidean covariant derivative is just the coordinatewise directional derivative, so for a tangent vector field $V = \left(V^1, \dots, V^{n+1} \right)$,

$$S(V)(p) = -\nabla_{V(p)}^{\mathbb{R}^{n+1}} U$$

$$= -\frac{1}{R} \left(V(p)[x^1], \dots, V(p)[x^{n+1}] \right)$$

$$= -\frac{1}{R} \left(V^1(p), \dots, V^{n+1}(p) \right)$$

$$= -\frac{V(p)}{R}$$

where the third line follows since the x^i are coordinate functions and we have

$$V[x^i] = \sum_j \frac{\partial x^i}{\partial x^j} V^j = V^j \delta_i^j = V^i.$$

Hence, the shape operator of the n-sphere is completely analogous to that of the 2-sphere — it is simply scalar multiplication by the negative reciprocal of the radius of the sphere.

While the shape operator for hypersurfaces plays a role similar to that for surfaces in $\mathbb{R}^3$, it must be realized that it can never assume as important a position in general for higher dimensional manifolds. This is simply because hypersurfaces are rather rare among manifolds in higher dimensions whereas surfaces in $\mathbb{R}^3$ are, by definition, hypersurfaces. For manifolds in $\mathbb{R}^{n+1}$ which are not hypersurfaces, we may replace the shape operator, or more precisely $\langle S(V), Z \rangle U$, by the more general *second fundamental form*. When we write, for tangent vector fields V and Z,

$$\nabla_V Z = \text{proj}_{T_p M} \nabla_V^{\mathbb{R}^{n+1}} \bar{Z}$$

we are ignoring the normal component of $\nabla^{\mathbb{R}^{n+1}}$. Let us now put it back in by writing superscripts T and N to denote the tangential and normal projections respectively;

$$\nabla_V^{\mathbb{R}^{n+1}} Z = (\nabla_V^{\mathbb{R}^{n+1}} Z)^T + (\nabla_V^{\mathbb{R}^{n+1}} Z)^N$$
$$\stackrel{\text{def}}{=} \nabla_V Z + B(V, Z).$$

$B(V, Z) = (\nabla_V^{\mathbb{R}^{n+1}} Z)^N$ is called the *second fundamental form* of $M^k \subset \mathbb{R}^{n+1}$.

Proposition 3.4. $B(\cdot, \cdot)$ *satisfies the following properties.*

(1) $B(fV, Z) = f B(V, Z)$
(2) $B(V, fZ) = f B(V, Z)$
(3) $B(V, Z) = B(Z, V)$

The first two properties are included in the statement that B is bilinear; that is, linear in each variable separately. The last property is what is meant by saying that B is symmetric.

Proof. We shall prove (2) and leave (1) and (3) to the following exercise. By definition,

$$B(V, fZ) = \nabla_V^{\mathbb{R}^{n+1}} fZ - \nabla_V fZ$$
$$= V[f] Z + f \nabla_V^{\mathbb{R}^{n+1}} Z - V[f] Z - f \nabla_V Z$$

by property (iii) of Proposition 3.1. Hence,

$$= f\nabla_V^{\mathbb{R}^{n+1}} Z - f\nabla_V Z$$
$$= f\left(\nabla_V^{\mathbb{R}^{n+1}} Z - \nabla_V Z\right)$$
$$= fB(V, Z).$$

$\square$

EXERCISE 3.5. Show that properties (1) and (3) hold above. Hints: for (1), use property (ii) of Proposition 3.1. For (3), use property (v) of Proposition 3.1 and the fact that $[V, Z]$ is a tangent vector field, so $[V, Z]^N = 0$.

Now, by Proposition 3.4, at each point $p \in M$, $B(p)$ is a symmetric bilinear mapping $T_pM \times T_pM \to N_pM$, where $N_pM = \langle U_1, \dots, U_{n+1-k}\rangle$ is the *normal space* to M in $\mathbb{R}^{n+1}$. From B we can define the *mean curvature vector field* to be

$$H(p) = \sum_{s=1}^{k}(\nabla_{\partial_s}^{\mathbb{R}^{n+1}} \partial_s)^N$$
$$= \sum_{s=1}^{k} B(p)(\partial_s, \partial_s)$$

where $\partial_1, \dots, \partial_k$ are assumed to be orthonormal at p.

EXERCISE 3.6. Show that this definition of H, in the case of a two dimensional patch $\mathbf{x}(u, v)$ in $\mathbb{R}^3$, is just twice our old notion of mean curvature.

We will talk more about the second fundamental form after we discuss Christoffel symbols in the next section. A good general reference is [DoC2]. It is important to note that what we have done in this section is not the most general definition of a covariant derivative. We have already said that manifolds may be defined without reference to an ambient Euclidean space, but it is also the case that covariant derivatives ∇ (or *connections* as they are also called) may be defined without reference to an ambient Euclidean covariant derivative $\nabla^{\mathbb{R}^{n+1}}$. The essential properties of an abstract covariant derivative are properties (i), (ii) and (iii) of Proposition 3.1, while properties (iv) and (v) of that result endow an abstract covariant derivative with the special names *Levi-Civita connection* or *Riemannian connection*. Every manifold (with given metric) has a unique Riemannian connection and, for our manifolds $M^k \subset \mathbb{R}^{n+1}$ with the induced metric, the connection we have defined is plainly that Riemannian connection.

9.4 Christoffel Symbols

In Chapter 3.4, we introduced Christoffel symbols as the coefficients of a basis expansion for the second partials $\mathbf{x}_{uu}$ and $\mathbf{x}_{vv}$ in terms of $\mathbf{x}_u$, $\mathbf{x}_v$ and U. Recall that we assumed that the basis had $F = \mathbf{x}_u \cdot \mathbf{x}_v = 0$ so that we obtained

$$\mathbf{x}_{uu} = \Gamma^u_{uu}\mathbf{x}_u + \Gamma^v_{uu}\mathbf{x}_v + lU$$
$$\mathbf{x}_{uv} = \Gamma^u_{uv}\mathbf{x}_u + \Gamma^v_{uv}\mathbf{x}_v + mU$$
$$\mathbf{x}_{vv} = \Gamma^u_{vv}\mathbf{x}_u + \Gamma^v_{vv}\mathbf{x}_v + nU.$$

We were able to identify the Christoffel symbols in terms of the metric E and G (with $F = 0$ remember);

$$\mathbf{x}_{uu} = \frac{E_u}{2E}\mathbf{x}_u - \frac{E_v}{2G}\mathbf{x}_v + lU$$
$$\mathbf{x}_{uv} = \frac{E_v}{2E}\mathbf{x}_u + \frac{G_u}{2G}\mathbf{x}_v + mU$$
$$\mathbf{x}_{vv} = -\frac{G_u}{2E}\mathbf{x}_u + \frac{G_v}{2G}\mathbf{x}_v + nU.$$

Now, $\mathbf{x}_{uu}$ is nothing more than $\nabla^{\mathbb{R}^{n+1}}_{\mathbf{x}_u}\mathbf{x}_u$ and similarly for $\mathbf{x}_{uv}$ and $\mathbf{x}_{vv}$. Also, removing the U-components lU, mU and nU from the expressions above then produces $\nabla_{\mathbf{x}_u}\mathbf{x}_u$, $\nabla_{\mathbf{x}_u}\mathbf{x}_v$ and $\nabla_{\mathbf{x}_v}\mathbf{x}_v$ respectively. To generalize this to k dimensions, we simply write

$$\nabla^{\mathbb{R}^{n+1}}_{\boldsymbol{\partial}_i}\boldsymbol{\partial}_j = \sum_{s=1}^{k}\Gamma^s_{ij}\boldsymbol{\partial}_s + \sum_{t}b^t_{ij}U_t$$

and, taking tangential components,

$$\nabla_{\boldsymbol{\partial}_i}\boldsymbol{\partial}_j = \sum_{s=1}^{k}\Gamma^s_{ij}\boldsymbol{\partial}_s.$$

EXERCISE 4.1. Use Lemma 3.2 to prove that $\Gamma^s_{ij} = \Gamma^s_{ji}$ for all i, j, s.

Just as Christoffel symbols in the two dimensional case may be expressed in terms of the metric, so too can higher dimensional Christoffel symbols be expressed in terms of the metric (in traditional higher dimensional notation)

$$g_{ij} = \langle \boldsymbol{\partial}_i, \boldsymbol{\partial}_j \rangle \qquad \text{with} \qquad g = \det(g_{ij})$$

where (g_{ij}) is the matrix of metric coefficients. Of course, we have $g_{ij} = g_{ji}$ so the matrix is symmetric as well as being invertible. We also have need of the inverse of the matrix (g_{ij}) which is denoted by (g^{ij}). This inverse will allow us to isolate certain quantities as exemplified by the proof below. In particular, note that we always have the relation

$$\sum_j g^{mj} g_{js} = \delta_s^m$$

where δ_s^m is the Kronecker delta.

Proposition 4.1. *The Christoffel symbols are determined by the metric as*

$$\Gamma_{is}^m = \frac{1}{2} \sum_j g^{mj} \left(\partial_i g_{sj} + \partial_s g_{ij} - \partial_j g_{si} \right).$$

Proof. Property (iv) of Proposition 3.1 gives

$$\begin{aligned}
\partial_s g_{ij} &= \partial_s \langle \partial_i, \partial_j \rangle \\
&= \langle \nabla_{\partial_s} \partial_i, \partial_j \rangle + \langle \partial_i, \nabla_{\partial_s} \partial_j \rangle \\
&= \langle \sum_l \Gamma_{si}^l \partial_l, \partial_j \rangle + \langle \partial_i, \sum_l \Gamma_{sj}^l \partial_l \rangle \\
&= \sum_l \left(\Gamma_{si}^l g_{lj} + \Gamma_{sj}^l g_{il} \right).
\end{aligned}$$

Similarly,

$$\partial_i g_{js} = \sum_l \left(\Gamma_{ij}^l g_{ls} + \Gamma_{is}^l g_{jl} \right) \qquad \text{and} \qquad \partial_j g_{si} = \sum_l \left(\Gamma_{js}^l g_{li} + \Gamma_{ji}^l g_{sl} \right)$$

with

$$\partial_s g_{ij} + \partial_i g_{js} - \partial_j g_{si} = \sum_l \Gamma_{is}^l g_{jl}.$$

Now, multiplying by the mth row of the inverse metric matrix, we obtain

$$\frac{1}{2} \sum_j g^{mj} \left(\partial_i g_{sj} + \partial_s g_{ij} - \partial_j g_{si} \right) = \sum_l \Gamma_{is}^l \sum_j g^{mj} g_{jl} = \Gamma_{is}^m$$

since $\sum_j g^{mj} g_{jl} = \delta_l^m$. $\square$

Corollary 4.2. *For $M = \mathbb{R}^{n+1}$, the Christoffel symbols are all zero.*

Proof. The metric coefficients g_{ij} are all constants, so their partials are zero. By Proposition 4.1, the Christoffel symbols vanish. $\square$

Example 4.1. Consider a two dimensional patch $\mathbf{x}(u,v) = \mathbf{x}(u^1, u^2)$ with $F = \mathbf{x}_u \cdot \mathbf{x}_v = 0$. The metric matrix and its inverse are then

$$(g_{ij}) = \begin{pmatrix} E & 0 \\ 0 & G \end{pmatrix} \qquad (g^{ij}) = \begin{pmatrix} 1/E & 0 \\ 0 & 1/G \end{pmatrix}$$

with determinant $g = \det(g_{ij}) = EG$. Then, for example,

$$\Gamma_{uu}^u = \Gamma_{11}^1 = \frac{1}{2}\frac{1}{E}(E_u + E_u - E_u) + \frac{1}{2}\cdot 0 = \frac{E_u}{2E}$$

$$\Gamma_{uv}^u = \Gamma_{12}^1 = \frac{1}{2}\frac{1}{E}(0 + E_v - 0) + \frac{1}{2}\cdot 0 = \frac{E_v}{2E}.$$

EXERCISE 4.2. Show that $\Gamma_{vv}^u = \Gamma_{22}^1 = -G_u/2E$, $\Gamma_{uu}^v = \Gamma_{11}^2 = -E_v/2G$, $\Gamma_{uv}^v = \Gamma_{12}^2 = G_u/2G$ and $\Gamma_{vv}^v = \Gamma_{22}^2 = G_v/2G$.

Since the formulas for the Γ's above occur in the geodesic equations in two dimensions, it should be no surprise that they occur as well in higher dimensions. More generally, a tangent vector field V is said to be *parallel along a curve* $\alpha: I \to M^k = \mathbf{x}(u^1, \dots, u^k)$ if $\nabla_{\alpha'}V = 0$ at every point on the curve. This condition may be translated into local coordinates as follows. Let $V = \sum_j v^j \boldsymbol{\partial}_j$.

$$\nabla_{\alpha'}V = \sum_i \frac{du^i}{dt}\nabla_{\boldsymbol{\partial}_i}V$$

$$= \sum_i \frac{du^i}{dt}\left(\sum_j \frac{\partial v^j}{\partial u^i}\boldsymbol{\partial}_j + \sum_j v^j \sum_s \Gamma_{ij}^s \boldsymbol{\partial}_s\right)$$

$$= \sum_i \sum_s \frac{du^i}{dt}\frac{\partial v^s}{\partial u^i}\boldsymbol{\partial}_s + \sum_i \frac{du^i}{dt}\sum_j \sum_s v^j \sum_s \Gamma_{ij}^s \boldsymbol{\partial}_s$$

$$= \sum_s \left(\sum_i \left(\frac{du^i}{dt}\frac{\partial v^s}{\partial u^i} + \frac{du^i}{dt}\sum_j v^j \Gamma_{ij}^s\right)\right)\boldsymbol{\partial}_s.$$

This expression is zero exactly when each component is zero. Hence, for each s we have

$$\sum_i \left(\frac{du^i}{dt}\frac{\partial v^s}{\partial u^i} + \frac{du^i}{dt}\sum_j v^j \Gamma_{ij}^s\right) = 0.$$

Now, the first term $\sum_i \left(du^i/dt \cdot \partial v^s \partial u^i \right)$ is precisely dv^s/dt by the chain rule. Therefore,

Proposition 4.3. *V is parallel if and only if, for each s,*

$$\frac{dv^s}{dt} + \sum_i \sum_j \Gamma^s_{ij} \frac{du^i}{dt} v^j = 0.$$

Corollary 4.4. *A curve α is a geodesic on M^k if and only if*

$$\frac{d^2 u^s}{dt^2} + \sum_i \sum_j \Gamma^s_{ij} \frac{du^i}{dt} \frac{du^j}{dt} = 0.$$

Proof. We have $\alpha(t) = \mathbf{x}(u^1(t), \dots, u^k(t))$, so that $\alpha' = \sum_j (du^j/dt)\partial_j$. Thus, the geodesic condition $\nabla_{\alpha'} \alpha' = 0$ translates into the equation above by taking the formula of Proposition 4.3 and substituting $du^j/dt = v^j$. $\qquad\square$

Parallel vector fields always exist and are unique because the condition of Proposition 4.3 is a linear first order differential equation. Here however, we cannot determine one specific angle function which describes the parallel vector rotation. The reason for this is apparent. It is only in the plane that rotations are determined by angles. In higher dimensions, it is the *special orthogonal group* $SO(n)$, consisting of matrices A with $\det(A) = 1$ and $AA^t = I$, which describes rotations.

EXERCISE 4.3. Define *parallel translation* as follows. Given V_0 at $p \in M$ and a curve α on M with $\alpha(0) = p$. Let V be the unique parallel vector field along α with $V(0) = V_0$. Then $V(t)$ is the parallel translate of V_0. Show that this association is a linear transformation from $T_p M$ to $T_{\alpha(t)} M$. Further, this linear transformation is an isometry. That is, for parallel vector fields V and W along α, the inner product $\langle V, W \rangle$ is constant along α. It may also be shown that parallel translation preserves orientation and it is known that orientation preserving isometries are in $SO(n)$. Hint: to show an isometry, use property (iv) of Proposition 3.1.

In Chapter 8 we showed that geodesics arise as extrema for the energy integral $\int T\, dt$ where the kinetic energy is given by $T = 1/2 \, |\alpha'|^2$. In higher dimensions $\alpha' = \sum_i (du^i/dt)\, \partial_i$ and, since we do not assume $\partial_i \cdot \partial_j = 0$, we then have $|\alpha'|^2 = \sum_{i,j} g_{ij} \dot{u}^i \dot{u}^j$. The appropriate energy integral we wish to extremize is then

$$\int L(t, u^1(t), \dots, u^k(t), \dot{u}^1(t), \dots, \dot{u}^k(t))\, dt \overset{\text{def}}{=} \int \sum_{i,j} g_{ij} \dot{u}^i \dot{u}^j \, dt.$$

Here we have dropped the superfluous $1/2$. Generalizing Exercise 8.1.2 to k functions, for each s we obtain an Euler-Lagrange equation

$$\frac{d}{dt}\left(\frac{\partial L}{\partial \dot{u}^s}\right) - \frac{\partial L}{\partial u^s} = 0.$$

We can compute each piece directly as follows.

$$\frac{\partial L}{\partial u^s} = \sum_{i,j} \frac{\partial g_{ij}}{\partial u^s} \dot{u}^i \dot{u}^j \qquad \text{and} \qquad \frac{\partial L}{\partial \dot{u}^s} = \sum_j g_{sj} \dot{u}^j + \sum_i g_{is} \dot{u}^i$$

with

$$\frac{d}{dt}\left(\frac{\partial L}{\partial \dot{u}^s}\right) = \sum_j \left(g_{sj} \ddot{u}^j + \sum_i \frac{\partial g_{sj}}{\partial u^i} \dot{u}^i \dot{u}^j \right) + \sum_i \left(g_{is} \ddot{u}^i + \sum_i \frac{\partial g_{is}}{\partial u^j} \dot{u}^i \dot{u}^j \right)$$

$$= 2 \sum_i g_{is} \ddot{u}^i + \sum_{i,j} \left(\frac{\partial g_{sj}}{\partial u^i} + \frac{\partial g_{is}}{\partial u^j} \right) \dot{u}^i \dot{u}^j.$$

Hence, the Euler-Lagrange equation becomes

$$\frac{d}{dt}\left(\frac{\partial L}{\partial \dot{u}^s}\right) - \frac{\partial L}{\partial u^s} = \sum_i g_{is} \ddot{u}^i + \frac{1}{2} \sum_{i,j} \left(\frac{\partial g_{sj}}{\partial u^i} + \frac{\partial g_{is}}{\partial u^j} - \frac{\partial g_{ij}}{\partial u^s} \right) \dot{u}^i \dot{u}^j = 0.$$

Now here is where we can use the identity $\sum_s g^{ms} g_{si} = \delta_i^m$ to isolate the second derivative $\ddot{u}^m = \sum_i \sum_s g^{ms} g_{si} \ddot{u}^i$ above. Sum the equation above by $\sum_s g^{ms}$ to obtain

$$\sum_s g^{ms} \left(\sum_i g_{is} \ddot{u}^i + \frac{1}{2} \sum_{i,j} \left(\frac{\partial g_{sj}}{\partial u^i} + \frac{\partial g_{is}}{\partial u^j} - \frac{\partial g_{ij}}{\partial u^s} \right) \dot{u}^i \dot{u}^j \right) = 0$$

$$\sum_s g^{ms} \sum_i g_{is} \ddot{u}^i + \sum_{i,j} \frac{1}{2} \sum_s g^{ms} \left(\frac{\partial g_{sj}}{\partial u^i} + \frac{\partial g_{is}}{\partial u^j} - \frac{\partial g_{ij}}{\partial u^s} \right) \dot{u}^i \dot{u}^j = 0,$$

which then reduces to

$$\ddot{u}^m + \sum_{ij} \Gamma_{ij}^m \dot{u}^i \dot{u}^j = 0.$$

These are, of course, the geodesic equations. Therefore, in higher dimensions as well, the geodesic equations arise from a variational principle.

EXERCISE 4.4 Show that, for an orthogonal two dimensional patch $\mathbf{x}(u, v)$, the geodesic equations above are those of Chapter 5.

We now wish to show that certain *normal coordinates* may be chosen for which the covariant derivative is quite well behaved. This choice of coordinates will simplify our discussion of curvature in the next section greatly.

Theorem 4.5. *For a manifold M^k and $p \in M$, a patch $\mathbf{x}(u^1, \dots, u^k)$ may be chosen about p with $\mathbf{x}(0, \dots, 0) = p$ so that*

$$g_{ij}(p) = \langle \boldsymbol{\partial}_i(p), \boldsymbol{\partial}_j(p) \rangle = \delta_i^j \qquad \text{and} \qquad \nabla_{\boldsymbol{\partial}_i} \boldsymbol{\partial}_j(p) = 0.$$

These coordinates $u^1, \dots, u^k$ are called normal coordinates.

Sketch of Proof. The details of a proof of this result require knowledge of the exponential map. Since we have not talked about this subject, we can only outline the ideas involved here. Take the tangent space at p and, for any vector $\mathbf{v} \in T_p M$, map the line through the origin of $T_p M$ in the direction $\mathbf{v}$ to the unit speed geodesic $\alpha_{\mathbf{v}}$ through p in the $\mathbf{v}$ direction,

$$t\mathbf{v} \mapsto \alpha_{\mathbf{v}}(t).$$

For simplicity, assume M is complete, so that geodesics run forever. This definition then defines a k-patch from $T_p M$ to M. Choose an orthogonal coordinate system $u^1, \dots, u^k$ for $T_p M$ as coordinates for the patch $\mathbf{x}$. By definition, $g_{ij} = \delta_i^j$ at p since the coordinates are orthogonal. Also, for a tangent vector $\mathbf{v} = (v^1, \dots, v^k)$, we have,

$$\mathbf{x}(u^1(t), \dots, u^k(t)) = \alpha_{\mathbf{v}}(t) = \mathbf{x}(tv^1, \dots, tv^k)$$

so that $\ddot{u}^i = d^2(tv^i)/dt^2 = 0$ for all i. The geodesic equations, which of course hold on α, become

$$\sum_{i,j} \Gamma_{ij}^m(\alpha_{\mathbf{v}}(t))\dot{u}^i\dot{u}^j = \sum_{i,j} \Gamma_{ij}^m(\alpha_{\mathbf{v}}(t))v^i v^j = 0,$$

and this is true for all $\mathbf{v} \in T_p M$. But this can only happen if all the Γ_{ij}^m are zero. Hence,

$$\nabla_{\boldsymbol{\partial}_i} \boldsymbol{\partial}_j(p) = \sum_m \Gamma_{ij}^m(p)\boldsymbol{\partial}_m = 0.$$

$\square$

With the preliminaries above, we can end this section by giving a brief (and somewhat simplistic) sketch of the relationship between the higher dimensional mean curvature vector field and higher dimensional "area" minimization. Note how the use of normal coordinates makes the argument much simpler than it otherwise might be. The reader is recommended to [Law, Theorem 1.4] and [GHL, Theorem 5.19] for rigorous details.

Suppose we have a manifold M (which we assume is rather local in the sense of being a single patch, say) and we vary it as in Theorem 4.3.2. Namely, we take $M_t = M + tU$, where we assume U is a vector field which is always normal to M. For a set of coordinates $(u^1, \dots, u^k)$, the *area element* is given by

$$dA = \sqrt{g}\, du^1 \dots du^k$$

where $g = \det(g_{ij})$ is the determinant of the metric coefficient matrix. (In two dimensions recall that $dA = \sqrt{EG - F^2}\, du\, dv$.) Similarly, for $g_t = \det(g_{t\,ij})$ on M_t, we have area element

$$dA_t = \sqrt{g_t}\, du^1 \dots du^k$$

$$= \frac{\sqrt{g_t}\, dA}{\sqrt{g}}$$

$$= \frac{\sqrt{g_t}}{\sqrt{g}}\, dA$$

and the higher dimensional area is then

$$A(t) = \int_{M_t} \frac{\sqrt{g_t}}{\sqrt{g}}\, dA.$$

As in Chapter 4, we want to find critical points for this area functional. Note that the integral above depends on the point about which we take our coordinate system, so let us fix $p \in M$ and let us take a normal coordinate system about p. Therefore, for these normal coordinates, *at* p, the tangent vector fields $\partial_1, \dots, \partial_k$ are orthonormal and all the covariant derivatives $\nabla_{\partial_i} \partial_j$ vanish. We can consider the normal vector field U as another local coordinate, so we have, for all i and j,

$$[\partial_i, \partial_j] = 0 \qquad \text{and} \qquad [U, \partial_i] = 0.$$

Now we wish to calculate

$$\frac{dA(t)}{dt}\bigg|_{t=0} = \int \frac{d}{dt}\left(\frac{\sqrt{g_t}}{\sqrt{g}}\right)\bigg|_{t=0} dA.$$

Because the coordinate vector fields are orthonormal at p, we have $g = 1$ there. Hence,

$$\frac{d}{dt}\left(\frac{\sqrt{g_t}}{\sqrt{g}}\right)\bigg|_{t=0} = \frac{1}{2}\frac{dg_t/dt}{\sqrt{g}}\bigg|_{t=0}$$

$$= \frac{1}{2}\frac{dg_t}{dt}\bigg|_{t=0}.$$

EXERCISE 4.5. Let $Q(t) = (q_{ij}(t))$ be a matrix whose entries depend on t and for which $Q(0) = I$, the identity matrix. Show that

$$\frac{d}{dt} \det Q(t)\Big|_{t=0} = \operatorname{tr}\left(\frac{dq_{ij}(0)}{dt}\right).$$

Hints: write out the definition of the determinant of $Q(t)$,

$$\det Q = \sum_{\sigma \in S_k} (-1)^{\operatorname{sgn}\sigma} q_{1\sigma(1)} \cdots q_{k\sigma(k)}$$

where σ is a permutation in the permutation group on k letters S_k, and use the product rule to find dQ/dt. Then evaluate at $t = 0$ to make all terms in the determinant zero except for those having each factor as $q_{rr}(0) = 1$ together with a single $dq_{ii}(0)/dt$. Thus, the determinant sum reduces to the sum of $dq_{ii}(0)/dt$'s and this is $\operatorname{tr} Q'(0)$.

Since $g = \det(g_{ij})$, the exercise may be applied to the result above to produce

$$\frac{d}{dt}\left(\frac{\sqrt{g_t}}{\sqrt{g}}\right)\Big|_{t=0} = \frac{1}{2} \sum_{i=1}^{k} \frac{\partial g_{t\,ii}}{\partial t}\Big|_{t=0}$$

$$= \frac{1}{2} \sum_{i=1}^{k} \frac{\partial}{\partial t}\langle \tilde{\partial}_i, \tilde{\partial}_i \rangle\Big|_{t=0}$$

where $\tilde{\partial}_i$ is the vector field at level t determined by the local diffeomorphism $m \mapsto m + tU$. Then,

$$\frac{d}{dt}\left(\frac{\sqrt{g_t}}{\sqrt{g}}\right)\Big|_{t=0} = \frac{1}{2} \sum_{i=1}^{k} U\langle \tilde{\partial}_i, \tilde{\partial}_i \rangle\Big|_{t=0}$$

since U *is* the t-direction. By property (iv) of Proposition 3.1, we then have

$$\frac{d}{dt}\left(\frac{\sqrt{g_t}}{\sqrt{g}}\right)\Big|_{t=0} = \frac{1}{2} \sum_{i=1}^{k} \langle \nabla_U^{\mathbb{R}^{n+1}} \tilde{\partial}_i, \tilde{\partial}_i \rangle\Big|_{t=0} + \langle \tilde{\partial}_i, \nabla_U^{\mathbb{R}^{n+1}} \tilde{\partial}_i \rangle\Big|_{t=0}$$

$$= \sum_{i=1}^{k} \langle \nabla_U^{\mathbb{R}^{n+1}} \tilde{\partial}_i, \tilde{\partial}_i \rangle\Big|_{t=0}$$

$$= \sum_{i=1}^{k} \langle \nabla_{\tilde{\partial}_i}^{\mathbb{R}^{n+1}} U, \tilde{\partial}_i \rangle\Big|_{t=0}$$

by (v) of Proposition 3.1 and the vanishing of the brackets mentioned above. Hence, again by (iv),

$$\frac{d}{dt}\left(\frac{\sqrt{g_t}}{\sqrt{g}}\right)\bigg|_{t=0} = \sum_{i=1}^{k} \tilde{\partial}_i \langle U, \tilde{\partial}_i \rangle \big|_{t=0} - \langle U, \nabla_{\tilde{\partial}_i}^{\mathbb{R}^{n+1}} \tilde{\partial}_i \rangle \big|_{t=0}$$

$$= \sum_{i=1}^{k} \partial_i \langle U, \partial_i \rangle - \langle U, \nabla_{\partial_i} \partial_i \rangle - \langle U, \left(\nabla_{\partial_i}^{\mathbb{R}^{n+1}} \partial_i\right)^N \rangle$$

since $\nabla^{\mathbb{R}^{n+1}} = \nabla + (\nabla^{\mathbb{R}^{n+1}})^N$ and, at $t = 0$, we are on M. Also, for $t = 0$, U and the ∂_i are orthogonal and the coordinates are normal, so

$$\frac{d}{dt}\left(\frac{\sqrt{g_t}}{\sqrt{g}}\right)\bigg|_{t=0} = \sum_{i=1}^{k} -\langle U, \left(\nabla_{\partial_i}^{\mathbb{R}^{n+1}} \partial_i\right)^N \rangle$$

$$= -\langle U, \sum_{i=1}^{k} \left(\nabla_{\partial_i}^{\mathbb{R}^{n+1}} \partial_i\right)^N \rangle$$

$$= -\langle U, H \rangle$$

where H is the mean curvature vector field of M in $\mathbb{R}^{n+1}$. After integration, we have the

Theorem 4.6. *The derivative of the area functional at zero is given by*

$$\frac{dA(t)}{dt}\bigg|_{t=0} = \int_M -\langle U, H \rangle \, dA$$

where H is the mean curvature vector field and U is the normal variational direction.

If M (i.e. $t = 0$) is a critical point for all variations, then we must have $H = 0$. This follows because any nonzero part of H would produce a nonzero $\langle U, H \rangle$ for some U and this then could be localized by a bump function to produce a nonzero integral, contradicting the assumption that M is a critical point of $A(t)$. Thus, we have the higher dimensional version of Theorem 4.3.2,

Corollary 4.7. *If M is least area, then the mean curvature vector field is zero.*

9.5 Curvatures

In Chapter 3.4 we developed a formula for Gauss curvature which depended only on the metric. This led of course to Gauss's Theorem Egregium. If we look closely at how the formula was derived, we see that we used the fact that $\mathbf{x}_{uuv} - \mathbf{x}_{uvu} = 0$. Written in our "covariant" notation (and noting that the covariant derivative with respect to a parameter curve is simply partial differentiation), we have

$$\nabla_{\mathbf{x}_v}^{\mathbb{R}^{n+1}} \nabla_{\mathbf{x}_u}^{\mathbb{R}^{n+1}} \mathbf{x}_u - \nabla_{\mathbf{x}_u}^{\mathbb{R}^{n+1}} \nabla_{\mathbf{x}_v}^{\mathbb{R}^{n+1}} \mathbf{x}_u = 0.$$

Furthermore, the derivation also reveals that we made essential use of the basis descriptions of U_u and U_v, where U is the unit normal of the surface. In fact, it was these terms which provided the $(ln - m^2)/EG = K$ in the formula. Now, however, we must use M^k's covariant derivative ∇ without the benefit of a unit normal. Therefore, we should not expect the corresponding quantity

$$\nabla_{\partial_j} \nabla_{\partial_i} \partial_s - \nabla_{\partial_i} \nabla_{\partial_j} \partial_s$$

to be zero. Indeed, analogous to the derivation of the two dimensional formula (but without the ingredients U_u and U_v), we may define a vector field $R(\partial_i, \partial_j)\partial_s = \sum_l R_{ijs}^l \partial_l$ by

$$R(\partial_i, \partial_j)\partial_s = \nabla_{\partial_j} \nabla_{\partial_i} \partial_s - \nabla_{\partial_i} \nabla_{\partial_j} \partial_s.$$

The object $R(\cdot, \cdot)$ is called the *Riemann curvature*. We want R to possess certain qualities. Namely, we want R to be bilinear with respect to smooth functions. By this we mean that $R(fX, Y) = fR(X, Y)$, $R(X + Z, Y) = R(X, Y) + R(Z, Y)$, $R(X, fY) = fR(X, Y)$ and $R(X, Y + Z) = R(X, Y) + R(X, Z)$ for any smooth function $f: M \to \mathbb{R}$. But, as the reader is asked to check in the following exercise, as yet, this is *not the case*! Moreover, we want $R(\cdot, \cdot)$ to be a linear map on vector fields as well.

EXERCISE 5.1. Use the properties of covariant derivatives to show that

(1) $$R(\partial_i, \partial_j)f\partial_s = fR(\partial_i, \partial_j)\partial_s.$$

(2) $$R(f\partial_i, \partial_j)\partial_s = \partial_j f \nabla_{\partial_i} \partial_s + fR(\partial_i, \partial_j)\partial_s$$
$$= -\nabla_{[f\partial_i, \partial_j]}\partial_s + fR(\partial_i, \partial_j)\partial_s.$$

where the last line follows from Proposition 3.1 (ii) and Exercise 3.3 (4). Hence, R is not linear in the first (or second) variable.

The last formula of the exercise gives a hint as to how to redefine R so as to achieve bilinearity. Namely, for tangent vector fields X, Y and Z, we have

Definition 5.1. The *Riemann curvature* is defined to be

$$R(X,Y)Z = \nabla_{[X,Y]}Z + \nabla_Y\nabla_X Z - \nabla_X\nabla_Y Z.$$

EXERCISE 5.2. Show that

$$R(\partial_i, \partial_j)\partial_s = \sum_l R^l_{ijs}\partial_l$$

$$= \sum_l \left(\frac{\partial\Gamma^l_{is}}{\partial u^j} - \frac{\partial\Gamma^l_{js}}{\partial u^i} + \sum_m \Gamma^m_{is}\Gamma^l_{jm} - \Gamma^m_{js}\Gamma^l_{im} \right)\partial_l.$$

Since, as we showed previously, all Christoffel symbols vanish in $\mathbb{R}^{n+1}$, the expression above says that the Riemann curvature of $\mathbb{R}^{n+1}$ is zero everywhere. We say that $\mathbb{R}^{n+1}$ is **flat**. Further, show that

$$R_{ijsr} \overset{\text{def}}{=} \sum_l R^l_{ijs}g_{lr}$$

$$= \sum_l \left(\frac{\partial\Gamma^l_{is}}{\partial u^j} - \frac{\partial\Gamma^l_{js}}{\partial u^i} + \sum_m \Gamma^m_{is}\Gamma^l_{jm} - \Gamma^m_{js}\Gamma^l_{im} \right)g_{lr}$$

and calculate R_{1212} for a two dimensional patch $\mathbf{x}(u^1, u^2) = \mathbf{x}(u, v)$ with $g_{11} = E$, $g_{12} = g_{21} = 0$ and $g_{22} = G$. Show that

$$R_{1212} = EG\left[\frac{E_u G_u}{4E^2 G} - \frac{1}{E}\left(\frac{E_v}{2G}\right)_v - \frac{E_v G_v}{4EG^2} + \frac{E_v E_v}{4E^2 G} - \frac{1}{E}\left(\frac{G_u}{2G}\right)_u - \frac{G_u G_u}{4EG^2} \right]$$

and compare this with the formula above Exercise 3.4.3 to see that

$$K = \frac{R_{1212}}{g}$$

where K is Gauss curvature and $g = \det(g_{ij}) = EG$. Thus the Riemann curvature generalizes the Gaussian curvature.

EXERCISE 5.3. Suppose $M^n \subset \mathbb{R}^{n+1}$ is a hypersurface with unit normal U and shape operator S. Show that, for tangent vector fields X, Y and Z,

$$R(X,Y)Z = \langle S(X), Z\rangle\, S(Y) - \langle S(Y), Z\rangle\, S(X).$$

Hints: (1) write

$$R(X,Y) = (\nabla^{\mathbb{R}^{n+1}}_{[X,Y]} - S_{[X,Y]}) + (\nabla^{\mathbb{R}^{n+1}}_Y - S_Y)(\nabla^{\mathbb{R}^{n+1}}_X - S_X)$$
$$- (\nabla^{\mathbb{R}^{n+1}}_X - S_X)(\nabla^{\mathbb{R}^{n+1}}_Y - S_Y)$$

where $\mathcal{S}_Y(Z) = \langle S(Y), Z \rangle$. (2) Use the facts that

$$\nabla^{\mathbb{R}^{n+1}}_{[X,Y]} + \nabla^{\mathbb{R}^{n+1}}_Y \nabla^{\mathbb{R}^{n+1}}_X - \nabla^{\mathbb{R}^{n+1}}_X \nabla^{\mathbb{R}^{n+1}}_Y = 0$$

(since $\mathbb{R}^{n+1}$ is flat) and $\langle \nabla^{\mathbb{R}^{n+1}}_Y U, U \rangle = 1/2\, Y\langle U, U \rangle = 0$ (since $\langle U, U \rangle = 1$ is constant).

It is clear that Riemann curvature is a complicated object to deal with. We can gain some intuition about it however by considering vector fields in pairs. First, let us introduce the notation

$$R(X, Y, Z, W) \overset{\text{def}}{=} \langle R(X, Y)Z, W \rangle.$$

Then we define the *sectional curvature* of the two dimensional subspace of T_pM spanned by $X(p)$ and $Y(p)$ to be

$$K(X, Y) = \frac{R(X, Y, X, Y)}{\langle X, X \rangle \langle Y, Y \rangle - \langle X, Y \rangle^2}.$$

As the notation suggests, the sectional curvature is a Gauss curvature of a two dimensional submanifold of M through p with tangent plane $\{X, Y\}$. This generalizes Exercise 5.2 above where $K = R_{1212}/g = R(\partial_1, \partial_2, \partial_1, \partial_2)/g$. Although we will not prove it, it is a fact that *the sectional curvature actually determines the Riemann curvature.* Of course the reverse is true by definition. Also note that if X and Y are orthonormal, then $K(X, Y) = R(X, Y, X, Y)$.

Example 5.2: The sphere S^n. In Example 3.1 we saw that the shape operator of S^n is given by $S(X) = -X/R$, where R is the radius of the sphere. From the formula of Exercise 5.3, we obtain

$$R(X, Y)Z = \langle S(X), Z \rangle S(Y) - \langle S(Y), Z \rangle S(X)$$
$$= \frac{1}{R^2}(\langle X, Z \rangle Y - \langle Y, Z \rangle X).$$

The sectional curvature has an even more perspicacious appearance — especially when we take X and Y to be orthonormal. For then we have

$$K(X, Y) = R(X, Y, X, Y)$$
$$= \frac{1}{R^2}(\langle X, X \rangle Y - \langle Y, X \rangle X, Y)$$
$$= \frac{1}{R^2}$$

since $\langle Y, X \rangle = 0 = \langle X, Y \rangle$ and $\langle X, X \rangle = 1 = \langle Y, Y \rangle$. Thus $K = 1/R^2$ is the *constant* sectional curvature of S^n, completely analogous to the constant Gauss curvature of S^2.

EXERCISE 5.4. Show that the Riemann curvature obeys the following symmetry relations.

(1) $R(X, Y, Z, W) = -R(Y, X, Z, W)$
(2) $R(X, Y, Z, W) = -R(X, Y, W, Z)$
(3) $R(X, Y, Z, W) = R(Z, W, X, Y)$.

Hints: (1) $\langle \nabla_Y \nabla_X V, V \rangle = Y \langle \nabla_X V, V \rangle - \langle \nabla_X V, \nabla_Y V \rangle$ and (2) $\langle \nabla_{[X,Y]} V, V \rangle = 1/2 [X, Y] \langle V, V \rangle$, both by (iv) of Proposition 3.1. For part (3), use the first Bianchi identity below.

EXERCISE 5.5. Prove the first Bianchi identity:

$$R(X, Y, Z, W) + R(Y, Z, X, W) + R(Z, X, Y, W) = 0.$$

Hints: W is superfluous, so drop it from the notation. Write out the definitions of the terms and group them to make use of the identity $\nabla_X Y - \nabla_Y X = [X, Y]$ etc. Then group the terms according to the respective Lie brackets which arise and apply the identity again to end up with only brackets in the formula. Apply the Jacobi identity (Exercise 3.3 (3)).

There is a *second Bianchi identity* which will prove very convenient a bit later. This identity involves covariant derivatives of the Riemann curvature, so we must understand how these are defined. The covariant derivative of R is given by

$$
\begin{aligned}
\nabla R(X, Y, Z, W, V) &= \nabla_V R(X, Y, Z, W) \\
&= V[R(X, Y, Z, W)] - R(\nabla_V X, Y, Z, W) \\
&\quad - R(X, \nabla_V Y, Z, W) - R(X, Y, \nabla_V Z, W) \\
&\quad - R(X, Y, Z, \nabla_V W).
\end{aligned}
$$

Proposition 5.1: The second Bianchi identity.

$$\nabla R(X, Y, Z, W, V) + \nabla R(X, Y, W, V, Z) + \nabla R(X, Y, V, Z, W) = 0.$$

Proof. Choose normal coordinates (Theorem 4.5) about a point $p \in M$ and test the identity at this arbitrary point. Let X, Y, Z, W and V be in the orthonormal coordinate basis with $\nabla_X Y(p) = 0$ etc. Also note that brackets of these vector fields vanish as well by $[X, Y] = \nabla_X Y - \nabla_Y X = 0 - 0 = 0$. Then

$$
\begin{aligned}
\nabla R(X, Y, Z, W, V) &= \nabla_V R(X, Y, Z, W) \\
&= \nabla_V R(Z, W, X, Y) \\
&= V[R(Z, W, X, Y)] \quad\quad \text{since } \nabla_V(\cdot) = 0 \\
&= \langle \nabla_V R(Z, W)X, Y \rangle \quad \begin{array}{l} \text{by Proposition 3.1 (iv)} \\ \text{and } \nabla_V Y = 0 \end{array}
\end{aligned}
$$

and similarly, $\nabla R(X, Y, W, V, Z) = \langle \nabla_Z R(W, V) X, Y \rangle$ and $\nabla R(X, Y, V, Z, W)$
$= \langle \nabla_W R(V, Z) X, Y \rangle$. Hence, X and Y are superfluous, so we remove them
from the notation. Then

$$\nabla_V R(Z, W) + \nabla_Z R(W, V) + \nabla_W R(V, Z) = (\nabla_V \nabla_W - \nabla_W \nabla_V) \nabla_Z$$
$$+ (\nabla_Z \nabla_V - \nabla_V \nabla_Z) \nabla_W$$
$$+ (\nabla_W \nabla_Z - \nabla_Z \nabla_W) \nabla_V$$

and, evaluating on X, we obtain $(\nabla_V R(Z, W) + \nabla_Z R(W, V) + \nabla_W R(V, Z))(X)$
$= R(W, V, \nabla_Z X) + R(V, Z, \nabla_W X) + R(Z, W, \nabla_V X)$ since brackets are zero.
But then, since $\nabla_Z X = 0$ etc., we have

$$(\nabla_V R(Z, W) + \nabla_Z R(W, V) + \nabla_W R(V, Z))(X) = 0.$$

Hence, the result is proved.   □

While the Bianchi identities may seem rather esoteric, they *do* simplify various
calculations considerably as we will see below.

Higher dimensional differential geometry offers us both the challenge of
understanding nonvisualizable geometric phenomena and the opportunity to
create new tools with which to study such phenomena. Some of these new
tools are new types of curvatures which, in two dimensions, become the Gauss
curvature. We have already seen that sectional curvature can tell us information somewhat obscured by the Riemann curvature. Sectional curvature,
however, cannot be a panacea since, as we mentioned previously, it determines
Riemann curvature. Instead, there is a general method called *contraction*
which is available to us in higher dimensions and which allows us to isolate
more tractable portions of the Riemann curvature. In order to do this, we
will generalize the notion of a *frame field* which was used in Chapter 6. If
$\mathcal{E}_1, \ldots, \mathcal{E}_k$ are vector fields defined on a neighborhood of a point $p \in M^k$ with

$$\langle \mathcal{E}_i, \mathcal{E}_j \rangle = \delta_i^j$$

at every point in the neighborhood, then the collection $\{\mathcal{E}_1, \ldots, \mathcal{E}_k\}$ is called
a *frame* about p. One way to obtain a frame is to choose normal coordinates
around p (Theorem 4.5) and then parallel translate the corresponding vector
fields along geodesics passing through p. By Exercise 4.3, parallel translation
is an isometry, so the orthonormality of the frame is preserved as it is translated to other points in the normal coordinate patch. The one bothersome
thing about using a frame is that we lose many of the coordinate formulas
we obtained earlier. Nevertheless, we shall see that the advantages of a frame
outweigh the disadvantages, so assume that we have a frame about p in what
follows.

Definition 5.3. The *Ricci curvature* is defined to be

$$\mathrm{Ric}(X, Y) = \sum_{i=1}^{k} \langle R(X, \mathcal{E}_i)Y, \mathcal{E}_i \rangle$$

where X and Y are tangent vector fields on M^k. The *scalar curvature* is defined to be

$$\kappa = \sum_{j=1}^{k} \mathrm{Ric}(\mathcal{E}_j, \mathcal{E}_j)$$

$$= \sum_{i,j=1}^{k} \langle R(\mathcal{E}_j, \mathcal{E}_i)\mathcal{E}_j, \mathcal{E}_i \rangle.$$

The first thing we should note is that these definitions are independent of the frame field we choose. The reason is this. Any two frames are related at a point by an orthogonal matrix; that is, a matrix A with $A^t = A^{-1}$. This is equivalent to saying that the rows of A form an orthonormal set of vectors. This, together with the symmetry relations of Exercise 5.4 suffice to prove the invariance of the definition. Rather than give a proof of this in general, we will simply concentrate on the $k = 2$ case. So, suppose that $\{\mathcal{F}_j\}$ is another frame field which is related to $\{\mathcal{E}_i\}$ by a 2×2 orthogonal matrix $A = (a_{st})$ as follows:

$$\mathcal{E}_i = a_{1i}\mathcal{F}_1 + a_{2i}\mathcal{F}_2.$$

Now we can just compute using the fact that the rows of A are orthonormal. That is, $a_{11}a_{21} + a_{12}a_{22} = 0$, $a_{11}^2 + a_{12}^2 = 1$ and $a_{21}^2 + a_{22}^2 = 1$.

$$
\begin{aligned}
\mathrm{Ric}(X, Y) &= \langle R(X, \mathcal{E}_1)Y, \mathcal{E}_1 \rangle + \langle R(X, \mathcal{E}_2)Y, \mathcal{E}_2 \rangle \\
&= \langle R(X, a_{11}\mathcal{F}_1 + a_{21}\mathcal{F}_2)Y, a_{11}\mathcal{F}_1 + a_{21}\mathcal{F}_2 \rangle \\
&\quad + \langle R(X, a_{12}\mathcal{F}_1 + a_{22}\mathcal{F}_2)Y, a_{12}\mathcal{F}_1 + a_{22}\mathcal{F}_2 \rangle \\
&= a_{11}^2 \langle R(X, \mathcal{F}_1)Y, \mathcal{F}_1 \rangle + a_{11}a_{21} \langle R(X, \mathcal{F}_1)Y, \mathcal{F}_2 \rangle \\
&\quad + a_{21}a_{11} \langle R(X, \mathcal{F}_2)Y, \mathcal{F}_1 \rangle + a_{21}^2 \langle R(X, \mathcal{F}_2)Y, \mathcal{F}_2 \rangle \\
&\quad + a_{12}^2 \langle R(X, \mathcal{F}_1)Y, \mathcal{F}_1 \rangle + a_{12}a_{22} \langle R(X, \mathcal{F}_1)Y, \mathcal{F}_2 \rangle \\
&\quad + a_{22}a_{12} \langle R(X, \mathcal{F}_2)Y, \mathcal{F}_1 \rangle + a_{22}^2 \langle R(X, \mathcal{F}_2)Y, \mathcal{F}_2 \rangle \\
&= a_{11}^2 \langle R(X, \mathcal{F}_1)Y, \mathcal{F}_1 \rangle + 2a_{11}a_{21} \langle R(X, \mathcal{F}_1)Y, \mathcal{F}_2 \rangle \\
&\quad + a_{21}^2 \langle R(X, \mathcal{F}_2)Y, \mathcal{F}_2 \rangle \\
&\quad + a_{12}^2 \langle R(X, \mathcal{F}_1)Y, \mathcal{F}_1 \rangle + 2a_{12}a_{22} \langle R(X, \mathcal{F}_1)Y, \mathcal{F}_2 \rangle \\
&\quad + a_{22}^2 \langle R(X, \mathcal{F}_2)Y, \mathcal{F}_2 \rangle
\end{aligned}
$$

using the symmetry relations of Exercise 5.4. Then,

$$
\begin{aligned}
\mathrm{Ric}(X,Y) &= (a_{11}^2 + a_{12}^2)\langle R(X,\mathcal{F}_1)Y,\mathcal{F}_1\rangle \\
&\quad + 2(a_{11}a_{21} + a_{12}a_{22})\langle R(X,\mathcal{F}_1)Y,\mathcal{F}_2\rangle \\
&\quad + (a_{21}^2 + a_{22}^2)\langle R(X,\mathcal{F}_2)Y,\mathcal{F}_2\rangle \\
&= \langle R(X,\mathcal{F}_1)Y,\mathcal{F}_1\rangle + \langle R(X,\mathcal{F}_2)Y,\mathcal{F}_2\rangle
\end{aligned}
$$

by the equalities $a_{11}a_{21} + a_{12}a_{22} = 0$, $a_{11}^2 + a_{12}^2 = 1$ and $a_{21}^2 + a_{22}^2 = 1$. Hence, the Ricci curvature is well defined.

The second thing to note is that both Ricci and scalar curvatures come about through a "trace-like" process. Of course, to take the trace of a matrix $A = (a_{ij})$, we form

$$
\mathrm{tr}\, A = \sum_i a_{ii} = \sum_i \langle A(\mathcal{E}_i), \mathcal{E}_i\rangle
$$

for an orthonormal basis $\{\mathcal{E}_1, \dots, \mathcal{E}_k\}$. Here, for Ric, we take two slots of the Riemann curvature and sum over the same elements of the frame. Similarly, and even more closely analogous to trace, to form κ, we sum over the only available slots in Ric. This process of summing over two slots with the same frame elements is called *contraction* and is sometimes denoted $\mathcal{C}$. Thus, $\mathcal{C}_{24}R = \mathrm{Ric}$ and $\mathcal{C}_{12}\,\mathrm{Ric} = \kappa$ since, in the first case, we sum over the second and fourth slots and, in the second case, we sum over the first and second slots. We note here that some authors order the subscripts on R_{ijlr} differently, so contraction subscripts may have to be adjusted to correspond to ours. As the reader might expect, there is a more formal and precise notion of "contraction," but we shall have no need of it here.

Example 5.4. (i) For a k-manifold M^k, let us compute the (only available) contraction of the metric $\langle \cdot, \cdot \rangle$. For a frame as above,

$$
\begin{aligned}
\mathcal{C}(\langle\cdot,\cdot\rangle) &= \sum_{i=1}^k \langle \mathcal{E}_i, \mathcal{E}_i\rangle \\
&= k \\
&= \dim M^k
\end{aligned}
$$

since $\langle \mathcal{E}_i, \mathcal{E}_i\rangle = \delta_i^j$.

(ii) Contraction can also be done by coordinates. Without going into details, for coordinates $u^1, \dots, u^k$ and basis tangent vector fields $\partial_1, \dots, \partial_k$ with $\langle R(\partial_i, \partial_j)\partial_s, \partial_r\rangle = R_{ijsr}$, let

$$
R_{is} = \sum_{j,r} R_{ijsr} g^{jr}.
$$

These are then the components of Ricci curvature. Thus, by Exercise 5.2,

$$
\begin{aligned}
R_{is} &= \sum_{j,r} \sum_{l} \left(\frac{\partial \Gamma_{is}^l}{\partial u^j} - \frac{\partial \Gamma_{js}^l}{\partial u^i} + \sum_{m} \Gamma_{is}^m \Gamma_{jm}^l - \Gamma_{js}^m \Gamma_{im}^l \right) g_{lr} g^{jr} \\
&= \sum_{j,l} \left(\frac{\partial \Gamma_{is}^l}{\partial u^j} - \frac{\partial \Gamma_{js}^l}{\partial u^i} + \sum_{m} \Gamma_{is}^m \Gamma_{jm}^l - \Gamma_{js}^m \Gamma_{im}^l \right) \sum_{r} g_{lr} g^{jr} \\
&= \sum_{l} \left(\frac{\partial \Gamma_{is}^l}{\partial u^l} - \frac{\partial \Gamma_{ls}^l}{\partial u^i} + \sum_{m} \Gamma_{is}^m \Gamma_{lm}^l - \Gamma_{ls}^m \Gamma_{im}^l \right)
\end{aligned}
$$

since $\sum_r g_{lr} g^{jr} = \delta_j^l$.

In order to see the power of these curvatures, we offer the following theorems without proof. For a proof of the the first, see [GHL] for example.

Theorem 5.2: Myers' Theorem. *Let M^k be a complete k-manifold and suppose that the Ricci curvature on all of M^k is strictly bounded away from zero; that is,*

$$
\mathrm{Ric}(X, X) \geq \frac{k-1}{r^2} > 0.
$$

Then M^k is compact and the diameter of M^k is less than or equal to πr.

Compare this result with Bonnet's result, Theorem 6.7.6. Scalar curvature also has the power to constrain the type of a manifold. Compare the following result (due to Ros [Ros]) to Alexandrov's result, Theorem 4.4.3.

Theorem 5.3. *A compact hypersurface $M^n \subset \mathbb{R}^{n+1}$ of constant scalar curvature is a sphere S^n with metric induced by that of $\mathbb{R}^{n+1}$.*

EXERCISE 5.6. For a surface defined by an orthogonal patch in $\mathbb{R}^3$, show that Ricci and scalar curvatures are given by

$$
\mathrm{Ric}(X, Y) = K \langle X, Y \rangle \qquad \text{and} \qquad \kappa = 2K
$$

where K is Gaussian curvature. Hint: use Exercise 5.2. The orthogonal patch condition is, in fact, unnecessary, but Exercise 5.2 assumed it for simplicity.

EXERCISE 5.7. Show that the following formula holds relating the covariant derivative, the metric and the bracket.

$$
\begin{aligned}
2\langle X, \nabla_Z Y \rangle = Z \langle X, Y \rangle &+ Y \langle X, Z \rangle - X \langle Y, Z \rangle \\
&+ \langle Y, [X, Z] \rangle + \langle Z, [X, Y] \rangle - \langle X, [Y, Z] \rangle.
\end{aligned}
$$

EXERCISE 5.8. Here is a first small step toward getting away from the induced metric of $\mathbb{R}^{n+1}$. A metric $\langle \cdot, \widetilde{\cdot} \rangle$ is a **multiple** of another metric $\langle \cdot, \cdot \rangle$ if

$$\langle X, \widetilde{Y} \rangle = \lambda \langle X, Y \rangle$$

for all X and Y, where λ is a constant. If λ is allowed to be a smooth function on M, then the metrics are **conformal**. Denoting the covariant derivative and curvatures of the associated metric $\langle \cdot, \widetilde{\cdot} \rangle$ by $\widetilde{\nabla}$, $\tilde{R}$, $\widetilde{\mathrm{Ric}}$, $\tilde{\kappa}$, show that the following properties hold for a "multiple" metric.

(1) $\widetilde{\nabla}_X Y = \nabla_X Y$. (Use the formula from Exercise 5.7.)

(2) $\tilde{R}(X, Y, Z, W) = \lambda R(X, Y, Z, W)$. (Use (1).)

(3) $\tilde{K}(X, Y) = \dfrac{1}{\lambda} K(X, Y)$. (Use (2).)

(4) $\widetilde{\mathrm{Ric}}(X, Y) = \mathrm{Ric}(X, Y)$. (Use (2).)

(5) $\tilde{\kappa} = \dfrac{1}{\lambda} \kappa$. (Use (4).)

EXERCISE 5.9. Let $P^k = \{(u^1, \dots, u^k) \in \mathbb{R}^k \mid u^k > 0\}$ be the "upper half space" analogous to the two dimensional Poincaré plane. Define an analogous metric on P^k by

$$\langle \mathbf{v}, \mathbf{w} \rangle = \frac{\mathbf{v} \cdot \mathbf{w}}{(u^k)^2}.$$

In our present language, this means that

$$g_{ij} = \langle \mathbf{e}_i, \mathbf{e}_j \rangle = \frac{\mathbf{e}_i \cdot \mathbf{e}_j}{(u^k)^2} = \frac{\delta^i_j}{(u^k)^2}.$$

where the $\mathbf{e}_i$ are the standard orthonormal $\mathbb{R}^k$ basis vectors. Hence, the metric is diagonal. The sectional curvature of P^k is known to be constant at -1. The following steps lead to a partial understanding of this result. For further and complete details, including formulas for general conformal metrics, see [DoC2, Chapter 8.3].

(1) Show that the Christoffel symbols are

$$\Gamma^r_{ij} = \frac{1}{2} \sum_l g^{lr} \left(\partial_i g_{jl} + \partial_j g_{li} - \partial_l g_{ij} \right)$$

$$= \frac{1}{2} g^{rr} \left(\partial_i g_{jr} + \partial_j g_{ri} - \partial_r g_{ij} \right)$$

and that the only nonzero ones are

$$\Gamma^j_{kj} = -\frac{1}{u^k} \quad \text{for } j \neq k, \qquad \Gamma^k_{ii} = \frac{1}{u^k} \quad \text{for } i \neq k, \qquad \Gamma^k_{kk} = -\frac{1}{u^k}.$$

(2) Now show that the components of Riemann curvature relevant to sectional curvature are

$$R_{ijij} = \sum_l \left(\frac{\partial \Gamma_{ii}^l}{\partial u^j} - \frac{\partial \Gamma_{ji}^l}{\partial u^i} + \sum_m \Gamma_{ii}^m \Gamma_{jm}^l - \Gamma_{ji}^m \Gamma_{im}^l \right) g_{lj}$$

$$= \left(\frac{\partial \Gamma_{ii}^j}{\partial u^j} - \frac{\partial \Gamma_{ji}^j}{\partial u^i} + \sum_m \Gamma_{ii}^m \Gamma_{jm}^j - \Gamma_{ji}^m \Gamma_{im}^j \right) \left(\frac{1}{(u^k)^2} \right)$$

and the only nonzero components are $R_{kjkj} = R_{ikik} = R_{ijij} = -1/(u^k)^4$. Here we assume $k \neq j$ in R_{kjkj}, $i \neq k$ in R_{ikik} and $i \neq j$, $i \neq k$ and $j \neq k$ in R_{ijij}.

(3) Using the calculation of (2), show that the components of sectional curvature $K_{ij} = R_{ijij}/(g_{ii} \cdot g_{jj})$ for the cases above are constantly equal to -1. This exercise indicates how the conformal metrics of Chapter 5 may be generalized to higher dimensions.

EXERCISE 5.10. A manifold M^k is said to have nonnegative sectional curvature if $K(X, Y) \geq 0$ for all X and Y. The Ricci curvature is said to be nonnegative when $\mathrm{Ric}(X, X) \geq 0$ for all X. Show that nonnegative sectional curvature on M implies nonnegative Ricci curvature on M and, in turn, this implies nonnegative scalar curvature on M. Hint: First prove the formula

$$\mathrm{Ric}(X, X) = \langle X, X \rangle \sum_{j=1}^{k-1} K(X, \mathcal{E}_j)$$

for a frame $\{\mathcal{E}_j\}$ spanning $X^{\perp}$, the space of vector fields orthogonal to X.

EXERCISE 5.11. Here is an exercise which at once goes beyond our definitions and tests them. A **Lie Group** G is a smooth manifold with a smooth associative multiplication $G \times G \to G$ having an identity e and inverses. Take G to be compact and connected. For example, the special orthogonal group $SO(n)$ is a compact connected Lie group. We may define certain vector fields on G by fixing a vector $X(e) \in T_e G$ and then taking $X(g) = g_*(X(e)) \in T_g G$. Here, g_* is the map induced on tangent vectors by the multiplication $(g, h) \mapsto g \cdot h$ for all $h \in G$. These types of vector fields are called **left invariant vector fields**. Lie groups are quite symmetric in the sense that the covariant derivative (which may be defined without reference to an ambient Euclidean space, but which has all the properties of Proposition 3.1) and curvature at the identity determine the covariant derivatives and curvatures at all other points of G. Furthermore, it is known that $\nabla_X X = 0$ for X any left invariant vector field. Thus, curves whose tangent vectors belong to left invariant vector fields are geodesics. Also, for G as above, the following identity is known to hold: $\langle [X, Y], Z \rangle = -\langle Y, [X, Z] \rangle$ for left invariant vector fields X, Y and Z. Here, $\langle \cdot, \cdot \rangle$ is a metric which is first defined on $T_e G$ and then transported around G by the Lie group multiplication.

Now, for a compact, connected Lie group G with left invariant vector fields X, Y and Z, show that

(1) $\nabla_X Y = \frac{1}{2}[X,Y]$. (Consider $\nabla_{X-Y}(X-Y)$ and use Proposition 3.1 (v).)

(2) $R(X,Y)Z = \frac{1}{4}[[X,Y],Z]$.

   (Start with the definition of R, then use (1) to express R completely in terms of brackets. Finally, use the Jacobi identity for the bracket (Exercise 3.3 (3)).)

(3) If X and Y are orthonormal, then $K(X,Y) = \frac{1}{4}\left\|[X,Y]\right\|^2$.

   Thus, sectional curvature, and hence, Ricci and scalar curvatures are non-negative for a compact, connected Lie group. (Use the definition of K, (1), Proposition 3.1 (iv) and the identity $\langle[X,Y],Z\rangle = -\langle Y,[X,Z]\rangle$ with $Z=Y$.)

The **Killing form** of a Lie group G is defined to be $b(X,Y) = \text{tr}(\text{ad}_X \, \text{ad}_Y)$, where $\text{ad}_Z(W) = [Z,W]$ is a linear transformation of the vector space of left invariant vector fields. It can be shown that b is symmetric, bilinear and invariant under all automorphisms of G. If b is also nondegenerate, in the sense that $b(V,W) = 0$ for all W implies $V = 0$, then b is actually the (bi-invariant) metric $\langle \cdot, \cdot \rangle$ on G. That is, $\langle X,Y\rangle = b(X,Y)$. A Lie group with nondegenerate Killing form is said to be **semisimple**. For instance, $SO(n)$ is semisimple. Using (2) above and the definition of ad_Z, show that a semisimple Lie group with metric b has Ricci curvature

$$\text{Ric}(X,Y) = -\frac{1}{4}b(X,Y).$$

To end this section, we wish to make several calculations using the definitions above to try to understand a tiny bit of modern physics. Einstein's general theory of relativity is based on a principle that energy and momentum distort the geometry of spacetime and, as a consequence, are responsible for various physical phenomena such as light "bending" around the Sun (along a spacetime geodesic) and the precession of the perihelion of Mercury. Although this principle is simple to state in our naive discussion, the precise form of the relationship is far from clear. If we are to believe in the relationship at all, then perhaps the simplest such formula is

$$G = cT$$

where G is some type of curvature (called the *Einstein curvature*) and T is a quantity (called the *stress-energy tensor*) which depends on the amount of energy-momentum in a particular region of spacetime. In particular, in a vacuum we have $T = 0$ — but interesting physics (such as the bending of light) can still occur. If G were Riemann curvature R, then the formula $G = cT$

together with the assumption $T = 0$ would imply the vanishing of Riemann curvature. But, as we have said earlier, $R = 0$ implies that a manifold is flat. Hence, spacetime would be forced to be flat and this would contradict the interesting geometrically induced physics observed. So, G cannot then be the Riemann curvature. The next logical choice to try is the Ricci curvature, so we take $G = \mathrm{Ric}$ provisionally. But we must ask, what else do we desire in a physical quantity? Certainly, the law of conservation of energy-momentum must be obeyed, so we obtain the vanishing of some sort of divergence of T. This means that we should also require G to be divergence free as well. Let us consider this now.

If $V = (V^1, V^2, V^3)$ is a vector field in $\mathbb{R}^3$ with coordinates x^1, x^2 and x^3, then we know

$$\mathrm{div}\, V = \frac{\partial V^1}{\partial x^1} + \frac{\partial V^2}{\partial x^2} + \frac{\partial V^3}{\partial x^3} = \sum_i \langle \nabla^{\mathbb{R}^3}_{\mathbf{e}_i} V, \mathbf{e}_i \rangle$$

where the $\mathbf{e}_i$ are the usual Euclidean orthonormal basis vectors.

EXERCISE 5.12. Verify the last equality

$$\frac{\partial V^1}{\partial x^1} + \frac{\partial V^2}{\partial x^2} + \frac{\partial V^3}{\partial x^3} = \sum_i \langle \nabla^{\mathbb{R}^3}_{\mathbf{e}_i} V, \mathbf{e}_i \rangle.$$

In the same way, on a manifold M^k, we can define the divergence of vector fields and bilinear quantities with two slots (i.e. 2-tensors) such as the metric $\langle \cdot, \cdot \rangle$ and Ric. Let $A(\cdot, \cdot)$ be such a quantity and define the *divergence of A*, $\mathrm{div}\, A(\cdot)$, to be a linear quantity with one slot (i.e. a 1-form; see section 6),

$$\mathrm{div}\, A(X) \overset{\mathrm{def}}{=} \sum_i \nabla_{\mathcal{E}_i} A(\mathcal{E}_i, X)$$

$$= \sum_i \mathcal{E}_i A(\mathcal{E}_i, X) - A(\nabla_{\mathcal{E}_i} \mathcal{E}_i, X) - A(\mathcal{E}_i, \nabla_{\mathcal{E}_i} X)$$

where $\{\mathcal{E}_i\}_{i=1}^k$ is a frame and we use a definition for ∇A similar to that for ∇R in the second Bianchi identity Proposition 5.1. There are several things to note here. First, the divergence produces a "tensor" which is completely determined at each point by the values of its constituents at that point. Also, although we have used a frame to define the divergence, this is in fact only a convenient way to compute — the divergence may be defined without reference to a particular frame used for computation. With this in mind, consider

Example 5.5: The metric $\langle \cdot, \cdot \rangle$ has zero divergence. To see this, take normal coordinates about an arbitrary point $p \in M$ and extend to a frame $\{\mathcal{E}_1, \ldots, \mathcal{E}_k\}$ by parallel translation along geodesics emanating from p. In particular, we have $\nabla_{\mathcal{E}_i}\mathcal{E}_j = 0$ at p. Now, because the divergence is determined by the values of the frame etc. at p, the same calculation as below may be made for a normal coordinate frame about every point in M. At p we have

$$\text{div}\,\langle \cdot, \cdot \rangle(\mathcal{E}_i) = \sum_j \nabla_{\mathcal{E}_j}\langle \mathcal{E}_j, \mathcal{E}_i \rangle$$

$$= \sum_j \mathcal{E}_j\langle \mathcal{E}_j, \mathcal{E}_i \rangle - \langle \nabla_{\mathcal{E}_j}\mathcal{E}_j, \mathcal{E}_i \rangle - \langle \mathcal{E}_j, \nabla_{\mathcal{E}_j}\mathcal{E}_i \rangle$$

$$= \sum_j \mathcal{E}_j\langle \mathcal{E}_j, \mathcal{E}_i \rangle - 0 - 0$$

since we have normal coordinates at p. But the first term vanishes as well because $\langle \mathcal{E}_i, \mathcal{E}_j \rangle = \delta_i^j$ on the coordinate neighborhood and the $\mathcal{E}_j$-derivative of a constant is zero. Therefore,

$$\text{div}\,\langle \cdot, \cdot \rangle(\mathcal{E}_i) = 0.$$

Since this is true for all $\mathcal{E}_i$ and the process is linear, then it holds for all vector fields as well.

EXERCISE 5.13. For a smooth function $f \colon M \to \mathbb{R}$, show that

$$\text{div}(f\langle \cdot, \cdot \rangle) = df$$

where df is the differential of f defined by $df(X) = X[f]$ for a vector field X. Thus, the divergence of conformal metrics (i.e. $f > 0$ at every p) depends on the conformality factor alone. Note that this makes sense in light of Example 5.4 since the divergence depends on the frame chosen. Verify that, for metric $f\langle \cdot, \cdot \rangle$, the frame $\{(1/\sqrt{f})\,\mathcal{E}_i\}$ gives $\text{div}(f\langle \cdot, \cdot \rangle) = 0$.

We will now apply the divergence to Ricci curvature to find a remarkable relationship.

Theorem 5.4. *For Ricci curvature* Ric *and scalar curvature* κ,

$$d\kappa = 2\,\text{div}(\text{Ric}).$$

Proof. Again choose a frame derived from parallel translation along geodesics of normal coordinates about an arbitrary point $p \in M$. Hence, at p, $\nabla_{\mathcal{E}_i}\mathcal{E}_j = 0$ for all i and j. Then

$$2\operatorname{div}(\operatorname{Ric})(\mathcal{E}_s) = 2\sum_j \nabla_{\mathcal{E}_j}\operatorname{Ric}(\mathcal{E}_j,\mathcal{E}_s)$$

$$= 2\sum_j \mathcal{E}_j \operatorname{Ric}(\mathcal{E}_j,\mathcal{E}_s)$$

$$= 2\sum_{i,j} \mathcal{E}_j \langle R(\mathcal{E}_j,\mathcal{E}_i)\mathcal{E}_s,\mathcal{E}_i\rangle$$

$$= \sum_{i,j} \mathcal{E}_j \langle R(\mathcal{E}_j,\mathcal{E}_i)\mathcal{E}_s,\mathcal{E}_i\rangle + \sum_{i,j} \mathcal{E}_i \langle R(\mathcal{E}_i,\mathcal{E}_j)\mathcal{E}_s,\mathcal{E}_j\rangle$$

where the second term is obtained by switching i and j. Then, the symmetry relations for R imply

$$2\operatorname{div}(\operatorname{Ric})(\mathcal{E}_s) = \sum_{i,j} \mathcal{E}_j \langle R(\mathcal{E}_j,\mathcal{E}_i)\mathcal{E}_s,\mathcal{E}_i\rangle + \sum_{i,j} \mathcal{E}_i \langle R(\mathcal{E}_j,\mathcal{E}_i)\mathcal{E}_j,\mathcal{E}_s\rangle$$

$$= \sum_{i,j} \nabla R(\mathcal{E}_j,\mathcal{E}_i,\mathcal{E}_s,\mathcal{E}_i,\mathcal{E}_j) + \nabla R(\mathcal{E}_j,\mathcal{E}_i,\mathcal{E}_j,\mathcal{E}_s,\mathcal{E}_i)$$

$$= -\sum_{i,j} \nabla R(\mathcal{E}_j,\mathcal{E}_i,\mathcal{E}_i,\mathcal{E}_j,\mathcal{E}_s)$$

by the second Bianchi identity. Then, interchanging the first two coordinates,

$$2\operatorname{div}(\operatorname{Ric})(\mathcal{E}_s) = \sum_{i,j} \nabla R(\mathcal{E}_i,\mathcal{E}_j,\mathcal{E}_i,\mathcal{E}_j,\mathcal{E}_s)$$

$$= \sum_{i,j} \mathcal{E}_s R(\mathcal{E}_i,\mathcal{E}_j,\mathcal{E}_i,\mathcal{E}_j)$$

$$= \mathcal{E}_s \sum_{i,j} R(\mathcal{E}_i,\mathcal{E}_j,\mathcal{E}_i,\mathcal{E}_j)$$

$$= \mathcal{E}_s[\kappa]$$

$$= d\kappa(\mathcal{E}_s).$$

Since this is true on a basis, it is true in general. $\qquad\square$

EXERCISE 5.14. The manifold $(M^k, \langle \cdot, \cdot \rangle)$ with metric $\langle \cdot, \cdot \rangle$ is an **Einstein manifold** if $\mathrm{Ric}(X, Y) = \lambda \langle X, Y \rangle$ for a fixed constant λ and all vector fields X and Y.

(1) Show that, if $\mathrm{Ric}(X, Y)(p) = \lambda(p) \langle X, Y \rangle(p)$ for a smooth function $\lambda \colon M^k \to \mathbb{R}$ with $k \geq 3$, then M^k is Einstein. Hints: (1) first compute κ and (2) then use Theorem 5.4.

(2) Show that a semisimple Lie group with Killing form metric b is an Einstein manifold. Hint: use the last part of Exercise 5.11.

Now, what has been the point of all this? The argument we presented earlier suggested that the Einstein and Ricci curvatures should be identified and that the Einstein curvature should be divergence free. But, as Theorem 5.4 shows, in general, the Ricci curvature has a non-zero divergence. So what is to be done to rescue the idea? By Exercise 5.13, we know

$$\mathrm{div}(\kappa \langle \cdot, \cdot \rangle) = d\kappa$$

so that $\mathrm{div}(\kappa \langle \cdot, \cdot \rangle) = 2 \, \mathrm{div}(\mathrm{Ric})$ and, therefore,

$$\mathrm{div}\left(\mathrm{Ric} - \frac{1}{2} \kappa \langle \cdot, \cdot \rangle\right) = 0.$$

This calculation then says that, if we define the *Einstein curvature* to be $G = \mathrm{Ric} - \frac{1}{2} \kappa \langle \cdot, \cdot \rangle$, then we obtain a type of curvature which is divergence free. Also, we have

Theorem 5.5. *The Einstein and Ricci curvatures determine each other. In particular, $G = 0$ if and only if $\mathrm{Ric} = 0$.*

Proof. Denoting contraction by $\mathcal{C}$, we may write

$$G = \mathrm{Ric} - \frac{1}{2} \kappa \, \langle \cdot, \cdot \rangle$$

$$= \mathrm{Ric} - \frac{1}{2} \mathcal{C}(\mathrm{Ric}) \, \langle \cdot, \cdot \rangle$$

since the contraction of the Ricci curvature is scalar curvature by the discussion preceding Example 5.4. Of course, Example 5.4 itself shows that $\mathcal{C}(\langle \cdot, \cdot \rangle) = \dim M$, so we also have

$$\mathcal{C}(G) = \mathcal{C}(\mathrm{Ric}) - \frac{1}{2} \kappa \, \mathcal{C}(\langle \cdot, \cdot \rangle)$$

$$= \kappa - \frac{1}{2} \kappa \dim M$$

$$= \left(1 - \frac{\dim M}{2}\right) \kappa.$$

Now, spacetime has four dimensions — time and $\mathbb{R}^3$ — so dim $M = 4$. Hence, $\mathcal{C}(G) = -\kappa$ and

$$\mathrm{Ric} = G + \frac{1}{2}\kappa\langle\cdot,\cdot\rangle$$

$$= G - \frac{1}{2}\mathcal{C}(G)\langle\cdot,\cdot\rangle.$$

Thus, Ric and G determine each other. $\square$

Therefore, it seems that the Einstein curvature is a good choice for the Einstein field equation $G = 8\pi T$ (where the constant 8π is determined by taking the Newtonian limit). Indeed, in a vacuum, where $T = 0$, it is generally more convenient to solve the equations Ric $= 0$ rather than $G = 0$, so the relationship with Ricci curvature is important. Perhaps we should mention what the word "solve" means here. What is known and what is unknown? Generally, we try to determine the geometric structure of spacetime from a hypothesis on the Ricci curvature and a knowledge of some facet of the relevant geometry. In Example 5.4 we showed that the Ricci curvature is given in terms of Christoffel symbols and their derivatives. These, in turn, are given in terms of derivatives of the metric. Hence, the expressions

$$R_{is} = 0$$

form a system of second order partial differential equations *in the metric*. Therefore, when we solve this system, we are *determining the metric in that region of spacetime*. Since the metric determines everything, it can be safely said then that we understand the geometry of spacetime in that particular region.

Example 5.6: The Schwarzschild Solution. The solution (i.e. the metric coefficients) for the vacuum field equations $R_{is} = 0$ outside a spherically symmetric body of mass M is given by

$$g_{00} = 1 - \frac{2M}{r} \qquad g_{11} = -\left(1 - \frac{2M}{r}\right)^{-1} \qquad g_{22} = -r^2 \qquad g_{33} = -r^2\sin^2\phi$$

where time t is usually given the index 0. This is the Schwarzschild solution discovered by Schwarzschild in 1916. Note that this metric is different from our usual ones in the sense that it is not positive definite. Spacetime is, in fact, based on the flat (Minkowski) metric of special relativity which has the form $g_{00} = 1$, $g_{11} = -1$, $g_{22} = -1$ and $g_{33} = -1$. These metrics are said to be *semi-riemannian metrics* and much of the theory of ordinary Riemannian geometry carries over to these metrics as well.

There are many good texts which discuss the material of this section in great detail. In particular, for general discussions of higher dimensional geometry, see [Spi], [GHL] and [DoC2]. For a particularly elementary discussion of the relativistic bending of light and the precession of Mercury's perihelion, see [Fab]. For general relativity itself, see the classic [MTW] and for a beautiful combination of differential geometry and relativity, see [O'N2].

9.6 THE CHARMING DOUBLENESS

But the beauty here lay in the duality, in the charming doubleness ...

— Thomas Mann (Felix Krull, p. 77)

Everything that we have done above has a dual formulation which is often powerful and quite beautiful. First, recall that any (real, say) vector space V has a dual vector space $V^* = \{f \colon V \to \mathbb{R} \mid f \text{ is linear}\}$ consisting of linear functionals on V. In terms of a basis for V, $\{\mathbf{v}_1, \ldots, \mathbf{v}_k\}$, the dual vector space has basis $\{\bar{\mathbf{v}}^1, \ldots, \bar{\mathbf{v}}^k\}$ with the defining property

$$\bar{\mathbf{v}}^i(\mathbf{v}_j) = \delta_i^j.$$

In the same way, if the $\mathbf{v}_i$ are tangent vector fields which form a basis at each point $p \in M$, then their *duals* are called 1-*forms* and are denoted by θ^i. Of course, the dual 1-forms have the defining property $\theta^i(\boldsymbol{\partial}_j) = \delta_j^i$ as well. If we take the vector field basis to be $\{\boldsymbol{\partial}_i\}$, then a dual 1-form θ^i is usually denoted by du^i to indicate that coordinate vector fields $\boldsymbol{\partial}_i = \partial/\partial u^i$ are in use.

Now, a linear transformation $f \colon V \otimes V \to V$ with $f(\mathbf{v}_i \otimes \mathbf{v}_s) = \sum_r \tilde{\Gamma}_{si}^r \mathbf{v}_r$ may be described by writing $\tilde{f} \colon V \to V^* \otimes V$ with $\tilde{f}(\mathbf{v}_i)(\mathbf{v}_s) = f(\mathbf{v}_i \otimes \mathbf{v}_s)$. Perhaps we should remind the reader here that the symbol $\otimes$ stands for *tensor product*. The tensor product of two vector spaces V and W with bases $\{\mathbf{v}_1, \ldots, \mathbf{v}_m\}$ and $\{\mathbf{w}_1, \ldots, \mathbf{w}_n\}$ is formed by taking the vector space $V \otimes W$ with basis

$$\{\mathbf{v}_i \otimes \mathbf{w}_j\}$$

where $i = 1, \ldots, m$ and $j = 1, \ldots, n$. Of course we may also write

$$\tilde{f}(\mathbf{v}_i) = \sum_{j,r} a_{ji}^r \bar{\mathbf{v}}^j \otimes \mathbf{v}_r$$

where the $a_{ji}^r \in \mathbb{R}$ and, since there is a dual basis element in the formula, we may evaluate at $\mathbf{v}_s$ to get

$$\tilde{f}(\mathbf{v}_i)(\mathbf{v}_s) = \sum_{j,r} a_{ji}^r \bar{\mathbf{v}}^j(\mathbf{v}_s)\, \mathbf{v}_r \, \mathbf{v}^j$$

$$= \sum_r a_{si}^r \mathbf{v}_r.$$

But, $\tilde{f}(\mathbf{v}_i)(\mathbf{v}_s) = f(\mathbf{v}_i \otimes \mathbf{v}_s) = \sum_r \tilde{\Gamma}^r_{si} \mathbf{v}_r$ and, comparing the two formulas, we see that

$$a^r_{si} = \tilde{\Gamma}^r_{si}$$

and this relation holds for all i and s. Hence, $\tilde{f}(\mathbf{v}_i) = \sum_{j,r} \tilde{\Gamma}^r_{si} \bar{\mathbf{v}}^j \otimes \mathbf{v}_r$.

We can now do the same thing for the covariant derivative. More precisely, if $\mathcal{V}$ denotes tangent vector fields on M, then the covariant derivative is an $\mathbb{R}$-linear map $\nabla \colon \mathcal{V} \otimes \mathcal{V} \to \mathcal{V}$. Just as above, this map may be rewritten in a dual form as $\tilde{\nabla} \colon \mathcal{V} \to \mathcal{V}^* \otimes \mathcal{V}$ with

$$\tilde{\nabla}(\mathcal{E}_i) = \sum_r \omega^r_i \mathcal{E}_r$$

where $\{\mathcal{E}_1, \ldots, \mathcal{E}_k\}$ is a frame on M^k and the ω^r_i are 1-forms. Note that we omit writing the tensor product symbol $\otimes$ for convenience. Also, now that we see that this is simply a different description of the same covariant derivative, we can dispense with $\tilde{\nabla}$ and simply write ∇. Let us identify the 1-form coefficients in terms of something we know. Let $\{\theta_1, \ldots, \theta_k\}$ denote the dual 1-forms to the frame $\{\mathcal{E}_1, \ldots, \mathcal{E}_k\}$ and consider the following calculation analogous to the vector space calculation above. (We use symbols $\tilde{\Gamma}$ to denote coefficients in a basis decomposition even though the chosen frame may not be a coordinate frame. Therefore, the $\tilde{\Gamma}$'s are not the usual Christoffel symbols in general, but, in case $\mathcal{E}_i = \partial_i$, then $\tilde{\Gamma} = \Gamma$.)

$$\sum_r \tilde{\Gamma}^r_{si} \mathcal{E}_r = \nabla_{\mathcal{E}_s} \mathcal{E}_i$$

$$= \nabla(\mathcal{E}_i)(\mathcal{E}_s)$$

$$= \sum_{j,r} \tilde{\Gamma}^r_{ji} \theta^j(\mathcal{E}_s) \mathcal{E}_r$$

by the vector space formula above. By the 1-form formula, however, we have

$$\nabla(\mathcal{E}_i)(\mathcal{E}_s) = \sum_r \omega^r_i(\mathcal{E}_s) \mathcal{E}_r$$

so

$$\sum_r \left(\sum_j \tilde{\Gamma}^r_{ji} \theta^j(\mathcal{E}_s) \right) \mathcal{E}_r = \sum_r \omega^r_i(\mathcal{E}_s) \mathcal{E}_r.$$

Since this holds for all s,

$$\sum_j \tilde{\Gamma}^r_{ji} \theta^j = \omega^r_i.$$

Therefore, the *connection 1-forms* ω^r_i may be written in terms of the dual frame and the coefficients of the covariant derivative's basis expansion.

EXERCISE 6.1. Show that

$$\sum_l (\tilde{\Gamma}^l_{ji} - \tilde{\Gamma}^l_{ij})\mathcal{E}_l = [\mathcal{E}_i, \mathcal{E}_j]$$

so that there is no reason, in general, that $\tilde{\Gamma}^l_{ji} = \tilde{\Gamma}^l_{ij}$. Hint: Proposition 3.1 (v). Also, show that $\tilde{\Gamma}^l_{ji} = -\tilde{\Gamma}^i_{jl}$. Hint: Proposition 3.1 (iv).

EXERCISE 6.2. Show that

$$\omega^j_i = -\omega^i_j.$$

Hint: Proposition 3.1 (iv).

Riemann curvature is defined in terms of "second covariant derivatives," so if we want to have a dual version of this as well, then we must know how to differentiate 1-forms. For a moment, consider a function f of k-variables $u^1, \ldots, u^k$ and compute its differential to be

$$df = \sum_i \frac{\partial f}{\partial u^i} \, du^i.$$

The du^i are 1-forms dual to the vector fields ∂_i, so df is itself a 1-form. This calculation from calculus tells us that the derivative of a 1-form should be a 2-form. Now, just as a 1-form acts on vector fields, a 2-form should act on pairs of vector fields.

Example 6.1. Let θ and ϕ be 1-forms. Define the 2-form $\theta \wedge \phi$ by how it acts on a pair of vector fields V and W:

$$\theta \wedge \phi(V, W) = \theta(V) \cdot \phi(W) - \theta(W) \cdot \phi(V).$$

This "product" $\theta \wedge \phi$ is called the *wedge product of 1-forms*. Note that 1-forms have the property that $\theta \wedge \phi = -\phi \wedge \theta$. In particular, this means that $\theta \wedge \theta = 0$ for all 1-forms θ.

To conform with the definition of df above, we define the *exterior derivative* of a 1-form $f \, du^s$ to be

$$d(f \, du^s) = \sum_j \frac{\partial f}{\partial u^j} \, du^j \wedge du^s$$

where the u^i are coordinates. This is equivalent to defining d by the formula

$$d\theta(V, W) = V[\theta(W)] - W[\theta(V)] - \theta([V, W])$$

for any two vector fields V and W.

EXERCISE 6.3. For coordinate vector fields $V = \partial_j$ and $W = \partial_r$, show that the formula $d\theta(V, W) = V[\theta(W)] - W[\theta(V)] - \theta([V, W])$ gives

$$d(f\, du^i) = \partial_j f \delta_i^r - \partial_r f \delta_i^j$$

where we may assume $j < r$. Explain why this is the same as the coordinate formula for d above.

Theorem 6.1: First Structure Equation. *Let $\{\mathcal{E}_i\}$ be a frame on M^k with dual frame $\{\theta^i\}$ and connection coefficients $\{\omega_s^r\}$. Then,*

$$d\theta^i = -\sum_l \omega_l^i \wedge \theta^l.$$

Proof. We will compute both sides of the equation on vector fields in the frame.

$$d\theta^i(\mathcal{E}_j, \mathcal{E}_r) = -\theta^i([\mathcal{E}_j, \mathcal{E}_r])$$

since $\theta^i(\mathcal{E}_j) = \delta_j^i$ has zero derivative. Then, by Proposition 3.1 (v),

$$
\begin{aligned}
d\theta^i(\mathcal{E}_j, \mathcal{E}_r) &= -\theta^i\left(\nabla_{\mathcal{E}_j}\mathcal{E}_r - \nabla_{\mathcal{E}_r}\mathcal{E}_j\right) \\
&= -\theta^i\left(\nabla \mathcal{E}_r(\mathcal{E}_j) - \nabla \mathcal{E}_j(\mathcal{E}_r)\right) \\
&= -\theta^i\left(\sum_l \omega_r^l(\mathcal{E}_j)\,\mathcal{E}_l - \sum_l \omega_j^l(\mathcal{E}_r)\,\mathcal{E}_l\right) \\
&= \omega_i^j(\mathcal{E}_r) - \omega_r^i(\mathcal{E}_j)
\end{aligned}
$$

since $\theta^i(\mathcal{E}_l) = \delta_l^i$. Now we compute the sum of wedge products.

$$
\begin{aligned}
\sum_l \omega_l^i \wedge \theta^l(\mathcal{E}_j, \mathcal{E}_r) &= \sum_l \left(\omega_l^i(\mathcal{E}_j)\theta^l(\mathcal{E}_r) - \omega_l^i(\mathcal{E}_r)\theta^l(\mathcal{E}_j)\right) \\
&= \omega_r^i(\mathcal{E}_j) - \omega_i^j(\mathcal{E}_r).
\end{aligned}
$$

Because these calculations are the same on a basis, the general formula follows. $\square$

Now let's relate these ideas to Riemann curvature. We write, as usual, $R(\mathcal{E}_i, \mathcal{E}_j)\mathcal{E}_m = \sum_r R^r_{ijm}\mathcal{E}_r$ and we compute

$$R(\mathcal{E}_i, \mathcal{E}_j)\mathcal{E}_m = \nabla_{[\mathcal{E}_i,\mathcal{E}_j]}\mathcal{E}_m + \nabla_{\mathcal{E}_j}\nabla_{\mathcal{E}_i}\mathcal{E}_m - \nabla_{\mathcal{E}_i}\nabla_{\mathcal{E}_j}\mathcal{E}_m$$

$$= \nabla\mathcal{E}_m([\mathcal{E}_i, \mathcal{E}_j]) + \nabla(\nabla\mathcal{E}_m(\mathcal{E}_i))(\mathcal{E}_j) - \nabla(\nabla\mathcal{E}_m(\mathcal{E}_j))(\mathcal{E}_i)$$

$$= \sum_l \omega^l_m([\mathcal{E}_i, \mathcal{E}_j])\mathcal{E}_l + \nabla(\sum_s \omega^s_m(\mathcal{E}_i)\mathcal{E}_s)(\mathcal{E}_j)$$

$$- \nabla(\sum_s \omega^s_m(\mathcal{E}_j)\mathcal{E}_s)(\mathcal{E}_i)$$

$$= \sum_l \omega^l_m([\mathcal{E}_i, \mathcal{E}_j])\mathcal{E}_l + \sum_s \mathcal{E}_j\omega^s_m(\mathcal{E}_i)\mathcal{E}_s + \sum_s \omega^s_m(\mathcal{E}_i)\sum_r \omega^r_s(\mathcal{E}_j)\mathcal{E}_r$$

$$- \sum_s \mathcal{E}_i\omega^s_m(\mathcal{E}_j)\mathcal{E}_s - \sum_s \omega^s_m(\mathcal{E}_j)\sum_r \omega^r_s(\mathcal{E}_i)\mathcal{E}_r$$

$$= -\sum_s \left[\mathcal{E}_i\omega^s_m(\mathcal{E}_j) - \mathcal{E}_j\omega^s_m(\mathcal{E}_i) - \omega^s_m([\mathcal{E}_i, \mathcal{E}_j])\right]\mathcal{E}_s$$

$$+ \sum_r \left[\sum_s \omega^s_m(\mathcal{E}_i)\omega^r_s(\mathcal{E}_j) - \omega^s_m(\mathcal{E}_j)\omega^r_s(\mathcal{E}_i)\right]\mathcal{E}_r$$

$$= -\sum_r \left(d\omega^r_m(\mathcal{E}_i, \mathcal{E}_j) - \sum_s \omega^s_m \wedge \omega^r_s(\mathcal{E}_i, \mathcal{E}_j)\right)\mathcal{E}_r$$

$$= -\sum_r ((d\omega^r_m - \sum_s \omega^s_m \wedge \omega^r_s)(\mathcal{E}_i, \mathcal{E}_j))\mathcal{E}_r.$$

If we let $\Omega^r_m = d\omega^r_m - \sum_s \omega^s_m \wedge \omega^r_s$, then we obtain

$$R^r_{ijm} = -\Omega^r_m(\mathcal{E}_i, \mathcal{E}_j).$$

The matrix of 2-forms $\Omega = (\Omega^r_m)$ is known as *the curvature 2-form*. Now, we could have defined Ω by the formula $R^r_{ijm} = -\Omega^r_m(\mathcal{E}_i, \mathcal{E}_j)$ and then we would have

Theorem 6.2: The Second Structure Equation. If Ω denotes the curvature 2-form and ω^r_m are the connection 1-forms determined by ∇, then

$$\Omega^r_m = d\omega^r_m - \sum_s \omega^s_m \wedge \omega^r_s.$$

Whether we define Ω in terms of R or derive the relationship, we see that we can determine curvature by combining the exterior derivative and wedge product of the connection 1-forms.

Example 6.2. Suppose $M: \mathbf{x}(u, v) \subset \mathbb{R}^3$ is a surface patch with $\mathbf{x}_u \cdot \mathbf{x}_u = E$, $\mathbf{x}_u \cdot \mathbf{x}_v = 0$ and $\mathbf{x}_u \cdot \mathbf{x}_v = G$. Define a frame by

$$\mathcal{E}_1 = \frac{\mathbf{x}_u}{\sqrt{E}} \qquad \text{and} \qquad \mathcal{E}_2 = \frac{\mathbf{x}_v}{\sqrt{G}}.$$

Let θ^1 and θ^2 be the dual frame having $\theta^i(\mathcal{E}_j) = \delta^i_j$. Hence, in terms of the coordinate 1-forms, $\theta^1 = \sqrt{E}\, d\mathbf{x}_u$ and $\theta^2 = \sqrt{G}\, d\mathbf{x}_v$. Then,

$$d\theta^1 = \frac{E_u}{2\sqrt{E}}\, d\mathbf{x}_u \wedge d\mathbf{x}_u + \frac{E_v}{2\sqrt{E}}\, d\mathbf{x}_v \wedge d\mathbf{x}_u$$

$$= -\frac{E_v}{2\sqrt{EG}}\, d\mathbf{x}_u \wedge \theta^2$$

since $d\mathbf{x}_u \wedge d\mathbf{x}_u = 0$ and $d\mathbf{x}_v = \theta^2/\sqrt{G}$. Similarly,

$$d\theta^2 = -\frac{G_u}{2\sqrt{EG}}\, d\mathbf{x}_v \wedge \theta^1.$$

Now, if we write $\omega_2^1 = a\, d\mathbf{x}_u + b\, d\mathbf{x}_v$, then $\omega_1^2 = -a\, d\mathbf{x}_u - b\, d\mathbf{x}_v$ and $d\theta^1 = a\, d\mathbf{x}_u \wedge \theta^2$ and $d\theta^2 = -b\, d\mathbf{x}_v \wedge \theta^1$. This means that

$$a = -\frac{E_v}{2\sqrt{EG}} \qquad \text{and} \qquad b = \frac{G_u}{2\sqrt{EG}}.$$

Hence, $\omega_2^1 = -\frac{E_v}{2\sqrt{EG}}\, d\mathbf{x}_u + \frac{G_u}{2\sqrt{EG}}\, d\mathbf{x}_v$ and

$$d\omega_2^1 = \frac{1}{2}\left(\left(-\frac{E_v}{\sqrt{EG}}\right)_v d\mathbf{x}_v \wedge d\mathbf{x}_u + \left(\frac{G_u}{\sqrt{EG}}\right)_u d\mathbf{x}_u \wedge d\mathbf{x}_v\right)$$

$$= \frac{1}{2\sqrt{EG}}\left(\left(\frac{E_v}{\sqrt{EG}}\right)_v + \left(\frac{G_u}{\sqrt{EG}}\right)_u\right)\theta^1 \wedge \theta^2$$

since $\theta^1/\sqrt{E} = d\mathbf{x}_u$ and $\theta^2/\sqrt{G} = d\mathbf{x}_v$. Therefore,

$$d\omega_2^1 = -K\,\theta^1 \wedge \theta^2,$$

where K is Gauss curvature, by Theorem 3.4.1. Of course, we have met the 1-form ω_2^1 before in Chapter 6, but there we confined it to be along a curve so that it could be thought of as a function. Here its true nature as a 1-form is revealed. The formula above, which is surely one of the most beautiful in all of Mathematics, also explains results such as Exercise 6.3.7. (Why?)

Example 6.3: The sphere S^2. We have $E = R^2 \cos^2 v$, $F = 0$ and $G = R^2$ for the usual patch on S^2. A dual frame is then given by

$$\theta^1 = R \cos v \, d\mathbf{x}_u \qquad \text{and} \qquad \theta^2 = R \, d\mathbf{x}_v.$$

The exterior derivatives are

$$d\theta^1 = \sin v \, d\mathbf{x}_u \wedge \theta^2 \qquad \text{and} \qquad d\theta^2 = 0.$$

Hence, $\omega_2^1 = \sin v \, d\mathbf{x}_u$ and we have

$$d\omega_2^1 = \cos v \, d\mathbf{x}_v \wedge d\mathbf{x}_u$$

$$= -\frac{\cos v}{R^2 \cos v} \, \theta^1 \wedge \theta^2$$

$$= -\frac{1}{R^2} \, \theta^1 \wedge \theta^2.$$

Therefore, comparing this result with the formula $d\omega_2^1 = -K \, \theta^1 \wedge \theta^2$, we obtain $K = 1/R^2$ just as we should.

EXERCISE 6.4. Compute K for other surfaces using the formula $d\omega_2^1 = -K \, \theta^1 \wedge \theta^2$.

In modern geometry and physics, differential forms play an important role, both conceptually and in terms of calculation. For a classical approach to forms, see [Fla]. The reader will also find many applications to physics there. A newer treatment is given in [Dar], where forms are applied to understand modern gauge theories. For an approach to surface geometry via forms, see [O'N1].

LIST OF EXAMPLES, DEFINITIONS, AND REMARKS

Chapter 3: Curvature(s)

Chapter 4: Constant Mean Curvature Surfaces

Chapter 5: Geodesics, Metrics and Isometries

ANSWERS AND HINTS
TO SELECTED EXERCISES

CHAPTER 1: THE GEOMETRY OF CURVES

1.1 Let $p = (-1, 0, 5)$ and $q = (3, -1, -2)$ and substitute in the equation $\alpha(t) = p + t(q - p)$.

1.2 $\alpha(t) = (-1 + t, 6 - 5t, 5 - 8t, 9t)$.

1.3 The vector $(v_1, v_2, 0)$ is the hypotenuse of the right triangle whose sides are the vector $(v_1, 0, 0)$ and a translation of the vector $(0, v_2, 0)$. The vector (v_1, v_2, v_3) is the hypotenuse of the right triangle whose sides are the vector $(v_1, v_2, 0)$ and a translation of the vector $(0, 0, v_3)$.

1.4 $135°$ intersection.

1.11 Consider $\alpha(0)$, $\alpha(\frac{\pi}{2})$, $\alpha(\pi)$, and $\alpha(\frac{3\pi}{2})$. What is the distance of each of these points from the origin? Also note that $\alpha(t) \cdot \alpha'(t) = 0$ and that $\alpha''(t) = -\alpha(t)$.

1.12 Start with the basic parametrization $\alpha(t) = (r\cos t, r\sin t)$ and recognize that $t = 0$ should correspond to the point $(r + a, b)$ in the xy-plane. Similarly, $t = \frac{\pi}{2}$ should correspond to the point $(a, r + b)$.

1.13 $\alpha(t) = (a\cos(t), b\sin(t))$.

2.1 Use the chain rule on $\beta(s) = \alpha(h(s))$.

2.2 Show that $|\alpha'(t)| = r$; $s(t) = rt$; and $t(s) = \frac{s}{r}$. Use the definition $\beta(s) = \alpha(t(s))$ to obtain $\beta(s) = (r\cos\frac{s}{r}, r\sin\frac{s}{r})$.

2.3 Show that $|\alpha'(t)| = \sqrt{a^2\sin^2 t + b^2\cos^2 t}$. Is $s(t) = \int_0^t |\alpha'(u)|du$ integrable in closed form?

2.3 Using the fact that $T \cdot e_1 = |T||e_1|\cos\theta$ and taking derivatives of both sides, first show that $\kappa N \cdot e_1 = -\sin\theta\frac{d\theta}{ds}$. There are two normals to β at any point on the curve, so we have two cases: (1) $N = N_1$, where the angle between N_1 and e_1 is $\theta + \frac{\pi}{2}$; and (2) $N = N_2$, where the angle between N_2 and e_1 is $\frac{\pi}{2} - \theta$. Use $\kappa N \cdot e_1 = -\sin\theta\frac{d\theta}{ds}$ and the definition of the dot product to show that, for $N = N_1$, $\kappa = \frac{d\theta}{ds}$, and for $N = N_2$, $\kappa = -\frac{d\theta}{ds}$.

3.2 Recall that cofactor expansion gives

$$\begin{bmatrix} i & j & k \\ v_1 & v_2 & v_3 \\ w_1 & w_2 & w_3 \end{bmatrix} = \begin{bmatrix} v_2 & v_3 \\ w_2 & w_3 \end{bmatrix} i - \begin{bmatrix} v_1 & v_3 \\ w_1 & w_3 \end{bmatrix} j + \begin{bmatrix} v_1 & v_2 \\ w_1 & w_2 \end{bmatrix} k.$$

3.3 Recall that interchanging two rows of a determinant changes the sign of the determinant. What does this imply about $w \times v$ and $v \times w$?

3.4 Try Maple.

3.5 Rewrite Lagrange's Identity as

$$|v \times w|^2 = |v|^2|w|^2 - |v|^2|w|^2 \cos^2 \theta$$

and simplify the right-hand side to obtain the desired result.

3.7 The area of a parallelogram is given by bh, where b is the length of the base and h is the altitude. In the parallelogram spanned by v and w, h is equal to $|v| \sin \theta$, where θ is the angle between v and w.

3.8 (1) Show that $\beta'(s) = \left(\frac{\sqrt{1+s}}{2}, \frac{-\sqrt{1-s}}{2}, \frac{\sqrt{2}}{2} \right)$.

(2) Note that $T' = \beta'' = \left(\frac{1}{4\sqrt{1+s}}, \frac{1}{4\sqrt{1-s}}, 0 \right)$. Also, since $T' = \kappa N$, $|T'| = |\kappa||N| = \kappa$.

(3) Since $T' = \kappa N$, $N = \frac{T'}{\kappa}$.

(4) $B' = -\tau N$, so $|B'| = |-\tau||N|$, or $|B'| = |-\tau| = \tau$.

3.16 Use the fact that $p = \beta(s) + r(s)\beta'(s)$ for some function $r(s)$. Differentiate both sides to obtain

$$(1 + r'(s))T + r(s)(\kappa N) = 0.$$

Take the dot products of both sides with N to establish that $\kappa r(s) = 0$ and with T to establish that $r(s) \neq 0$.

3.17 Let $\alpha(s) - p = aT + bN + cB$ so that $T \cdot (\alpha - p) = a$, $N \cdot (\alpha - p) = b$, and $B \cdot (\alpha - p) = c$. Recognize that $(\alpha - p) \cdot (\alpha - p) = R^2$ since α lies on a sphere of center p and radius R. Take derivatives of both sides of this equation to obtain an expression for $T \cdot (\alpha - p)$. Then take derivatives of both sides of $T \cdot (\alpha - p) = a$ to obtain an expression for $N \cdot (\alpha - p)$. Finally, take derivatives of both sides of $N \cdot (\alpha - p) = b$ to obtain an expression for $B \cdot (\alpha - p)$.

3.18 Use the previous problem. Let the constant be R^2 and show that $\alpha + \frac{1}{\kappa}N + \frac{1}{\tau}(\frac{1}{\kappa})'B$ is a constant.

4.1 If the road is *not* banked, $\alpha''(t)$ can be resolved into two components: (1) tangential acceleration $= \frac{dv}{dt}T(t) = 0$ since the car is traveling at a constant speed; and (2) centripetal acceleration $= \kappa v^2 N(t)$. By Newton's Law, the magnitude of the force due to centripetal acceleration is $|m\kappa v^2 N(t)| = m\kappa v^2$, which must be balanced by the force due to friction, μmg.

If the road *is* banked, there are three primary forces acting on the car: (1) a downward force mg due to gravity; (2) a corresponding normal force exerted by the road; and (3) a kinetic frictional force preventing the car from flying off the road. The static frictional force preventing the car from sliding downward is negligible.

Recall from physics that $|f| = \mu|N|$, yielding $f_x = \mu N_y$ and $f_y = \mu N_x$. Summing the vertical forces on the car yields

$$N_y = mg + f_y = mg + \mu N_x.$$

The total of the horizontal forces, $f_x + N_x$, produces the centripetal acceleration, so we have

$$m\kappa v^2 = f_x + N_x = \mu N_y + N_x.$$

Solve simultaneously for N_x and N_y to obtain

$$N_x = \frac{m}{1+\mu^2}[\kappa\nu^2 - \mu g]$$

$$N_y = \frac{m}{1+\mu^2}[g + \mu\kappa\nu^2].$$

An expression for $\tan\theta$ may now be obtained. Solve this expression for ν and obtain

$$\nu \le \sqrt{\frac{g(\tan\theta + \mu)}{\kappa(1 - \mu\tan\theta)}}.$$

4.2 Use $\kappa = \frac{|\alpha' \times \alpha''|}{|\alpha'|^3}$ for both parts of this problem. In the case of the general plane curve $\alpha(t) = (x(t), y(t))$, we have $\alpha'(t) = (x'(t), y'(t))$ and $\alpha''(t) = (x''(t), y''(t))$, leading to:

$$\kappa = \frac{x'(t)y''(t) - x''(t)y'(t)}{[(x'(t))^2 + (y'(t))^2]^{\frac{3}{2}}}.$$

4.5 As part of Exercise 4.4, we have $|\alpha'(t)| = \sqrt{2}\cosh t$.

5.1 Since $\gamma(s)$ is not necessarily a unit speed parametrization, use $\kappa_\gamma = \frac{|\gamma' \times \gamma''|}{|\gamma'|^3}$.

    (1) Show that $\gamma' \times \gamma'' = \kappa B - \kappa\cos\theta(u \times N)$.
    (2) Show that $u \times N = \cos\theta B - \sin\theta T$.
    (3) Show that $|\gamma' \times \gamma''| = \kappa\sin\theta$.
    (4) Show that $|\gamma'| = \sin\theta$.

5.2 ($\Rightarrow$) β a circular helix $\Rightarrow$ γ a circle $\Rightarrow$ κ_γ is constant. A circular helix is a special case of a cylindrical helix. Thus, $T \cdot u = \cos\theta$ is constant. What do these results imply about $\kappa = \kappa_\gamma \sin^2\theta$? Finally, use the fact that for a cylindrical helix, $\frac{\tau}{\kappa}$ is a constant.

    ($\Leftarrow$) τ and κ constant $\Rightarrow$ $\frac{\tau}{\kappa} = \cot\theta$ is constant. Use this to show that κ_γ is constant. Also show that $\tau_\gamma = 0$ and conclude that γ is a circle.

2.19 Use the fact that $\kappa = \tau = \frac{1}{\sqrt{8(1-s^2)}}$ (Exercise 2.10) to show that $\cot\theta = \frac{\tau}{\kappa}$ is a constant.

2.20 Use $\frac{|\beta' \times \beta''|}{|\beta'|^3}$ to compute $\kappa = \frac{1}{4}$. Use $\frac{(\beta' \times \beta'') \cdot \beta'''}{|\beta' \times \beta''|^2}$ to compute $\tau = -8$. What is true if both τ and κ are constants?

5.5 Prove that $\frac{\tau}{\kappa}$ constant $\Leftrightarrow 4b^4 = 9a^2$ by using MAPLE.

5.6 WLOG assume that β has unit speed. Show that $\frac{T'(s)}{\kappa} \cdot u = 0$ or, equivalently, $\frac{1}{\kappa}(T'(s) \cdot u) = 0$. Use the fact that $(T(s) \cdot u)' = T'(s) \cdot u + T(s) \cdot u' = T'(s) \cdot u$ since u is a constant vector.

CHAPTER 2: SURFACES

1.1 ($\Rightarrow$) If x_u and x_v are linearly dependent, then $x_u = cx_v$ where c is a scalar. Use the fact that $x_u \times x_v$ may be expressed as a determinant and use properties of determinants.

    ($\Leftarrow$) $x_u \times x_v = 0 \Rightarrow |x_u||x_v|\sin\theta = 0$. What does this imply about θ, the angle between x_u and x_v?

1.2 Use a Monge patch, $x(u,v) = (u, v, u^2 + v^2)$; determine the u-parameter curve, $x(u, v_0)$, and the v-parameter curve, $x(u_0, v)$. Note that each of the parameter curves lies in a coordinate plane of $\mathbb{R}^3$.

1.4 $\mathbf{x}(u,v) = (u, v, +\sqrt{1 - u^2 - v^2})$.

1.10 A ruling patch for a cone is of the form $x(u,v) = p + v\delta(u)$ where p is a fixed point. Let $p = (0,0,0)$ as the cone emanates from the origin. The line that is to sweep out the surface must thus extend from $(0,0,0)$ to a point on the circle $(a\cos u, a\sin u, a)$ lying parallel to and above the xy-plane. (The z-coordinate of any point on the circle is a because $z = \sqrt{x^2 + y^2} = \sqrt{a^2\cos^2 u + a^2\sin^2 u} = a$.) A ruling patch for a cylinder is of the form $x(u,v) = \beta(u) + vq$ where q is a fixed direction vector. The directrix $\beta(u)$ for a standard cylinder is the unit circle in the xy-plane $(\cos u, \sin u, 0)$. We want a standard, right circlular cylinder, so $q = (0,0,1)$.

1.11 To show that a surface is doubly ruled, we need to identify two ruling patches for the surface. Since $z = xy = f(x,y)$, we can use a Monge patch $x(u,v) = (u, v, uv)$ and write $x(u,v)$ in terms of $\beta(u) = (u, 0, 0)$ and $\delta(u) = (0, 1, u)$. Alternatively, $y(u,v) = (v, u, vu)$ is also a patch for the surface.

1.12 A patch for the helicoid is $x(u,v) = (0, 0, bu) + v(a\cos u, a\sin u)$.

1.13 A directrix for the hyperboloid of one sheet is the ellipse $\beta(u) = (a\cos u, b\sin u, 0)$. Let $\delta(u) = \beta'(u) + (0, 0, c)$. It can be shown that $x(u,v) = \beta(u) + v\delta(u)$ is indeed a patch for the hyperboloid. Alternatively, let $\beta(u)$ be as above and let $\delta(u) = \beta'(u) + (0, 0, -c)$ and verify that this, also, is a patch for the hyperboloid.

2.1 By definition,

$$v[fg] = \frac{d}{dt}(fg(\alpha(t)))\mid_{t=0} = \nabla fg(p) \cdot v.$$

Express ∇fg as $\left(\frac{\partial(fg)}{\partial x}, \frac{\partial(fg)}{\partial y}, \frac{\partial(fg)}{\partial z}\right)$ and recognize that $\frac{\partial(fg)}{\partial x} = \frac{\partial f}{\partial x}g + \frac{\partial g}{\partial x}f$. Finally, collect like terms to obtain the desired result, $v[fg] = v[f]g + fv[g]$.

2.2 $x = f(p_1, p_2, p_3)$ and $v = (v_1, v_2, v_3)$. Thus, by definition, $v[x] = \left(\frac{\partial x}{\partial p_1}, \frac{\partial x}{\partial p_2}, \frac{\partial x}{\partial p_3}\right) \cdot (v_1, v_2, v_3)$. But since $x(p_1, p_2, p_3) = p_1$, $\frac{\partial x}{\partial p_2} = \frac{\partial x}{\partial p_3} = 0$ and $\frac{\partial x}{\partial p_1} = 1$. A similar procedure may be used for $v[y]$ and $v[z]$.

2.4 Let $\alpha(t) = x(a_1(t), a_2(t))$, $\beta(t) = x(b_1(t), b_2(t))$ with $\alpha(0) = p = \beta(0)$ and $\alpha'(0) = v$, $\beta'(0) = w$. Then if $\gamma(t) = x(a_1(t) + b_1(t), a_2(t) + b_2(t))$, $\gamma'(t) = x_u\frac{du}{dt} + x_v\frac{dv}{dt} = x_u(a_1'(t) + b_1'(t)) + x_v(a_2'(t) + b_2'(t))$. Finding $\alpha'(t)$ and $\beta'(t)$ and substituting yields $\gamma'(0) = v + w$. Thus, $(v+w)[f] = \gamma'(0)[f] = \nabla f \cdot \gamma'(0)$. By using $v[f] + w[f] = \nabla f \cdot v + \nabla f \cdot w$, show that $(v+w)[f] = v[f] + w[f]$.

3.3 To compute the eigenvalues of a matrix S, we set $\det(\lambda I - S) = 0$. This yields, in the case of the 2×2 symmetric matrix

$$\begin{bmatrix} a & b \\ b & c \end{bmatrix},$$

the equation $\lambda^2 - (a+c)\lambda + ac - b^2 = 0$. Solve for λ_1 and λ_2 and show that they are both real.

4.2 For the first part, just use $S(\alpha') = -\nabla_{\alpha'}U$. For the second (which is also an if and only if), show that both $S(\alpha')$ and α' are in $P \cap T_{\alpha(t)}M$ for each t.

CHAPTER 3: CURVATURE(S)

1.1 Since $K = k_1 k_2$, k_1 and k_2 must be of opposite sign. Because $k_1(u)$ is defined to be the maximum curvature, $k_1(u) > k_2(u)$, so $k_1 > 0$ and $k_2 < 0$ is the case here.

1.4 (1) Euler's formula states that $k(u) = \cos^2 \theta k_1 + \sin^2 \theta k_2$ where $u = \cos \theta u_1 + \sin \theta u_2$ (i.e. u is a function of θ). Thus,

$$\frac{1}{2\pi} \int_0^{2\pi} k(\theta)d\theta = \frac{1}{2\pi} \int_0^{2\pi} (\cos^2 \theta k_1 + \sin^2 \theta k_2)d\theta.$$

Evaluate the integral, remembering that k_1 and k_2 are constants.
(2) Express v_1 and v_2 in terms of u_1 and u_2. That is, $v_1 = \cos \phi u_1 + \sin \phi u_2$ and $v_2 = \cos(\phi + \frac{\pi}{2})u_1 + \sin(\phi + \frac{\pi}{2})u_2$. Use Euler's formula to obtain $k(v_1)$ and $k(v_2)$.

1.7 M minimal $\Rightarrow H(p) = 0$ for every $p \in M$. What does this imply about k_1 and k_2 and, in turn, about K?

2.13 (a) Compute x_u, x_v, and $x_u \times x_v$ to obtain $U = \frac{\beta' \times \delta + v\delta' \times \delta}{W}$. Show that $(U \cdot x_{uv})^2 = \frac{(\beta' \cdot \delta \times \delta')^2}{W^2}$. To do so, you will need to recall that $a \cdot (b \times c) = -b \cdot (a \times c)$. Use the Lagrange Identity to show that $|x_u \times x_v|^2 = EG - F^2 = W^2$. Finally, show that $K = \frac{-(U \cdot x_{uv})^2}{EG - F^2}$ and combine with the above.
(b) From Exercise 3.11, a ruling patch for the saddle surface is $x(u, v) = (u, 0, 0) + v(0, 1, u)$. Then $\beta(u) = (u, 0, 0)$ and $\delta(u) = (0, 1, u)$. Use the results of (a) to obtain $K = \frac{-1}{(x^2 + y^2 + 1)^2}$.
(c) Note that $\beta(u) = (p_1, p_2, p_3)$ so that $\beta'(u) = (0, 0, 0)$.
(d) Note that $\delta(u) = (q_1, q_2, q_3)$ so that $\delta'(u) = (0, 0, 0)$.

2.15 For one direction, note that U cannot depend on v only when the term $v(\delta' \times \delta) = 0$. Thus, $\delta' \times \delta = 0$ and the formula for K of a ruled surface shows $K = 0$. For the other direction, note that $U_v = -S(x_v)$ is a tangent vector. Show that $U_v \cdot x_v = 0$ (automatically!) and $U_v \cdot x_u = 0$ by the hypothesis $K = 0$ (and the formula for K of a ruled surface).

2.18 If β is a line of curvature, then $\beta' \cdot U \times U' = \beta' \cdot U \times c\beta' = 0$ (why?). For the other way, show that developable implies that U' is perpendicular to $\beta' \times U$ which is also perpendicular to β'. Then note that all these vectors are in the tangent plane.

2.19 (a) and (b) are self-explanatory. In (c), use the results of part (b) in the expressions $K = \frac{ln - m^2}{EG - F^2}$ and $H = \frac{Gl + En - 2Fm}{2(EG - F^2)}$ and simplify. For part (d), recognize (using results of (c)), that D is the numerator of K evaluated at the critical point (u_0, v_0). Since the denominator of K is always positive, $D = 0 \Rightarrow K = 0$, $D < 0 \Rightarrow K < 0$, and $D > 0 \Rightarrow K > 0$. What must be true of k_1 and k_2 when $K = 0$? When $K < 0$? What do these results imply about the surface? Two sub-cases correspond to $K > 0$. If $f_{uu}(u_0, v_0)$ is positive, k_1 and k_2 must both be positive. If $f_{uu}(u_0, v_0)$ is negative, k_1 and k_2 must both be negative. What must be true of the surface in each of these sub-cases?

3.3 $F = m = 0$ for a surface of revolution. Thus, x_u and x_v are orthogonal and we can express $S(x_u)$ in terms of the basis vectors x_u and x_v. Thus, let $S(x_u) = ax_u + bx_v$. Compute $S(x_u) \cdot x_u$ and recognize that this is equal to l. Compute

$S(x_u) \cdot x_v$ and recognize that this is equal to m. Similarly, let $S(x_v) = cx_u + dx_v$ and take dot products with x_v and x_u.

3.6 (a) Derive

$$K = \frac{1 - u^2}{(1 + u^2 e^{-u^2})^2}$$

by using the expression for K for a surface of revolution. Algebraically determine when $K > 0$, $K = 0$, and $K < 0$. Part (b) is similar; parametrize the ellipse as $\alpha(u) = (R + a\cos u, b\sin u, 0)$ to produce the following patch for the elliptical torus: $x(u, v) = ((R + a\cos u)\sin v, b\sin u, (R + a\cos u)\cos v)$. Using the expression for K for a surface of revolution yields

$$K = \frac{ab^2 \cos u}{(R + a\cos u)(b^2 \cos^2 u + a^2 \sin^2 u)^2}.$$

3.9 By separating variables, we obtain the expression

$$u = \int \sqrt{\frac{1}{h^2} - 1}\, dh.$$

Make the substitution $h = 1/\cosh w$ to obtain $u = \int \tanh^2 w\, dw$. Integrate, then use the fact that $\cosh^{-1}\left(\frac{1}{h}\right) = \ln\left(\frac{1}{h} + \sqrt{\frac{1}{h^2} - 1}\right)$ and recall that $\tanh x = \frac{e^x - e^{-x}}{e^x + e^{-x}}$. Simplify to obtain $u = \ln\left|\frac{1}{h} + \frac{\sqrt{1 - h^2}}{h}\right| - \sqrt{1 - h^2} + C$.

4.2 To verify the expression for U_u, recall that a u-parameter velocity vector applied to a function of u and v takes the u-partial derivative of that function. Thus,

$$\nabla_{x_u} U = (x_u[u_1], x_u[u_2], x_u[u_3]) = U_u.$$

Then, since x_u and x_v form a basis for $T_p(M)$ and since ∇_{x_u} is in $T_p(M)$, we have $\nabla_{x_u} U = Ax_u + Bx_v$. Take the dot product of both sides to obtain $\nabla_{x_u} U \cdot x_u = Ax_u \cdot x_u = AE$. Also recognize that $0 = x_u[0] = x_u[U \cdot x_u] = \nabla_{x_u} U \cdot x_u + U \cdot x_{uu}$ and use this to show that $A = -l/E$. Proceed in a similar manner to find B and to obtain an expression for U_v.

4.3 By finding the two partial derivatives $\left(\frac{E_v}{2G}\right)_v$ and $\left(\frac{G_u}{2G}\right)_u$, we obtain, as an equivalent expression for the righthand side,

$$\frac{E_u G_u}{4E^2 G} - \frac{2GE_{vv}}{4G^2 E} + \frac{E_v G_v}{4G^2 E} + \frac{E_v E_v}{4E^2 G} - \frac{2GG_{uu}}{4G^2 E} + \frac{G_u G_u}{4G^2 E}.$$

Combine these terms over the common denominator $4E^2 G^2$. Next, compute

$$\frac{\partial}{\partial v}\left(\frac{E_v}{\sqrt{EG}}\right) = \frac{E_{vv}}{\sqrt{EG}} - \frac{E_v E_v G}{2(EG)^{\frac{3}{2}}} - \frac{EE_v G_v}{2(EG)^{\frac{3}{2}}}$$

and

$$\frac{\partial}{\partial u}\left(\frac{G_u}{\sqrt{EG}}\right) = \frac{G_{uu}}{\sqrt{EG}} - \frac{G_u E_u G}{2(EG)^{\frac{3}{2}}} - \frac{G_u E G_u}{2(EG)^{\frac{3}{2}}}.$$

Substitute into the given expression and write the result over the common denominator $4E^2 G^2$ to obtain the same expression as above.

4.4 A patch for a sphere of radius R is $x(u,v) = (R\cos u \cos v, R\sin u \cos v, R\sin v)$. Compute x_u, x_v, E, and G and substitute into the given expression to obtain $K = 1/R^2$.

5.1 ($\Rightarrow$) If two vectors v and w are equal and are written in terms of the same basis vectors x_u, x_v, and U, then their corresponding coefficients must be equal (*i.e.*,$v_1 = w_1$,$v_2 = w_2$, and $v_3 = w_3$). Compute $v \cdot x_u$ and $w \cdot x_u$ and use this fact to show that the two dot products are equal. Similarly, take dot products with x_v and U to obtain the other two equations.

($\Leftarrow$) Computing $v \cdot x_u$, $w \cdot x_u$, $v \cdot x_v$, and $w \cdot x_v$ and using the hypotheses $v \cdot x_u = w \cdot x_u$, $v \cdot x_v = w \cdot x_v$, we obtain the following two equations:

$$(v_1 - w_1)E + (v_2 - w_2)F = 0$$
$$(v_1 - w_1)F + (v_2 - w_2)G = 0.$$

Solve simulataneously and use the fact that, in $\mathbb{R}^3$, $EG - F^2 \neq 0$ to obtain $v_1 = w_1$ and $v_2 = w_2$. Use the hypothesis $v \cdot U = w \cdot U$ to show that $v_3 = w_3$.

5.2 Since S_p is a linear transformation from $T_p(M)$ to itself, we can write $S(x_u) = Ax_u + Bx_v$. But since p is an umbilic point, we have $S(x_u) = kx_u \Rightarrow B = 0$. Now use

$$l = S(x_u) \cdot x_u = AE + BF, \quad m = S(x_u) \cdot x_v = AF + BG$$

and solve for B to get $B = \frac{-Fl + Em}{EG - F^2} = 0$, so $-Fl + Em = 0$ or $l/E = m/F$. Do the same for $S(x_v)$.

5.4 A patch for a surface of revolution is given by $x(u,v) = (u, h(u)\cos v, h(u)\sin v)$. Then $K = \frac{-h''}{h(1+h'^2)^2} = 0 \Rightarrow -h'' = 0$. Thus, $h(u) = C_1 u + C_2$ (a line). Note that, if $C_1 = 0$, a cylinder is generated by revolving $h(u)$ about the x-axis; if $C_1 \neq 0$, a cone is generated.

CHAPTER 4: CONSTANT MEAN CURVATURE SURFACES

2.1 Use a Monge patch $x(u,v) = (u, v, f(u,v))$ to obtain $f_u = g'(u)$, $f_{uu} = g''(u)$, $f_v = h'(v)$, $f_{uv} = 0$, and $f_{vv} = h''(v)$. Then

$$H = 0 \Leftrightarrow (1 + h'^2(y))g''(x) + (1 + g'^2(x))h''(y) = 0.$$

Separate variables to obtain

$$\frac{-g''(x)}{1 + g'^2(x)} = \frac{h''(y)}{1 + h'^2(y)}.$$

Since x and y are independent, each side is constant relative to the other side. Thus, let

$$a = \frac{-g''(x)}{1 + g'^2(x)}.$$

Also let $w = g'(x)$ so that $g''(x) = \frac{dw}{dx}$ and integrate to obtain $w = \frac{dg}{dx} = -\tan ax$. Integrating again gives $g(x) = \frac{1}{a}\ln(\cos ax)$. Apply the same reasoning to the other side of the original differential equation to obtain $h(y) = -\frac{1}{a}\ln(\cos ay)$. Combining terms yields

$$f(x, y) = \frac{1}{a}\ln\left(\frac{\cos ax}{\cos ay}\right).$$

2.4 $\frac{d\tau_\beta}{du} = (\beta' \times \delta)' \cdot \delta' + (\beta' \times \delta) \cdot \delta''$. Both terms in this expression are zero.

3.2 Compute

$$\frac{\partial P}{\partial u} = \frac{(f_{uu}V + f_u V_u)(1 + f_u^2 + f_v^2) - V f_u^2 f_{uu} - V f_u f_v f_{uv}}{(1 + f_u^2 + f_v^2)^{\frac{3}{2}}}$$

and

$$\frac{\partial Q}{\partial v} = \frac{(f_{vv}V + f_v V_v)(1 + f_u^2 + f_v^2) - V f_v^2 f_{vv} - V f_u f_v f_{uv}}{(1 + f_u^2 + f_v^2)^{\frac{3}{2}}}.$$

Then apply Green's Theorem:

$$\int_v \int_u \frac{\partial P}{\partial u} + \frac{\partial Q}{\partial v}\, du dv = \int_C P\, dv - Q\, du.$$

4.1 Compute partial derivatives to obtain

$$\frac{\partial P}{\partial u} + \frac{\partial Q}{\partial v} = -V_u \cdot (U \times x_v) + V_v \cdot (U \times x_u) + V \cdot [U_v \times x_u - U_u \times x_v].$$

Since U is a function on M, we have $U_v = \nabla_{x_v}U = -S(x_v)$ and $U_u = \nabla_{x_u}U = -S(x_u)$. Substituting in the above equation and yields

$$\frac{\partial P}{\partial u} + \frac{\partial Q}{\partial v} = V_v \cdot (U \times x_u) + V_u \cdot (U \times x_v) + V \cdot (2Hx_u \times x_v).$$

Now apply Green's Theorem.

CHAPTER 5: GEODESICS, METRICS AND ISOMETRIES

1.7 A parametrization of a right circular cylinder is given by $x(u, v) = (R\cos u, R\sin u, bv)$. Then a curve on the surface is given by $\alpha(t) = (R\cos u(t), R\sin u(t), bv(t))$. Find α'' by differentiating twice and noting that the chain rule gives $\frac{d}{dt}\cos u(t) = -\sin u \frac{du}{dt}$. Now, $\alpha'' = \alpha''_{\tan} + (\alpha'' \cdot U)U$, where U in this case is $(\cos u, \sin u, 0)$. Taking the dot product of α'' with U yields $\alpha'' \cdot U = -R(\frac{du}{dt})^2$. From this and from $\alpha'' = \alpha''_{\tan} + (\alpha'' \cdot U)U$, we know that $\alpha_{\tan} = 0$ results in $\frac{d^2u}{dt^2} = 0$ and $\frac{d^2v}{dt^2} = 0$, yielding $u(t) = k_1 t + c_1$ and $v(t) = k_2 t + c_2$. Thus, $\alpha(t) = (R\cos(k_1 t + c_1), R\sin(k_1 t + c_1), b(k_2 t + c_2))$. Finally, consider the following cases: (1) $c_1 = c_2 = 0$, $k_1 = k_2 = 1$; (2) $k_1 = 1$, $k_2 = 0$, $c_1 = 0$, $c_2 \neq 0$; (3) $k_1 = 1$, $k_2 = 0$, $c_1 = 0$, $c_2 = \frac{1}{b}$.

2.6

$$\sqrt{G}\sin\phi = \sqrt{G}\cos(\pi/2 - \phi) = x_v \cdot \alpha' = Gv'$$

since $\alpha' = x_u u' + x_v v'$. Now use the relation $v' = c/G$ derived from the second geodesic equation.

2.7 In polar coordinates, a patch for the plane is given by $x(u,v) = (u\cos v, u\sin v)$. Compute $E = 1$, $F = 0$, and $G = u^2$, verifying that x is u-Clairut. Then we have

$$v(u) - v(u_0) = \int_{u_0}^{u} \frac{c\sqrt{E}}{\sqrt{G}\sqrt{G - c^2}} du = \int_{u_0}^{u} \frac{cdu}{u\sqrt{u^2 - c^2}}.$$

Integrate using the substitution $u = c\sec x \Rightarrow du = c\sec x\tan x dx$ to obtain $v(u) - v(u_0) = \pm\cos^{-1}\frac{c}{u}$, or $u\cos(v - v_0) = c$, the polar equation of a line.

2.8 Compute $E = 2$, $F = 0$, and $G = u^2$, verifying that the patch for the cone is u-Clairut. Then

$$v(u) - v(u_0) = \int_{u_0}^{u} \frac{c\sqrt{E}}{\sqrt{G}\sqrt{G - c^2}} du = \int_{u_0}^{u} \frac{c\sqrt{2}du}{u\sqrt{u^2 - c^2}}.$$

Integrate using the substitution $u = c\sec x$ to obtain $v(u) - v(u_0) = \sqrt{2}\sec^{-1}\frac{u}{c}$.

4.3 Compute $E = 1/(1 - u^2/4)^2$, $F = 0$, and $G = u^2/(1 - u^2/4)^2$. Then, $K =$

$$-\frac{1}{2\sqrt{EG}}\left(\frac{\partial}{\partial v}\left(\frac{E_v}{\sqrt{EG}}\right) + \frac{\partial}{\partial u}\left(\frac{G_u}{\sqrt{EG}}\right)\right) = -\frac{1}{2\sqrt{EG}}\left(\frac{\partial}{\partial u}\left(\frac{G_u}{\sqrt{EG}}\right)\right).$$

Find the required derivatives and work through the algebra to obtain $K = -1$.

CHAPTER 6: HOLONOMY AND THE GAUSS-BONNET THEOREM

1.3 A patch for the torus is given by $x(u,v) = ((R + r\cos u)\cos v, (R + r\cos u)\sin v, r\sin u)$. Find x_u, x_v, and compute $|x_u \times x_v| = r(R + r\cos u)$. Then

$$SA = \int_0^{2\pi}\int_0^{2\pi} r(R + r\cos u)dvdu = 4\pi^2 rR.$$

1.4 (1) A patch for a surface of revolution is given by $x(u,v) = (u, h(u)\cos v, h(u)\sin v)$. Find x_u, x_v, and compute $|x_u \times x_v| = h(u)(1 + h'^2(u))^{\frac{1}{2}}$. Recognize that for a surface of revolution, $h(u)$ is usually written as $f(x)$. Use the expression for surface area to complete the exercise. (2) Define a Monge patch $x(u,v) = (u, v, f(u,v))$. Then $|x_u \times x_v| = \sqrt{1 + f_u^2 + f_v^2}$.

1.5 A patch for the bugle surface (with $c = 1$) is given by

$$x(u,v) = (u - \tanh u, \operatorname{sech} u\cos v, \operatorname{sech} u\sin v).$$

Compute $|x_u \times x_v| = (\operatorname{sech}^4 u\tanh^2 u + \operatorname{sech}^2 u - 2\operatorname{sech}^4 u + \operatorname{sech}^6 u)^{\frac{1}{2}}$ and simplify using the identity $\tanh^2 u = 1 - \operatorname{sech}^2 u$, obtaining $|x_u \times x_v| = \operatorname{sech} u\tanh u$. Then

$$SA = \int_0^\infty\int_0^{2\pi} \operatorname{sech} u\tanh u\, dvdu = 2\pi.$$

1.6 (1) Total Gaussian curvature is given by

$$\int_M K = \int\int \frac{\cos u}{r(R + r\cos u)}|x_u \times x_v|\, dudv,$$

for the torus, where $|x_u \times x_v| = r(R + r\cos u)$. (2) A patch for the catenoid is given by $x(u,v) = (u, \cosh u\cos v, \cosh u\sin v)$, yielding $|x_u \times x_v| = \cosh^2 u$. Also,

$K = -1/\cosh^4 u$. Integrate $\int_M K$ to show that the total Gaussian curvature is -4π.

3.1 $\alpha'[V \cdot V] = 2\nabla_{\alpha'} V \cdot V = 0$ since V is parallel.

3.2 $\alpha'[V \cdot W] = \nabla_{\alpha'} V \cdot W + V \cdot \nabla_{\alpha'} W$. V and W parallel imply $\alpha'[V \cdot W] = 0$ and V parallel, $\alpha'[V \cdot W] = 0$, $\alpha'[W \cdot W] = 0$ imply $\nabla_{\alpha'} W$ is perpendicular to both V and W in a plane. Thus, $\nabla_{\alpha'} W = 0$ and W is parallel.

3.7 For the R-sphere, $K = 1/R^2$ and $|x_u \times x_v| = R^2 \cos v$. Integrate $\int_M K$ to show that the total Gaussian curvature above v_0 is $2\pi - 2\pi \sin v_0$. By Exercise 10.9, the holonomy around the v_0-latitude curve is $-2\pi \sin v_0$. Thus, the holonomy around a curve is equal to the total Gaussian curvature over the portion of the surface bounded by the curve (up to additions of multiples of 2π).

3.8 At the Equator, $v_0 = 0$. What is the holonomy along the Equator and what does this imply about the apparent angle of rotation of a vector moving along the Equator? What does this signify about the Equator?

4.1 But, of course, gravity really doesn't point that way on a planetary torus, does it?

5.3 The vector must come back to itself, so the total number of revolutions it makes is a multiple of 2π.

5.4 Note that the Gaussian curvature for both $\mathcal{H}$ and $\mathcal{P}$ is a constant $K = -1$. Thus we have

$$\int_\triangle K = -\int_\triangle = -\text{area of } \triangle.$$

But, since the sum of the interior angles of the triangle differs from π by $(+$ or $-)$ the total Gaussian curvature, we have

$$\Sigma i_j - \pi = -\text{area of } \triangle.$$

What does this imply about the sum of the angles, noting that area is a strictly positive quantity?

6.3 If $K \leq 0$ and $K < 0$ at even a single point, then the total Gauss curvature is negative. But the Euler characteristic of the torus is zero.

6.6 A disk has Euler characteristic 1.

7.8 Their curvatures are not bounded away from zero.

CHAPTER 7: MINIMAL SURFACES AND COMPLEX VARIABLES

1.2 For Cauchy-Riemann, $z^2 = x^2 - y^2 + i2xy$, so

$$\frac{\partial \phi}{\partial x} = 2x = \frac{\partial \psi}{\partial y} \qquad \frac{\partial \phi}{\partial y} = -2y = -\frac{\partial \psi}{\partial x}.$$

Thus, $f(z) = z^2$ is holomorphic and $f'(z) = 2x + i2y = 2z$ as it should.

1.8

$$\frac{\partial f}{\partial \bar{z}} = \frac{1}{2}\left(\frac{\partial \phi}{\partial u} + i\frac{\partial \psi}{\partial u} + i\frac{\partial \phi}{\partial v} + i^2\frac{\partial \psi}{\partial v}\right)$$

$$= \frac{\partial \phi}{\partial u} - \frac{\partial \psi}{\partial v} + i\frac{\partial \psi}{\partial u} + i\frac{\partial \phi}{\partial v}$$

$$= 0$$

by Cauchy-Riemann.

3.5 The calculations are exactly as in Example 3.1 except that an extra factor of i occurs in each term. This affects the real parts to produce $x^1 = \sinh u \sin v$, $x^2 = -\sinh u \cos v$ and $x^3 = v$; thus, a helicoid.

3.21 M is minimal with isothermal coordinates, so $l = -n$ and, consequently,

$$K = \frac{ln - m^2}{EG - F^2} = \frac{-l^2 - m^2}{E^2 - 0} = -\frac{l^2 + m^2}{E^2}.$$

3.22 A conformal Gauss map implies $U_u \cdot U_v = 0$. Plugging in the usual expressions for U_u and U_v, we get $0 = mH$. Now consider the two cases, $m = 0$ and $H = 0$.

4.7 Use Exercise 4.2 with $B = (0, 1, 0)$. For instance, $x^1 = \mathrm{Re}[\alpha^1(z) + i \int 0 \, dw] = \mathrm{Re}\,\beta(z)$, $x^3 = \mathrm{Re}[\alpha^3(z) + i \int 0 \, dw] = \mathrm{Re}\,\gamma(z)$, $x^2 = \mathrm{Re}[\alpha^2(z) + i \int 1 \cdot |\alpha'(w)| \, dw] = \mathrm{Re}[0 + i \int \sqrt{\beta'(w)^2 + \gamma'(w)^2} \, dw] = \mathrm{Im} \int \sqrt{\beta'(w)^2 + \gamma'(w)^2} \, dw$.

CHAPTER 8: THE CALCULUS OF VARIATIONS AND GEOMETRY

1.4

$$\frac{d}{dt}\left(f - \dot{x}\frac{\partial f}{\partial \dot{x}}\right) = \frac{\partial f}{\partial t} + \frac{\partial f}{\partial x}\dot{x} + \frac{\partial f}{\partial \dot{x}}\ddot{x} - \ddot{x}\frac{\partial f}{\partial \dot{x}} - \dot{x}\frac{d}{dt}\frac{\partial f}{\partial \dot{x}}$$

$$= 0 + 0 + \dot{x}\left(\frac{\partial f}{\partial x} - \frac{d}{dt}\frac{\partial f}{\partial \dot{x}}\right)$$

$$= 0$$

if and only if $\frac{\partial f}{\partial x} - \frac{d}{dt}\frac{\partial f}{\partial \dot{x}} = 0$.

1.8 $x(t) = t - \sin t$.

1.10 $x(t) = \sin t$.

2.1 The Euler-Lagrange equation for the time integral

$$T = \int \frac{\sqrt{1 + y'^2}}{ky} \, dx$$

is

$$\frac{\sqrt{1 + y'^2}}{ky} - y'\frac{y'}{ky\sqrt{1 + y'^2}} = c.$$

Then

$$\frac{1}{ky\sqrt{1 + y'^2}} = c$$

is separable with

$$\frac{y}{\sqrt{c^2 - y^2}} \, dy = dx.$$

The solutions are then $(x - a)^2 + y^2 = c^2$, circles centered on the x-axis — the geodesics of the Poincaré plane!

3.1 $x(t) = t - \sin t + b$.

3.3 $x(t) = \sin t + c$.

3.7

$$E = x \sin t + \frac{1}{2}\dot{x}^2 - x \sin t - \frac{1}{2}p^2 + x^2 - (\dot{x} - p)p$$
$$= \frac{1}{2}(\dot{x} - p)^2$$
$$\geq 0$$

3.9

$$E = \dot{x}^2 - x^2 - p^2 + x^2 - (\dot{x} - p)2p$$
$$= \dot{x}^2 - p^2 - 2p\dot{x} + 2p^2$$
$$= (\dot{x} - p)^2$$
$$\geq 0$$

4.2 The Euler-Lagrange equation is

$$\dot{x} - \lambda - \frac{d}{dt}(\dot{x} + x) = 0$$

with simplification $\ddot{x} = -\lambda$ and solution

$$x(t) = -\frac{\lambda}{2}t^2 + at + b.$$

The initial conditions give $a = \lambda/2$ and $b = 0$. Applying the constraint, we obtain

$$\frac{7}{12} = -\frac{\lambda}{2}\int_0^1 t^2 - t\,dt = \frac{\lambda}{12}$$

so that $\lambda = 7$. Finally, $x(t) = -\frac{7}{2}t^2 + \frac{7}{2}t$.

4.8 The equations of motion for the particle in the paraboloid are

$$(1 + 4u^2)\ddot{u} + 4u\dot{u}^2 + 2u - u\dot{v}^2 = 0 \qquad \dot{v} = \frac{1}{u^2}.$$

4.9 $T = m/2\,(E\dot{u}^2 + G\dot{v}^2)$, so (forgetting m)

$$E = 1/2(E\dot{u}^2 + G\dot{v}^2 - Ep_1^2 - Gp_2^2) - (\dot{u} - p_1)Ep_1 - (\dot{v} - p_2)Gp_2$$
$$= \frac{1}{2}(E(\dot{u} - p_1)^2 + G(\dot{v} - p_2)^2)$$
$$\geq 0$$

CHAPTER 9: A GLIMPSE AT HIGHER DIMENSIONS

3.3

$$[fV, gW] = fV[gW] - gW[fV]$$
$$= fV[g]W + fgVW - gW[f]V - gfWV$$
$$= fg[V, W] + fV[g]W - gW[f]V.$$

3.3 Assume $\mathbf{x}_u \cdot \mathbf{x}_v = 0$ and take an orthonormal basis $\bar{\mathbf{x}}_u = \mathbf{x}_u/\sqrt{E}$ and $\bar{\mathbf{x}}_v = \mathbf{x}_v/\sqrt{G}$. Then

$$(\nabla_{\bar{\mathbf{x}}_u} \bar{\mathbf{x}}_u)^N = \frac{1}{\sqrt{E}}\left(\left(\frac{\mathbf{x}_u}{\sqrt{E}}\right)_u\right)$$

$$= \left(\frac{\mathbf{x}_{uu}\sqrt{E} - \mathbf{x}_u(\sqrt{E})_u}{E^{3/2}}\right)^N$$

$$= \frac{l}{E}.$$

Similarly, $(\nabla_{\bar{\mathbf{x}}_v} \bar{\mathbf{x}}_v)^N = n/G$. Then, the sum is

$$\frac{Gl + En}{EG} = 2H$$

since $F = 0$.

5.10 Note that $\langle R(X, X)X, X\rangle = 0$, so that, if we take $X/|X|$ as $\mathcal{E}_k$, then $\mathcal{E}_1, \ldots, \mathcal{E}_k$ is a frame for M^k. By definition then,

$$\mathrm{Ric}(X, X) = \mathrm{Ric}(\mathcal{E}_k, \mathcal{E}_k) = \sum_{j-1}^{k-1} \langle R(\mathcal{E}_k, \mathcal{E}_j)\mathcal{E}_k, \mathcal{E}_j\rangle$$

since $\langle R(\mathcal{E}_k, \mathcal{E}_k)\mathcal{E}_k, \mathcal{E}_k\rangle = 0$. Then use the definition of sectional curvature and the definition of $\mathcal{E}_k = X/|X|$.

5.13 Let $X = \sum_j X^j \mathcal{E}_j$ and $df = \sum_j df_j \theta^j$ where θ^j is a dual vector space basis element to $\mathcal{E}_j$ (see §6) defined by $\theta^j(\mathcal{E}_i) = \delta_i^j$ (where δ_i^j is the Kronecker delta). Then

$$df(X) = df(\sum_j X^j \mathcal{E}_j) = \sum_j X^j df(\mathcal{E}_j) = \sum_j \sum_i X^j df_i \theta^i(\mathcal{E}_j) = \sum_j X^j df_j.$$

Also. we have by the definition of divergence,

$$\mathrm{div}(f\langle \cdot, \cdot\rangle)(X) = \sum_j \nabla_{\mathcal{E}_j}(f\langle \cdot, \cdot\rangle)(\mathcal{E}_j, X)$$

$$= \sum_j \mathcal{E}_j[f\langle \mathcal{E}_j, X\rangle] - f\langle \nabla_{\mathcal{E}_j}\mathcal{E}_j, X\rangle - f\langle \mathcal{E}_j, \nabla_{\mathcal{E}_j}X\rangle$$

$$= \sum_j \mathcal{E}_j[f]\langle \mathcal{E}_j, X\rangle + f\mathcal{E}_j\langle \mathcal{E}_j, X\rangle$$

$$- f\langle \nabla_{\mathcal{E}_j}\mathcal{E}_j, X\rangle - f\langle \mathcal{E}_j, \nabla_{\mathcal{E}_j}X\rangle$$

$$= \sum_j \mathcal{E}_j[f]\langle \mathcal{E}_j, X\rangle + f\langle \nabla_{\mathcal{E}_j}\mathcal{E}_j, X\rangle + f\langle \mathcal{E}_j, \nabla_{\mathcal{E}_j}X\rangle$$

$$- f\langle \nabla_{\mathcal{E}_j}\mathcal{E}_j, X\rangle - f\langle \mathcal{E}_j, \nabla_{\mathcal{E}_j}X\rangle$$

$$= \sum_j \mathcal{E}_j[f]\langle \mathcal{E}_j, X\rangle$$

$$= \sum_j df(\mathcal{E}_j)X^j$$

$$= \sum_j df_j X^j$$

$$= df(X)$$

5.14

$$\kappa = \sum_{i=1}^{k} \text{Ric}(\mathcal{E}_i, \mathcal{E}_i) = \sum_{i=1}^{k} \lambda \langle \mathcal{E}_i, \mathcal{E}_i \rangle = k\lambda.$$

Thus, $d\kappa = kd\lambda$ and we also have (by Exercise 5.13)

$$
\begin{aligned}
2d\lambda &= 2\,\text{div}\,f\langle \cdot, \cdot \rangle \\
&= 2\,\text{div}(\text{Ric}) \\
&= d\kappa \qquad \text{by Theorem 5.4} \\
&= kd\lambda.
\end{aligned}
$$

Thus, $(k-2)d\lambda = 0$ and, since $k \geq 3$, we must have $d\lambda = 0$. Hence, λ is a constant.

REFERENCES

[AM]. R. Abraham and J. Marsden, *Foundations of Mechanics*, Second Edition, Updated 1985 Printing, Addison-Wesley, 1978.

[Arn]. V. I. Arnol'd, *Mathematical Methods of Classical Mechanics*, Grad. Texts. in Math., vol. 60, Springer-Verlag.

[ATF]. V. Alekseev, V. Tikhomirov and S. Fomin, *Optimal Control Theory*, Consultants Bureau, 1987.

[BC]. J. Barbosa and A. Colares, *Minimal Surfaces in* $\mathbb{R}^3$, Lecture Notes in Math., vol. 1195, Springer-Verlag, 1986.

[Bli]. G. Bliss, *Lectures on the Calculus of Variations*, U. of Chicago Press, 1946.

[BM]. J. Boersma and J. Molenaar, *Geometry of the shoulder of a packaging machine*, SIAM Rev. **37 no. 3** (1995), 406-422.

[Bo]. C. V. Boys, *Soap Bubbles: Their Colors and the Forces which Mold Them*, Dover, 1959.

[Can]. P. B. Canham, *The minimum energy of bending as a possible explanation of the biconcave shape of the human red blood cell*, J. Theoret. Biol. **26** (1970), 61-81.

[Che]. S. S. Chern, *An Elementary Proof of the Existence of Isothermal Parameters on a Surface*, Proc. Amer. Math. Soc. **6** (1955), 771-782.

[Cou]. R. Courant, *Dirichlet's Principle, Conformal Mapping and Minimal Surfaces*, Interscience, 1950.

[CH]. R. Courant and D. Hilbert, *Methods of Mathematical Physics*, vol. 1, Interscience, 1953.

[CM]. W. Curtis and F. Miller, *Differential Manifolds and Theoretical Physics*, Academic Press, 1985.

[Cox]. H. Coxeter, *Introduction to Geometry*, 2nd Edition, Wiley, 1969.

[CR]. R. Courant and H. Robbins, *What is Mathematics*, Oxford U. Press, 1943.

[Dar]. R. W. R. Darling, *Differential Forms and Connections*, Cambridge U. Press, 1994.

[Del]. C. Delaunay, *Sur la Surface de Revolution dont la Courbure Moyenne est Constante*, J. de Math. Pure et Appl. **6 ser. 1** (1841), 309-320.

[DH1]. H. Deuling and W. Helfrich, *The curvature elasticity of fluid membranes: a catalogue of vesicle shapes*, Le Journal de Physique **37** (1976), 1335-1345.

[DH2]. H. Deuling and W. Helfrich, *Red blood cell shapes as explained on the basis of curvature elasticity*, Biophys. J. **16** (1976), 861-868.

[DHKW]. U. Dierkes, S. Hildebrandt, A. Küster and O. Wahlrab, *Minimal Surfaces I*, Grundlehren, vol. 295, Springer-Verlag, 1992.

[DoC1]. M. Do Carmo, *Differential Geometry of Curves and Surfaces*, Prentice-Hall, 1976.

[DoC2]. M. Do Carmo, *Riemannian Geometry*, Birkhäuser, 1992.

[Dou]. J. Douglas, *Solution of the Problem of Plateau*, Trans. Amer. Math. Soc. **33** (1931), 263-321.

[DWT]. D. W. Thompson, *On Growth and Form: The Complete Revised Edition*, Dover, 1992.

[Ee1]. J. Eells, *The Surfaces of Delaunay*, Math. Intell. **9 no. 1** (1987), 53-57.

[Ee2]. J. Eells, *On the Surfaces of Delaunay and Their Gauss Maps*, Proc. IV Int. Colloq. Diff. Geom. Santiago de Compostela, 1978, pp. 97-116.

[Eis]. L. Eisenhart, *A Treatise on the Differential Geometry of Curves and Surfaces*, Ginn and Company, 1909.

[Ew]. G. Ewing, *Calculus of Variations with Applications*, Dover, 1985.

[Fab]. R. Faber, *Differential Geometry and Relativity Theory: An Introduction*, Marcel Dekker, 1983.

[Finn]. R. Finn, *Equilibrium Capillary Surfaces*, Springer-Verlag, 1986.

[Fla]. H. Flanders, *Differential Forms*, Dover, 1989.

[FT]. A. Fomenko and A. Tuzhilin, *Elements of the Geometry and Topology of Minimal Surfaces in Three-Dimensional Space*, Transl. of Math. Mono., vol. 93, Amer. Math. Soc., 1991.

[G]. H. Grant, *Practical Descriptive Geometry*, McGraw-Hill, 1952.

[GHL]. S. Gallot, D. Hulin and J. Lafontaine, *Riemannian Geometry*, Universitext, Springer-Verlag, 1990.

[Go]. D. Gottlieb, *All the way with Gauss-Bonnet and the sociology of Mathematics*, to appear, Amer. Math Monthly.

[Gr]. A. Gray, *Modern Differential Geometry of Curves and Surfaces*, CRC Press, 1993.

[Grau]. W. Graustein, *Harmonic minimal surfaces*, Trans. Amer. Math. Soc. **47** (1940), 173-206.

[Gug]. H. Guggenheimer, *Differential Geometry*, McGraw-Hill, 1963.

[HK]. L. Haws and T. Kiser, *Exploring the Brachistochrone problem*, Amer. Math. Monthly **102 no.4** (1995), 328-336.

[Hi]. N. Hicks, *Notes on Differential Geometry*, Van Nostrand, 1965.

[HT]. S. Hildebrandt and A. Tromba, *Mathematics and Optimal Form*, Scientific American Books, 1985.

[Hoff]. D. Hoffman, *The Computer-Aided Discovery of New Embedded Minimal Surfaces*, Math. Intell. **9 no. 3** (1987).

[HM]. D. Hoffman and W. Meeks, *Minimal Surfaces Based on the Catenoid*, Amer. Math. Monthly **97 no. 8** (1990), 702-730.

[Hsi]. C. C. Hsiung, *A First Course in Differential Geometry*, John Wiley & Sons, 1981.

[Isen]. C. Isenberg, *The Science of Soap Films and Soap Bubbles*, Dover, 1992.

[Kir]. D. Kirk, *Optimal Control Theory: An Introduction*, Electrical Engineering Series, Prentice-Hall, 1970.

[Kl]. W. Klingenberg, *Lectures on Closed Geodesics*, Grund. Math. Wissen. vol. 230, Springer-Verlag, 1978.

[Kr]. E. Kreyszig, *Differential Geometry*, Dover, 1991.

[L]. G. Lawlor, *A New Minimization Proof for the Brachistochrone*, Amer. Math. Monthly **103 no. 3** (1996), 242-249.

[Lap]. P. S. Laplace, *Mechanique Celeste*, vol. IV, Little and Brown, 1839.

[Law]. H. B. Lawson, *Lectures on Minimal Submanifolds: Volume I*, Math. Lecture Series, vol. 9, Publish or Perish, 1980.

[LF]. L. Lusternik and A. Fet, *Variational problems on closed manifolds*, Dokl. Akad. Nauk. SSSR **81** (1951), 17-18.

[Lu]. M. Lunn, *A First Course in Mechanics*, Oxford U. Press, 1991.

[M]. J. Marsden, *Lectures on Mechanics*, Lecture Note Series 174, London Math. Soc., 1992.

[MH]. J. Marsden and M. Hoffman, *Basic Complex Analysis*, W. H. Freeman, 1987.

[MT]. J. Marsden and A. Tromba, *Vector Calculus*, W. H. Freeman, 1988.

[Mar]. J. Martin, *General Relativity*, Ellis Horwood Lmtd, 1988.

[Max]. J. Maxwell, *On the Theory of Rolling Curves*, Trans. Roy. Soc. Edinburgh **XVI Part V** (1849), 519-544.

[MP]. R. Millman and G. Parker, *Elements of Differential Geometry*, Prentice-Hall, 1977.

[MTW]. C. Misner, K. Thorne and J. Wheeler, *Gravitation*, W. H. Freeman, 1973.

[Mor]. F. Morgan, *Geometric Measure Theory: A Beginner's Guide*, Academic Press, 1988.

[Ni]. J. Nitsche, *Lectures on Minimal Surfaces: Volume I*, vol. 1, Cambridge U. Press, 1989.

[O'N1]. B. O'Neill, *Elementary Differential Geometry*, Academic Press, 1966.

[O'N2]. B. O'Neill, *Semi-Riemannian Geometry*, Academic Press, 1983.

[Op]. J. Oprea, *Geometry and the Foucault Pendulum*, Amer. Math. Monthly **102 no. 6** (1995), 515-522.

[Oss1]. R. Osserman, *A Survey of Minimal Surfaces*, Dover, 1986.

[Oss2]. R. Osserman, *Curvature in the Eighties*, Amer. Math. Monthly **97 no. 8** (1990), 731-756.

[P]. O. C. Pin, *Curvature and Mechanics*, Adv. in Math. **15** (1975), 269-311.

[Pau]. W. Paulsen, *What is the Shape of a Mylar Balloon*, Amer. Math. Monthly **101 no. 10** (1994), 953-958.

[Pin]. E. Pinch, *Optimal Control and the Calculus of Variations*, Oxford U. Press, 1993.

[Poh]. W. Pohl, *DNA and differential geometry*, Math. Intell. **3** (1980), 20-27.

[PoR]. W. Pohl and G. Roberts, *Topological considerations in the theory of replication of DNA*, J. Math. Biology **6** (1978), 383-402.

[PC]. J. Prussing and B. Conway, *Orbital Mechanics*, Oxford U. Press, 1993.

[R]. T. Radó, *On the Problem of Plateau/Subharmonic Functions*, Springer-Verlag, 1971.

[Red]. D. Redfern, *The Maple Handbook*, Springer-Verlag, 1993.

[Ros]. A. Ros, *Compact Surfaces with Constant Scalar Curvature and a Cogruence Theorem*, J. Diff. Geom. **27** (1988), 215-220.

[RW]. H. Resnikoff and R. Wells, *Mathematics in Civilization*, Dover, 1984.

[SS]. E. Saff and A. Snider, *Fundamentals of Complex Analysis for Mathematics, Science and Engineering*, Prentice-Hall, 1993.

[Sag]. H. Sagan, *Introduction to the Calculus of Variations*, Dover, 1992.

[Sco]. P. Scofield, *Curves of Constant precession*, Amer. Math. Monthly **102 no. 6** (1995), 531-537.

[Sol]. B. Solomon, *Tantrices of Spherical Curves*, Amer. Math. Monthly **103 no. 1** (1996), 30-39.

[SU]. J. Shigley and J. Uicker, *Theory of Machines and Mechanisms*, McGraw-Hill, 1980.

[Spi]. M. Spivak, *A Comprehensive Introduction to Differential Geometry: 5 Volumes*, Publish or Perish, 1979.

[St]. D. Struik, *Lectures on Classical Differential Geometry*, Dover, 1988.

[Sy]. K. Symon, *Mechanics*, 3rd Edition, Addison-Wesley, 1971.

[Tik]. V. M. Tikhomirov, *Stories about Maxima and Minima*, Math. World vol. 1, Amer. Math. Soc., 1990.

[Wei]. R. Weinstock, *Calculus of Variations*, Dover, 1974.

[We]. H. Wente, *Counter-example to the Hopf Conjecture*, Pac. J. Math. **121** (1986), 193-244.

[Whi]. L. Whitt, *The Standup Conic Presents: The Hyperbola and its Applications*, Umap Jour. **V no. 1** (1984), 9-21.

[WS]. F. Wilczek and A. Shapere, *Geometric Phases in Physics*, World Scientific, Singapore, 1989.

[Ya]. R. Yates, *Curves and their Properties*, Classics in Math. EducationL A Series, vol. 4, National Council of Teachers of Math., 1974.

[Y]. T. Young, *An Essay on the Cohesion of Fluids*, Phil. Trans. Roy. Soc. (London) **1** (1805), 65-87.

[Zw]. C. Zwikker, *The Advanced Geometry of Plane Curves and Their Applications*, Dover, 1963.

INDEX